Current Topics in Microbiology and Immunology

Volume 346

Current Topics in Microbiology and Immunology

Previously published volumes
Further volumes can be found at springer.com

Vol. 319: **Manser, Tim (Ed.):**
Specialization and Complementation of Humoral Immune Responses to Infection. 2008. ISBN 978-3-540-73899-2

Vol. 320: **Paddison, Patrick J.; Vogt, Peter K.(Eds.):** RNA Interference. 2008. ISBN 978-3-540-75156-4

Vol. 321: **Beutler, Bruce (Ed.):**
Immunology, Phenotype First: How Mutations Have Established New Principles and Pathways in Immunology. 2008. ISBN 978-3-540-75202-8

Vol. 322: **Romeo, Tony (Ed.):**
Bacterial Biofilms. 2008. ISBN 978-3-540-75417-6

Vol. 323: **Tracy, Steven; Oberste, M. Steven; Drescher, Kristen M. (Eds.):**
Group B Coxsackieviruses. 2008. ISBN 978-3-540-75545-6

Vol. 324: **Nomura, Tatsuji; Watanabe, Takeshi; Habu, Sonoko (Eds.):**
Humanized Mice. 2008. ISBN 978-3-540-75646-0

Vol. 325: **Shenk, Thomas E.; Stinski, Mark F. (Eds.):**
Human Cytomegalovirus. 2008. ISBN 978-3-540-77348-1

Vol. 326: **Reddy, Anireddy S.N; Golovkin, Maxim (Eds.):**
Nuclear pre-mRNA processing in plants. 2008. ISBN 978-3-540-76775-6

Vol. 327: **Manchester, Marianne; Steinmetz, Nicole F. (Eds.):**
Viruses and Nanotechnology. 2008. ISBN 978-3-540-69376-5

Vol. 328: **van Etten, (Ed.):**
Lesser Known Large dsDNA Viruses. 2008. ISBN 978-3-540-68617-0

Vol. 329: **Griffin, Diane E.; Oldstone, Michael B.A. (Eds.):** Measles. 2009. ISBN 978-3-540-70522-2

Vol. 330: **Griffin, Diane E.; Oldstone, Michael B.A. (Eds.):** Measles. 2009. ISBN 978-3-540-70616-8

Vol. 331: **Villiers, E. M. de (Ed.):**
TT Viruses. 2009. ISBN 978-3-540-70917-8

Vol. 332: **Karasev A. (Ed.):**
Plant produced Microbial Vaccines. 2009. ISBN 978-3-540-70857-5

Vol. 333: **Compans, Richard W.; Orenstein, Walter A. (Eds.):**
Vaccines for Pandemic Influenza. 2009. ISBN 978-3-540-92164-6

Vol. 334: **McGavern, Dorian; Dustin, Micheal (Eds.):**
Visualizing Immunity. 2009. ISBN 978-3-540-93862-0

Vol. 335: **Levine, Beth; Yoshimori, Tamotsu; Deretic, Vojo (Eds.):**
Autophagy in Infection and Immunity. 2009. ISBN 978-3-642-00301-1

Vol. 336: **Kielian, Tammy (Ed.):**
Toll-like Receptors: Roles in Infection and Neuropathology. 2009. ISBN 978-3-642-00548-0

Vol. 337: **Sasakawa, Chihiro (Ed.):**
Molecular Mechanisms of Bacterial Infection via the Gut. 2009. ISBN 978-3-642-01845-9

Vol. 338: **Rothman, Alan L. (Ed.):**
Dengue Virus. 2009. ISBN 978-3-642-02214-2

Vol. 339: **Spearman, Paul; Freed, Eric O. (Eds.):**
HIV Interactions with Host Cell Proteins. 2009. ISBN 978-3-642-02174-9

Vol. 340: **Saito, Takashi; Batista, Facundo D. (Eds.):**
Immunological Synapse. 2010. ISBN 978-3-642-03857-0

Vol. 341: **Bruserud, Øystein (Ed.):**
The Chemokine System in Clinical and Experimental Hematology. 2010. ISBN 978-3-642-12638-3

Vol. 342: **Arvin, Ann M. (Ed.):**
Varicella-zoster Virus. 2010. ISBN 978-3-642-12727-4

Vol. 343: **Johnson, John E. (Ed.):**
Cell Entry by Non-Enveloped Viruses. 2010. ISBN 978-3-642-13331-2

Vol. 345: **Simon, M. Celeste (Ed.):**
Diverse Effects of Hypoxia on Tumor Progression. 2010. ISBN 978-3-642-13328-2

Christian Rommel • Bart Vanhaesebroeck •
Peter K. Vogt
Editors

Phosphoinositide 3-kinase in Health and Disease

Volume 1

Springer

Editors
Dr. Christian Rommel
Intellikine
10931 North Torrey Pines Road
La Jolla, CA 92037
USA
christian@intellikine.com

Prof. Dr. Peter K. Vogt
The Scripps Research Institute
Dept. Molecular & Experimental Medicine
10550 North Torrey Pines Road
La Jolla, CA 92037
USA
pkvogt@scripps.edu

Prof. Dr. Bart Vanhaesebroeck
Queen Mary, University of London
Centre for Cell Signalling,
Institute of Cancer
Charterhouse Square
London EC1M 6BQ
United Kingdom
bart.vanh@qmul.ac.uk

ISSN 0070-217X
ISBN: 978-3-642-13662-7 e-ISBN: 978-3-642-13663-4
DOI 10.1007/978-3-642-13663-4
Springer Heidelberg Dordrecht London New York

Library of Congress Control Number: 2010936822

Cover design: WMXDesign GmbH, Heidelberg, Germany

Printed on acid-free paper

Springer is part of Springer Science+Business Media (www.springer.com)

Contents

Contributors

Jonathan M. Backer Department of Molecular Pharmacology, Albert Einstein College of Medicine, 1300 Morris Park, Avenue, Bronx, NY 10461, USA, jonathan.backer@einstein.yu.edu

José Ramón Bayascas Institut de Neurociències & Departament de Bioquímica i BiologiaMolecular, Universitat Autònoma de Barcelona, 08193 Barcelona, Spain, joseramon.bayascas@uab.cat

Lana Bozulic Friedrich Miescher Institute for Biomedical Research, Maulbeerstrasse 66, 4058 Basel, Switzerland

Lewis C. Cantley Department of Systems Biology and Division of Signal Transduction, Beth Israel Deaconess Medical Center, Harvard University, Boston, MA 02115, USA

Esther Castellano Signal Transduction Laboratory, Cancer Research UK London Research Institute, 44 Lincoln's Inn Fields, London WC2A 3PX, UK

Carlotta Costa Molecular Biotechnology Center, University of Torino, Via Nizza 52, 10126 Torino, Italy

Julian Downward Signal Transduction Laboratory, Cancer Research UK London Research Institute, 44 Lincoln's Inn Fields, London WC2A 3PX, UK, julian.downward@cancer.org.uk

Britta J. Eickholt MRC Centre for Developmental Neurobiology, King's College London, New Hunt's House, London SE1 1UL, UK, Britta.J.Eickholt@kcl.ac.uk

Elisabeth Fayard Friedrich Miescher Institute for Biomedical Research, Maulbeerstrasse 66, 4058 Basel, Switzerland

Lazaros C. Foukas Institute of Healthy Ageing, University College London, Gower Street, London WC1E 6BT, UK and Department of Genetics, Evolution and Environment, University College London, Gower Street, London WC1 6BT, UK, l.foukas@ucl.ac.uk

David A. Fruman Department of Molecular Biology and Biochemistry, Institute for Immunology, University of California-Irvine, Irvine, CA 92697-3900, USA, dfruman@uci.edu

David J. Glass Novartis Institute for Biomedical Research, 100 Technology Square, Cambridge, MA 02139, USA, david.glass@novartis.com

Phillip T. Hawkins The Babraham Institute, Babraham Research Campus, Cambridge CB22 3AT, UK, Philip.hawkins@bbsrc.ac.uk

Brian A. Hemmings Friedrich Miescher Institute for Biomedical Research, Maulbeerstrasse 66, 4058 Basel, Switzerland, brian.hemmings@fmi.ch

Emilio Hirsch Molecular Biotechnology Center, University of Torino, Via Nizza 52, 10126 Torino, Italy

Shaun P. Jackson Australian Centre for Blood Diseases, Monash University, 6th Level Burnet Building, Alfred Medical Research and Education Precinct (AMREP), 89 Commercial Road, Melbourne, VIC 3004, Australia, Shaun.Jackson@med.monash.edu.au

Klaus Okkenhaug Laboratory of Lymphocyte Signalling and Development, The Babraham Institute, Cambridge, UK, klaus.okkenhaug@bbsrc.ac.uk

Arnaud Parcellier Friedrich Miescher Institute for Biomedical Research, Maulbeerstrasse 66, 4058 Basel, Switzerland

Christian Rommel Intellikine Inc, 10931 North Torrey Pines Road, La Jolla, CA 92037, USA, christian@intellikine.com

Simone M. Schoenwaelder Australian Centre for Blood Diseases, Monash University, 6th Level Burnet Building, Alfred Medical Research and Education Precinct (AMREP), 89 Commercial Road, Melbourne, VIC 3004, Australia

Len R. Stephens The Babraham Institute, Babraham Research Campus, Cambridge CB22 3AT, UK

Sabine Suire The Babraham Institute, Babraham Research Campus, Cambridge CB22 3AT, UK

Kathryn Waite MRC Centre for Developmental Neurobiology, King's College London, New Hunt's House, London SE1 1UL, UK

Michael Wilson The Babraham Institute, Babraham Research Campus, Cambridge CB22 3AT, UK

Dominic J. Withers Metabolic Signalling Group, Medical Research Council Clinical Sciences Centre, Imperial College, Du Cane Road, London W12 0NN, UK, d.withers@imperial.ac.uk

Gongda Xue Friedrich Miescher Institute for Biomedical Research, Maulbeerstrasse 66, 4058 Basel, Switzerland

Tina L. Yuan Department of Systems Biology and Division of Signal Transduction, Beth Israel Deaconess Medical Center, Harvard University, Boston, MA 02115, USA

Introduction

Tina L. Yuan and Lewis C. Cantley

Contents

Abstract The phosphoinositide-3-kinase (PI3K) family of lipid kinases has been well conserved from yeast to mammals. In this evolutionary perspective on the PI3K family, we discuss the prototypical properties of PI3Ks: 1) the utilization of sparse but specifically localized lipid substrates; 2) the nucleation signaling complexes at membrane-targeted sites; and 3) the integration of intracellular signaling with extracellular cues. Together, these three core properties serve to establish order within the entropic environment of the cell. Many human diseases, including cancer and diabetes, are the direct result of loss or defects in one or more of these core properties, putting much hope in the clinical use of PI3K inhibitors singly and in combination to restore order within diseased tissues.

1 Establishing Order Within the Cell

The lipid kinase activity of phosphoinositide 3-kinase (PI3K) has been evolutionarily conserved from yeast to mammals and has evolved from a simple means of sorting vacuolar proteins to nucleating large signaling complexes that regulate

T.L. Yuan and L.C. Cantley
Harvard University, Department of Systems Biology and Beth Israel Deaconess Medical Center, Division of Signal Transduction, Boston, MA 02115

C. Rommel et al. (eds.). *Phosphoinositide 3-kinase in Health and Disease*, Volume 1
Current Topics in Microbiology and Immunology 346, DOI 10.1007/82_2010_55

growth, metabolism and survival (Engelman et al. 2006). Here, we reflect on the unique properties of PI3Ks that explain the diverse roles that these enzymes play in cellular regulation and their relevance in multiple human diseases.

A typical mammalian cell is composed of approximately 70% water and 20% proteins. In their textbook example, Lodish and colleagues estimate that for a hepatocyte this translates into roughly 8×10^9 protein molecules, most of which are randomly diffusing within a chaotic 15-μm^3 space (Lodish et al. 2000). In such a disordered environment, order and directionality must be established to successfully transmit growth and survival signals, for example from a membrane-anchored growth factor receptor to a transcription factor in the nucleus. Perhaps the most valuable and thus conserved property of PI3K is the ability to impose such order in a highly entropic environment.

The core properties that allow PI3K to carry out this function have been conserved from unicellular to multicellular organisms. These include (1) having low abundant but highly specific lipid substrates and products; (2) generating membrane-anchored products that nucleate signaling complexes at targeted sites; and (3) having the ability to associate with membrane-bound proteins that sense extracellular stimuli. Over the course of evolution, higher organisms have evolved several classes of PI3Ks that utilize these prototypical properties to regulate a wide range of functions ranging from directional motility to metabolism, growth, and survival. Importantly, it is also the loss of these core properties that result in aberrant signaling and disease.

2 Phosphatidylinositol and Phosphoinositides as Ideal Substrates

Evolving biological systems require simplicity that will not convolute cellular communication or waste resources. Yet there must be enough variability in the system to allow for diversification and selection. Following this model, PI3K has only three lipid substrates: phosphatidylinositol (PtdIns) and two of its phosphoinositide derivatives, PI-4-P and PI-4, 5-P_2. Additionally, these substrates are present at low levels within the cell. While only 5% of the mass of a mammalian cell is comprised of lipids, only 4% of total lipids are PtdIns and less than 1% of total PtnIns is phosphorylated. Importantly, the PI3K products make up only about 1% of the total phosphorylated forms of PtdIns (Mulgrew-Nesbitt et al. 2006). This extreme low abundance of PI3K lipid products ensures that PI3K signaling is deliberate, dynamic, non-promiscuous, and exquisitely localized.

Yet, despite the scarcity of PtdIns in the cell, the inositol head group contains five free hydroxyl groups that could potentially be phosphorylated to generate variability in the phosphoinositide pool. Three of the five hydroxyl groups (D3, D4, and D5 positions) are phosphorylated alone or in combination, yielding seven phosphoinositides, each with unique stereospecificity and charge. At least 10 discreet protein domains have independently evolved the ability to bind one or more phosphoinositides and have been identified in hundreds of proteins across numerous species (Lemmon 2008; DiNitto et al. 2003). Thus, by modifying a single lipid substrate, the phosphoinositide kinases have evolved the unique ability to regulate numerous proteins while carefully preserving specificity.

3 Nucleating a Protein Complex at a Target Location

The most ancient role of PI3K in unicellular organisms remains arguably its most relevant role in multicellular organisms. This is the role of nucleating a protein complex at a target location within the cell. *Saccharomyces cerevisiae* expresses the most primordial PI3K, the class III Vps34, which generates PI-3-P at sorting endosomes. Proteins containing FYVE domains bind to PI-3-P and form complexes that regulate vacuolar protein sorting (Burd and Emr 1998). The generation of PI-3-P specifically at sorting endosomes ensures that the protein-sorting complexes are carefully localized to this compartment. Proper localization of protein complexes is also critical for directional movement in another unicellular organism, *Dictyostelium discoideum*. The generation of PI-3, 4, 5-P_3 by class I PI3K at the cell's leading edge recruits PH domain containing proteins such as CRAC and AKT that rearrange the cytoskeleton for directed movement towards shallow chemoattractant gradients (Parent et al. 1998; Meili et al. 1999).

The need for proper protein localization for the most basic functions in unicellular organisms suggests that this was the original function of PI3K. Multicellular organisms have conserved this property by utilizing localized phosphoinositides to regulate cellular polarity and migration, particularly in epithelial cells, neutrophils, and macrophages (Gassama-Diagne et al. 2006; Fruman and Bismuth 2009). However, the utility of this enzyme in multicellular organisms extends into far more complex realms of signaling that nevertheless hinge on the ability to nucleate large signaling complexes at cellular membranes.

4 Coupling PI3K Activity to Extracellular Cues

The evolution towards multicellularity was accompanied by the emergence of two additional classes of PI3K. In addition to class III, class I and II PI3Ks are found in *Caenorhabditis elegans*, *Drosophila melanogaster*, and all vertebrates, suggesting that these later evolving classes specialize in mediating cell–cell communication. Extracellular sensing in unicellular organisms is essentially a survey of the local nutrient landscape, which informs the cell whether or not to grow and proliferate. In multicellular organisms, numerous extracellular stimuli instruct cells not just to grow and proliferate, but also to migrate to new location, to activate survival mechanisms, and to alter gene expression programs to control metabolic needs and other specialized functions of differentiated tissues.

Class I and II PI3Ks evolved to cope with these new and complex demands by targeting the assembly of various protein complexes directly downstream of membrane-bound receptors, thereby integrating intracellular signaling with extracellular cues. Class I PI3Ks are targeted to active receptor tyrosine kinases (RTKs) or G-protein coupled receptors (GPCRs). For class IA PI3Ks, localization to RTKs or adaptors downstream of RTKs is facilitated by conserved SH2 domains in the p85 family of regulatory subunits, which bind to Tyr-phosphorylated Y-X-X-M motifs on activated RTKs or adaptor molecules. The class IB PI3K, p110γ, localizes to GPCRs through the interaction of a regulatory subunit, p101 or p84, with G$\beta\gamma$ subunits. The class IA PI3K, p110β can bind to both RTKs and to G$\beta\gamma$ subunits of GPCRs and appears to act as an integrator of signaling through both pathways. Class II PI3Ks are less well studied, but can be activated by RTKs such as the epidermal growth receptor (EGFR) and insulin receptor and play an important role in clathrin-mediated vessel trafficking (Williams et al. 2009).

Due to their specialized role in interpreting extracellular cues, class I PI3Ks have been most extensively studied. It is the only class that generates PI-3, 4, 5-P_3, a potent second messenger, from PI-4, 5-P_2 in the membrane. High local concentrations of PI-3, 4, 5-P_3 in the membrane colocalize signaling molecules containing PH domains, which bind PI-3, 4, 5-P_3 with high specificity. Some of these molecules include guanine nucleotide exchange factors (GEF) such as GRP1, as well as kinases such as the Bruton tyrosine kinase (BTK), other members of the Tec family of non-receptor tyrosine kinases, and the serine/threonine kinases AKT (also known as PKB) and phosphoinositide-dependent kinase 1 (PDK1) (Cantley 2002). Importantly, localized pools of PI-3, 4, 5-P_3 are generated transiently, limited by the local availability of PI-4, 5-P_2 and by rapid degradation due to the presence of nearby phosphoinositide phosphatases. Three families of phosphoinositide phosphatases are important in modulating class I PI3K signaling: (1) PTEN dephosphorylates the 3$'$ position of PI-3, 4, 5-P_3 to regenerate PI-4, 5-P_2; (2) SHIP family members dephosphorylate the 5$'$ position of PI-3, 4, 5-P_3 to generate PI-3, 4-P_2; and (3) INPP4A/4B family members dephosphorylate the 4$'$ position of PI-3, 4-P_2 to generate PI-3-P. Ultimately, disengagement of PI3K from receptors due to various negative feedback loops terminates signaling. Temporal and spatial regulation of PI-3, 4, 5-P_3 (as well as PI-3, 4-P_2) production is unique to class I PI3Ks and highlights the importance of negatively regulating this complex pathway for receptor-mediated signaling (Lemmon 2008).

One PI-3, 4, 5-P_3 binder in particular, AKT, is responsible for much of this complexity. The PH domain of AKT can bind to both PI-3, 4, 5-P_3 and PI-3, 4-P_2, and both lipids appear to contribute to AKT recruitment to membranes (Franke et al. 1997), as well as recruitment of the upstream activating kinase, PDK1. There are more than 100 reported substrates for AKT, identified under varying degrees of stringency (Manning and Cantley 2007). Canonical AKT substrates contain R-X-R-X-X-S/T motifs, and phosphorylation on serine or threonine residues initiates a complex cascade of signaling events in the cytosol and nucleus. Briefly, AKT promotes proliferation through the inhibitory phosphorylation of FOXO transcription factors and GSK3, promotes protein synthesis by phosphorylating TSC2 and PRAS40, promotes cell survival by phosphorylating BAD and MDM2, and induces glucose uptake in GLUT4-containing cells through phosphorylation of AS160 (Manning and Cantley 2007). Interestingly, by comparing the number, specificity, and location of PI3K and AKT substrates, it is clear that these two well-conserved enzymes serve non-redundant functions as initiator and effector kinases, so to speak. Given the breadth of their cooperative actions, it is important to note that these two kinases are regulated by multiple feedback mechanisms to ensure that the system resets to basal levels following acute cell stimulation by growth factors (Engelman et al. 2006).

5 Disease Implications

As discussed above, though the responsibilities of PI3K in the cell have vastly increased over the course of evolution, its core properties have remained unchanged. Using low abundance, membrane-anchored lipids to organize signaling complexes downstream of extracellular cues persists as an optimal mode of transmitting signals within the cell. It is, thus, not surprising that alterations in these core properties result in evolutionarily unfavorable signaling and disease. Alterations in the PI3K pathway account for at least 30% of all human cancers and have been implicated in type-2 diabetes. To conclude, we briefly review how genetic alterations in PI3K pathway components unhinge the core properties of PI3K.

In normal cells, the low abundance of phosphoinositides ensures deliberate and targeted signaling downstream of PI3K activation. However, through loss of the PI-3, 4, 5-P_3 phosphatase, PTEN, or oncogenic hotspot mutations in p110α that confer constitutive kinase activity, PI-3, 4, 5-P_3 levels and cellular distribution increase dramatically, resulting in promiscuous and prolonged downstream signaling that contributes to tumorigenesis (Engelman et al. 2006). Recent work has also shown that other components of the PI3K signaling network that influence the phosphoinositide pool, including p85α and INPP4B, are frequently mutated or deleted in human cancers (Cancer Genome Atlas Research Network 2008; Gewinner et al. 2009).

The nucleation of protein complexes not only organizes signaling molecules to one location, but the systematic assembly of the complex ensures that signaling only occurs when all components are primed and ready. This system of checks and balances is disrupted, for example, in tumors with E17K mutations in AKT or p110α hotspot mutations (Carpten et al. 2007). Through enhanced enzymatic activity and membrane localization, cells bearing these mutations are no longer dependent on the step-wise assembly of the signaling complex, and can independently (over)activate downstream pathways that can lead to hyperproliferation.

Lastly, by coupling PI3K activity to activated extracellular receptors in normal cells, intracellular signaling is in harmony with extracellular conditions. However, in many tumors with RTK mutations or amplifications, such as in ERBB1 and ERBB2, extracellular conditions are exaggerated and can result in unsustainable PI3K signaling. In the other extreme, when PI3K activity is uncoupled from RTK signaling, such as in models of insulin resistance where PI3K is insensitive to insulin receptor activation, type-2 diabetes can arise (Engelman et al. 2006).

Given the large role PI3K plays in tumorigenesis and in the activation of macrophages and lymphocytes, there has been a profound effort by pharmaceutical companies to develop targeted inhibitors of this pathway for treating cancers and for suppressing immune responses. Several PI3K inhibitors, with varying specificities for submembers of the family, have shown great promise in pre-clinical cancer models and are currently in phase I/II clinical trials. AKT catalytic site inhibitors have also entered phase I clinical trials. It is now clear from extensive sequencing of genes from primary human cancers that mutations in the PI3K pathway are very frequent, but that they are invariably combined with mutations in other pathways. Thus, while there is much excitement about the PI3K and AKT inhibitors, it is likely that these compounds will need to be combined with other drugs that target other pathways activated in the same tumors in order to be effective. There is much hope that future "personalized" clinical trials that focus on matching drugs to mutational events in individual tumors will further validate targeted therapy and provide a logical path for conquering the myriad of cancers that have evaded more conventional cancer therapies over the past 40 years. It is likely that PI3K inhibitors will be a significant addition to the arsenal needed for this approach.

References

Burd CG, Emr SD (1998) Phosphatidylinositol(3)-phosphate signaling mediated by specific binding to RING FYVE domains. Mol Cell 2(1):157–162

Cancer Genome Atlas Research Network (2008) Comprehensive genomic characterization defines human glioblastoma genes and core pathways. Nature 455(7216):1061–1068

Cantley LC (2002) The phosphoinositide 3-kinase pathway. Science 296(5573):1655–1657

Carpten JD et al (2007) A transforming mutation in the pleckstrin homology domain of AKT1 in cancer. Nature 448(7152):439–444

DiNitto JP, Cronin TC, Lambright DG (2003) Membrane recognition and targeting by lipid-binding domains. Sci STKE 2003(213):re16

Engelman JA, Luo J, Cantley LC (2006) The evolution of phosphatidylinositol 3-kinases as regulators of growth and metabolism. Nat Rev Genet 7(8):606–619

Franke TF et al (1997) Direct regulation of the Akt proto-oncogene product by phosphatidylinositol-3, 4-bisphosphate. Science 275(5300):665–668

Fruman DA, Bismuth G (2009) Fine tuning the immune response with PI3K. Immunol Rev 228 (1):253–272

Gassama-Diagne A et al (2006) Phosphatidylinositol-3, 4, 5-trisphosphate regulates the formation of the basolateral plasma membrane in epithelial cells. Nat Cell Biol 8(9):963–970

Gewinner C et al (2009) Evidence that inositol polyphosphate 4-phosphatase type II is a tumor suppressor that inhibits PI3K signaling. Cancer Cell 16(2):115–125

Lemmon MA (2008) Membrane recognition by phospholipid-binding domains. Nat Rev Mol Cell Biol 9(2):99–111

Lodish H et al (2000) Molecular cell biology, 4th edn. Freeman & Co, New York, NY

Manning BD, Cantley LC (2007) AKT/PKB signaling: navigating downstream. Cell 129(7): 1261–1274

Meili R et al (1999) Chemoattractant-mediated transient activation and membrane localization of Akt/PKB is required for efficient chemotaxis to cAMP in Dictyostelium. EMBO J 18(8): 2092–2105

Mulgrew-Nesbitt A et al (2006) The role of electrostatics in protein-membrane interactions. Biochim Biophys Acta 1761(8):812–826

Parent CA et al (1998) G protein signaling events are activated at the leading edge of chemotactic cells. Cell 95(1):81–91

Williams R et al (2009) Form and flexibility in phosphoinositide 3-kinases. Biochem Soc Trans 37 (Pt 4):615–626

PDK1: The Major Transducer of PI 3-Kinase Actions

José Ramón Bayascas

Contents

Abstract Most of the cellular responses to phosphatidylinositol 3-kinase activation and phosphatidylinositol 3,4,5-trisphosphate production are mediated by the activation of a group of AGC kinases comprising PKB, S6K, RSK, SGK and PKC isoforms, which play essential roles in regulating physiological processes related to cell growth, proliferation, survival and metabolism. All these growth-factor-stimulated AGC kinases possess a common upstream activator, namely PDK1, a master kinase, which, being constitutively active, is still able to phosphorylate and activate its AGC substrates in response to rises in the levels of the PtdIns(3,4,5)P_3 second messenger. In this chapter, the biochemical, structural and genetic data on the mechanism of action and physiological roles of PDK1 are reviewed, and its

J.R. Bayascas
Institut de Neurociències & Departament de Bioquímica i Biologia Molecular, Universitat Autònoma de Barcelona, 08193 Barcelona, Spain
e-mail: joseramon.bayascas@uab.cat

C. Rommel et al. (eds.). *Phosphoinositide 3-kinase in Health and Disease*, Volume 1
Current Topics in Microbiology and Immunology 346, DOI 10.1007/82_2010_43

potential as a pharmaceutical target for the design of drugs therapeutically beneficial to treat human disease such us diabetes and cancer is discussed.

1 Introduction

The 3-phosphoinositide-dependent protein kinase-1 (PDK1) was first discovered in the field of the insulin signal transduction research as the protein kinase capable of phosphorylating and activating PKB in a PtdIns(3,4,5)P_3-dependent manner (Alessi et al. 1997b; Stokoe et al. 1997). In the same way as the activation of the phosphatidylinositol 3-kinase (PI 3-kinase) and subsequent production of phosphatidylinositol (3,4,5) trisphosphate, or PtdIns(3,4,5)P_3, is the major apical signalling event triggered by growth factors and hormones, PDK1 plays crucial roles in reading out the increases in PtdIns(3,4,5)P_3 levels and thereby in governing many of the cellular responses to this second messenger.

In fact, most of the physiological effects of PtdIns(3,4,5)P_3 rises in cells are mediated by a particular set of AGC protein kinase family members that controls cell growth, proliferation, survival as well as metabolic responses to insulin. These include protein kinase B (PKB)/akt isoforms, which regulate cell viability and proliferation as well as glucose homeostasis (Whiteman et al. 2002; Dummler and Hemmings 2007; Manning and Cantley 2007); p70 ribosomal S6 kinase isoforms (S6K), implicated in the regulation of protein synthesis and cell growth (Dann et al. 2007); the serum- and glucocorticoid-induced protein kinase isoforms (SGK) that play important roles in regulating ion transport, hormone release, neuroexcitability, cell proliferation and apoptosis (Lang et al. 2006); the p90 ribosomal S6 kinases (RSK) which control survival, proliferation, growth and motility (Hauge and Frodin 2006; Anjum and Blenis 2008); as well as several protein kinase C (PKC) isoforms (Newton 2003, 2010). PDK1, which itself belongs to the AGC family, is precisely the common upstream kinase phosphorylating and activating all these agonist-stimulated AGC kinases (Mora et al. 2004).

2 Mechanism of Activation of the AGC Kinases

AGC kinase family members share structural similarity and a common mechanism of activation relying on the dual phosphorylation of two residues that are each located in two highly conserved motifs: the T-loop or activation loop present in the catalytic domain of the majority of protein kinases and the hydrophobic motif, a structural signature of most AGC kinases that is positioned C-terminal to the kinase domain (Pearce et al. 2010). Phosphorylation of both residues is required for maximal catalytic activity. Some AGC kinases also contain a third phosphorylation site termed the *turn motif* or *zipper site* (Z/TM), which promotes their integrity both by stabilizing the active confirmation of the enzyme and by protecting the hydrophobic motif from dephosphorylation (Hauge et al. 2007). As mentioned before, PDK1 phosphorylates all these 23 agonist-stimulated AGC kinases at the serine/threonine residues in their T-loop (Mora et al. 2004). By contrast, the regulation of the hydrophobic motif phosphorylation is quite distinct among the different PDK1 targets. The mechanism of regulation of the turn motif phosphorylation, as well as the identity of the kinase/-s phosphorylating this motif, is just being characterized (Alessi et al. 2009).

2.1 PDK1, the Common T-Loop Kinase

PDK1 is expressed from a single-copy gene located on human chromosome 16p13.3, which produces a cytosolic protein of 556 amino acids (Alessi et al. 1997a; Stephens et al. 1998). PDK1 consists of two well-characterized functional domains, the N-terminal serine/threonine kinase domain of the AGC family and the C-terminal Pleckstrin homology (PH) domain that interacts with high affinity with both PtdIns(3,4,5)P_3 and PtdIns(3,4)P_2 as well as other phosphoinositides such as PtdIns(4,5)P_2 (Alessi et al. 1997a; Currie et al. 1999). Interestingly, the three PKB isoforms are, among all the PDK1 substrates characterized, the only ones possessing a PH domain, which in this case is located at the N-terminus of the protein and specifically interacts with PtdIns(3,4,5)P_3. The exclusive presence of a PH domain in the PKB isoforms entails quite a distinctive mechanism of activation when compared to the rest of PDK1 substrates.

After the discovery of PDK1 being the PKB T-loop kinase, many labs reasoned that the high degree of homology shared between different AGC kinase family members within the activation loop might indicate that PDK1 will perhaps also phosphorylate this site in other AGC kinases. This research area entered in the late 1990s onto a vibrating chase, which culminated in the confirmation of PDK1 being also the major T-loop kinase for S6K (Alessi et al. 1998; Pullen et al. 1998), RSK (Richards et al. 1999; Jensen et al. 1999), SGK (Park et al. 1999; Kobayashi and Cohen 1999) and many PKC isoforms (Dutil et al. 1998; Le Good et al. 1998; Chou et al. 1998). The final genetic confirmation that PDK1 was indeed the major T-loop kinase for all these PtdIns(3,4,5)P_3-regulated substrates in mammalian cells came from the finding that, in PDK1 knockout embryonic stem (ES) cells, agonist stimulation failed to activate PKB, S6K, RSK (Williams et al. 2000) and SGK1 (Collins et al. 2003), and the protein stability of a number of PKC isoforms was also compromised (Balendran et al. 2000a, b). All these studies further proved that the so-called "PDK2" hydrophobic motif kinase was distinct from PDK1.

PDK1 is ubiquitously expressed in cells and, surprisingly for an enzyme that regulates as much as 23 agonist-stimulated AGC kinases, its own catalytic activity is not stimulated by these agonists (Alessi et al. 1997a). Although several regulatory mechanisms and phosphorylation sites have been proposed to contribute to the regulation of PDK1 activity (Casamayor et al. 1999; Wick et al. 2003; Riojas et al. 2006; Yang et al. 2008), bacterially expressed PDK1 is fully active and autophosphorylated at its own Ser241 T-loop residue, perhaps explaining why it is constitutively active in mammalian cells. Much research has focussed on understanding how an enzyme that is always active is then capable of phosphorylating in an inducible manner its myriad of substrates in response to specific stimuli. Biochemical, genetic and structural data from many studies suggested that in the absence of stimuli, the catalytic activity of PDK1 is kept under control by limiting its access to substrates, and only after agonist stimulation, the different PDK1 substrates are converted into forms that can be recognized, phosphorylated and activated by PDK1. In this regard, phosphorylation of the hydrophobic motif becomes a rate-limiting step for the action of PDK1. Despite the importance of this second phosphorylation site for the understanding of the mechanism of activation of the AGC kinases, the identity of the kinase phosphorylating the hydrophobic motif remained elusive for years. Most evidence now demonstrates that at least two different complexes of mTOR function as the PDK2 kinase, as discussed in the next section.

2.2 mTORC1 and mTORC2, the Hydrophobic Motif Kinases

As PDK1 phosphorylated several growth-factor-stimulated AGC kinases at their T-loop, it was reasonable to propose that a common kinase might have also phosphorylated each of this group of AGC kinases at their hydrophobic motifs in an analogous manner. The mammalian target of rapamycin mTOR was the first characterized hydrophobic motif kinase that, in complex with Raptor (Hara et al. 2002; Kim et al. 2002) and mLST8, initially named GβL (Kim et al. 2003), was shown to phosphorylate the S6K hydrophobic motif, thereby contributing to the activation of this enzyme. Because the immunosuppressant rapamycin blocked the phosphorylation and activation of the S6K (Chung et al. 1992; Price et al. 1992) by inhibiting mTOR (Sabatini et al. 1994), while other AGC kinases such as PKB, RSK, SGK and most PKC isoforms were shown to be rapamycin insensitive, the notion of a common hydrophobic motif kinase fell out of favour.

The p90 ribosomal S6 kinase case is quite unusual, as it is a serine–threonine protein kinase containing two catalytic domains. The N-terminal kinase domain belongs to the AGC family and phosphorylates and regulates a number of cellular substrates mediating growth-factor-induced cell survival, proliferation, growth and motility. By contrast, the C-terminal catalytic domain belongs to the CAMK family and accomplishes a regulatory role. Following mitogen stimulation, ERK1/2 phosphorylates and activates the C-terminal catalytic domain of RSK. The activated C-terminal kinase domain then phosphorylates the hydrophobic motif, which in RSK is located in a linker region between the two kinase domains (Dalby et al. 1998). Phosphorylation of the hydrophobic motif is essential for PDK1-mediated activation of the N-terminal kinase domain (Frodin et al. 2000). Hence, the hydrophobic motif kinase activity for RSK relies on an autophosphorylation catalysed by the C-terminal kinase domain.

Regulation of the PKC hydrophobic motif phosphorylation encompasses diverse mechanisms among the members of this rather intricate family of AGC kinases. The conventional PKC isoforms (PKCα, β and γ) and the novel PKC isoforms (PKCδ, ε, η and θ) are primed for allosteric activation by both diacylglycerol and calcium (conventional PKCs) or diacylglycerol (novel PKCs) following phosphorylation of both the T-loop and the hydrophobic motif (Mellor and Parker 1998). For the conventional forms, phosphorylation of the T-loop occurs first by PDK1, followed by autophosphorylation of the hydrophobic motif, at least in PKC βII (Newton 2001). For the novel forms, a role of mTOR as the hydrophobic motif kinase was suggested by the observation that phosphorylation of the hydrophobic motif site in PKC δ and ε isoforms was shown to be sensitive to both rapamycin and nutrients (Parekh et al. 1999). The atypical isoforms (PKC ζ and λ) and the PKC-related kinases (PRK1 and PRK2) are not regulated by diacylglycerol or calcium. These members possess a negatively charged amino acid at the position equivalent to the hydrophobic motif phosphorylation site present in other AGC kinases (Balendran et al. 1999, 2000a) and, when expressed in cells, they are constitutively active. A distinct and major role of the turn motif phosphorylation in regulating the docking interaction with PDK1 and the activation of PRK2 has been recently demonstrated (Dettori et al. 2009). Hence, hydrophobic motif auto-phosphorylation, regulation by mTOR and presence of aspartic/glutamic acid residues mimicking the negative charge of a phosphate group in the hydrophobic motif are all mechanisms found to contribute to the activation of the different PKC isoforms.

The identity of the PKB hydrophobic motif kinase attracted for years the attention of many investigators, and several kinases were miscalled to mediate this phosphorylation (Bayascas and Alessi 2005). As mentioned before, since phosphorylation of PKB at Ser473 was not sensitive to rapamycin, mTOR was never considered a convincing candidate, until Sabatini and co-workers demonstrated in a seminal paper that mTOR, as part of a rapamycin insensitive complex with Rictor and mLST8, was indeed the PKB hydrophobic motif kinase (Sarbassov et al. 2005). This rapamycin-insensitive complex of mTOR, which also consisted of Sin1 (Frias et al. 2006; Jacinto et al. 2006; Yang et al. 2006) and Protor (Pearce et al. 2007), was termed mTORC2, whereas the original mTOR: raptor:mLST8 rapamycin-sensitive complex, which phosphorylated the hydrophobic motif site of S6K, has been thereafter termed mTORC1 (Guertin and Sabatini 2007). Ablation in mice of different mTORC2 components such as Rictor or mLST8 demonstrated that mTORC2 was, in addition, the hydrophobic motif kinase for PKCα (Sarbassov et al. 2004; Guertin et al. 2006). It was only very recently that mTORC2 was reportedly shown to phosphorylate the hydrophobic motif of the SGK isoforms (Garcia-Martinez and Alessi 2008), completing our understanding on the identity of the hydrophobic motif kinases for the PDK1-regulated kinases (Fig. 1). However, the existence of mTOR complexes others than mTORC1 and mTORC2 has been also suggested (Garcia-Martinez et al. 2009).

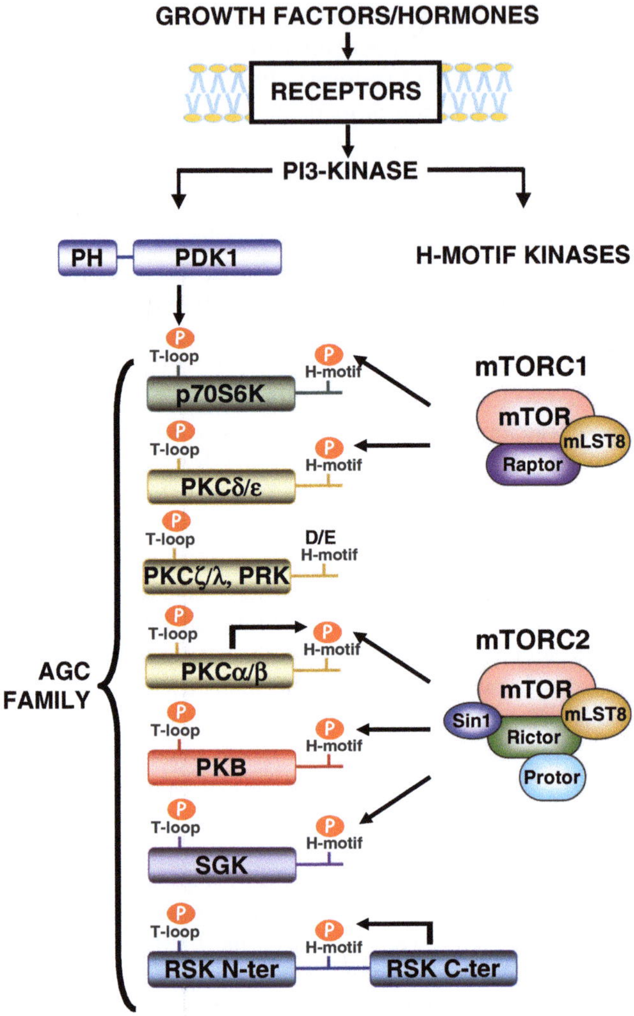

Fig. 1 Mechanism of activation of the agonist-stimulated AGC kinases by dual T-loop and hydrophobic motif phosphorylation. Following agonist stimulation and PI 3-kinase activation, the mTOR complexes 1 and 2 play a major role phosphorylating the hydrophobic motif, while PDK1 is the common kinase phosphorylating the T-loop, of all these kinases

2.3 Two Mechanisms of Regulation by PDK1

The phosphorylated hydrophobic motif is meant to play a dual role for the activation of many PDK1-regulated kinases. First, it functions as a docking site for the binding of PDK1, enabling the phosphorylation of the substrate at its T-loop by PDK1; in addition, once phosphorylated, the hydrophobic motif interacts with a groove in the catalytic domain of the same molecule, promoting in that way the transition of the corresponding AGC kinase to an active conformation. Interestingly, PDK1 itself does not possess any hydrophobic motif, but presents a hydrophobic motif binding pocket, termed the PIF pocket, which is precisely the domain in PDK1 that recognizes the docking site presented by the substrate. Moreover, interaction of the PIF pocket of PDK1 with the phosphorylated hydrophobic motif in the substrate induces the allosteric activation of PDK1 (Mora et al. 2004; Biondi 2004). The final demonstration that this was the mechanism that operates in vivo for the vast majority of PDK1 substrates came from the finding that in ES cells expressing PIF pocket mutant forms of PDK1 incapable of recognizing the phosphorylated hydrophobic motif, activation of S6K, RSK and SGK by growth factors was severely impaired, as it was the PDK1-dependent stabilization of a number of PKC isoforms (Collins et al. 2003, 2005). By contrast, activation of PKB isoforms proceeded normally, further highlighting the existence of a docking-site-independent mechanism regulating PKB isoforms activation.

Because phosphorylation of PKB by PDK1 is PtdIns(3,4,5)P_3 dependent (Alessi et al. 1997b; Stokoe et al. 1997), both PDK1 and PKB contain PtdIns(3,4,5)P_3 interacting PH domains (Alessi et al. 1997a) and agonist stimulation induces the recruitment of PKB to the plasma membrane (Andjelkovic et al. 1997), a model was proposed by which, upon growth factor stimulation and PtdIns(3,4,5)P_3 production, both PKB and PDK1 translocated to the plasma membrane via the specific interaction of their PH domains with newly generated PtdIns(3,4,5)P_3, where PDK1 could then readily phosphorylate and activate PKB (Lizcano and Alessi 2002). In vivo disruption of the interaction of the PDK1 PH domain with PtdIns(3,4,5)P_3 by knock-in mutation has been shown to severely affect the activation of PKB, without altering the intrinsic PDK1 catalytic activity, in ES cells (McManus et al. 2004). A pool of PDK1 constitutively associated with the plasma membrane has been previously observed (Currie et al. 1999) that could account for the marginal activation of PKB detected in mice expressing a mutant form of PDK1 incapable of interacting with PtdIns(3,4,5)P_3 (Bayascas et al. 2008). How PDK1 is constitutively anchored to the plasma membrane and how interaction of PtdIns(3,4,5)P_3 with the PH domain of PDK1 enhances the rate of phosphorylation and activation of PKB are aspects that still need to be clarified. Phosphorylation of PKB at Ser473 might as well positively influence phosphorylation of Thr308 by PDK1, as suggested by the fact that mTOR-specific inhibitors not only prevented Ser473 phosphorylation but also affected, although to a lesser extent, Thr308 phosphorylation (Garcia-Martinez et al. 2009). The interaction of PtdIns(3,4,5)P_3 with the PH domain of PKB not only triggers the translocation of PKB to the plasma membrane but also induces a large conformational change in the PKB protein, which is thought to create a PDK1 interacting site or to expose the T-loop phosphorylation residue (Thomas et al. 2002; Milburn et al. 2003). More recently, a comprehensive model accounting for the inactive conformer of PKB has been suggested, in which both the PH domain and the C-terminal hydrophobic motif are folded in together with the kinase domain, preventing in this way the phosphorylation of both the Thr308 by PDK1 and the Ser473 by mTORC2 (Calleja et al. 2007, 2009). Binding to PtdIns (3,4,5)P_3 would disrupt the interaction of the PKB PH domain with its own kinase domain, allowing the phosphorylation of the two regulatory residues. Moreover, in vivo analysis with fluorescent reporters provided strong evidence that PDK1 might form in cells a complex with the inactive form of its substrate PKB, and that this interaction might play a role in the recruitment of PDK1 to the plasma membrane by agonists (Calleja et al. 2007).

3 Structure of PDK1

High-resolution crystal structures of both the kinase domain (Biondi et al. 2002) and the Pleckstrin homology domain (Komander et al. 2004) of PDK1 have been reported and provided insightful information regarding the mechanism of action of this enzyme.

The catalytic domain of PDK1 assumes the classical bilobal kinase fold and is more similar to other AGC kinases crystallized, such as PKA or PKB. As in other kinases, the αC-helix is a key element in the kinase core, linking together the hydrophobic and phosphopeptide pockets as well as the serine 241 in the T-loop of PDK1 (Biondi et al. 2002). In contrast to PKBβ, in which phosphorylation of the T-loop is essential for the stabilization of both the αC-helix and the hydrophobic pocket (Yang et al. 2002a, b), T-loop phosphorylation of PDK1 is not required for the structural integrity of the αC-helix or the hydrophobic pocket, but contributes to the catalytic activity of PDK1 upon substrate binding (Komander et al. 2005).

A close examination of the small lobe of the catalytic domain of PDK1 confirmed the existence of an already postulated hydrophobic groove, named PIF pocket, involved in the interaction with the hydrophobic motif in the substrates (Biondi et al. 2000; 2001) and permitted the definition of the residues that form this domain. Among them, Leu155, the amino acid that was mutated to glutamic acid for knock-in analysis (Collins et al. 2003), was shown to be located at the core of that hydrophobic pocket. Interestingly, a second pocket occupied by a sulfate ion in the crystal was also observed next to the PIF pocket, which was hypothesized to be the phosphate-binding site for the phosphorylated residue in the hydrophobic motif. Mutation of Arg131 to Ala in this second pocket severely affected the ability of PDK1 to recognize the phosphorylated hydrophobic motif of S6K, and, as a consequence, the S6K T-loop was poorly phosphorylated (Biondi et al. 2002; Frodin et al. 2002). Accordingly, in knock-in ES cells expressing the Arg131Ala mutant form of the phosphate pocket of PDK1, agonist stimulation failed to efficiently activate S6K, RSK and SGK isoforms, whereas PKB was normally activated (Collins et al. 2005).

The crystal structure of the isolated PDK1 PH domain showed, when compared to the standard PH domain fold, an unusual N-terminal extension of unknown function. The phosphoinositide-binding pocket is a rather shallow and positively charged surface and is significantly more spacious than other PH domains, thus providing an explanation for the ability of PDK1 to efficiently bind different phosphoinositides. Furthermore, this larger ligand binding site might account for the ability of PDK1 to interact with Ins(1,3,4,5,6)P_5 and $InsP_6$ with high affinity, a mechanism that could serve to anchor PDK1 in the cytosol, thereby creating a complementary pool to the one interacting with phosphoinositides in the membrane. This second interaction might be involved in the regulation of substrates that do not translocate to the plasma membrane (Komander et al. 2004). In contrast to PKBα (Milburn et al. 2003), the PDK1 PH domain does not undergo any ligand-induced conformational change when it binds PtdIns(3,4,5)P_3, as suggested by the fact that both the non-complexed and phosphoinositide-complexed structures of the PDK1 PH domain superimposed well, and no remarkable conformational changes were detected. Mutational analysis of the different residues defining the phosphoinositide binding site revealed that Lys465, which established hydrogen bonds with the D3 and D4 phosphates, was an essential residue for phosphoinositide binding. Accordingly, mutation of this residue to glutamic acid completely abrogated binding of PDK1 to phosphoinositides as well as translocation of PDK1 to the plasma membrane in cells (Komander et al. 2004), and severely affected the activation of PKB, but not that of other AGC kinases, in mouse tissues (Bayascas et al. 2008).

4 Genetic Models and Disease

The first attempts to study the physiological roles of PDK1 in regulating the metabolic responses to insulin at the organism level came up against the early embryonic lethality described for the PDK1 knockout mice (Lawlor et al. 2002). Two PDK1 knock-in mice expressing mutant forms of PDK1 affecting either the function of the PIF pocket (PDK1 Leu155Glu) (Collins et al. 2003) or the PH domain (PDK1 Arg472,473,474Leu) (McManus et al. 2004) were also shown to be embryonic lethal, thus highlighting the essential roles that the PDK1-regulated signalling pathways play during development (McManus et al. 2004). As mentioned before, ES cells derived from all these three genetic models have been proved priceless biological tools to elucidate the physiological substrates of PDK1 and to investigate the mechanisms of action of this master regulatory kinase. To circumvent the prenatal lethality and explore the role of PDK1 in vivo in differentiated cells, tissue-specific conditional knockout and knock-in strategies were employed, which are summarized in Table 1.

4.1 PDK1 and Diabetes

Cre/Lox methodology was exploited to generate a series of tissue-specific conditional knockout mice lacking PDK1 in different insulin-responsive tissues. Mice lacking PDK1 specifically in muscle were first generated (Mora et al. 2003). These mice died from heart failure between 5 and 11 weeks of age as a consequence of the genetic deficiency affecting the cardiac muscle. PDK1-deficient cardiomyocytes

Table 1 PDK1 genetically modified mouse models

Mutation	Tissue	Viability	Other phenotypes	References
PDK1 −/−	Whole organism	Embryonic lethal E9.5	Lack of somites, forebrain and neural crest derived tissues	Lawlor et al. (2002)
PDK1 −/−	Muscle	Lethal at 5–11 weeks	Dilated cardiomyopathy	Mora et al. (2003)
PDK1 −/−	Liver	Lethal at 4–16 weeks	Defective postprandial glucose disposal and liver failure	Mora et al. (2005) and Okamoto et al. (2007)
PDK1 −/−	Pancreas	Viable	Diabetes resulting from loss of beta cell mass	Hashimoto et al. (2006)
PDK1 −/−	T-cell	Viable	Impaired T-cell differentiation	Hinton et al. (2004)
PDK1 −/−	Nervous system	Viable	Microcephaly	Chalhoub et al. (2009)
PDK1 fl/fl	Whole organism	Reduced viability	Reduced organ and body size	Lawlor et al. (2002)
			Protected from PTEN-induced tumourogenesis	Bayascas et al. (2005)
			Defective T-cell proliferative expansion	Kelly et al. (2006)
			Impaired electrolyte intestinal transport	Sandu et al. (2006)
			Impaired intestinal and renal amino acid absorption	Rexhepaj et al. (2006)
			Reduced erythrocyte cell death	Foller et al. (2008)
			Increased gastric acid secretion	Rotte et al. (2008)
			Defective phagocytosis of dendritic cells	Zaru et al. (2008)
PDK1 L155E	Whole organism	Embryonic lethal E12	Forebrain and body axis development defects	Collins et al. (2003)
PDK1 L155E	Muscle	Viable	Normal glucose homeostasis	Bayascas et al. (2006)
PDK1 L155E	T-cell	Viable	Defective T-cell proliferative expansion	Kelly et al. (2007)
PDK1 R131E	Whole organism	Embryonic lethal E19.5	Growth retardation and craniofacial defects	Collins et al. (2005)
PDK1 RRR/LLL	Whole organism	Embryonic lethal E10.5	Head blood vessel and placental defects	McManus et al. (2004)
PDK1 K465E	Whole organism	Viable	Reduced organ and body size, insulin resistance and hyperinsulinemia with normoglycemia	Bayascas et al. (2008)
			Deficient T-cell migration	Waugh et al. (2009)

obtained from these mice were shown to be more sensitive to hypoxia, which may have caused reduced cell viability and ultimately the organ failure (Mora et al. 2003). The PDK1 liver-specific knockout mice were next generated, which also died between 4 and 16 weeks of age as a result of liver failure accompanied by large-interstitial oedema. Before dying, they exhibited glucose intolerance caused by the inability of glucose to silence the expression of the gluconeogenic genes (Mora et al. 2005), a deficit that can be rescued by re-expression of liver glucokinase (Okamoto et al. 2007). The pancreas-specific PDK1 knockout mice were shown to be diabetic due to a deficit in β-cell mass than can be restored by Foxo1 haploinsufficiency (Hashimoto et al. 2006). Moreover, PDK1 PH domain knock-in mice carrying an improved version of the original triple arginine to leucine mutation, namely PDK1 Lys465Glu, were shown to be viable and exhibited some hallmarks of diabetes such us glucose intolerance, insulin resistance and hyperinsulinemia (Bayascas et al. 2008). By contrast, tissue-specific conditional knock-in mice expressing the PIF pocket PDK1 Leu155Glu mutation in muscle were shown to be of normal phenotype regarding glucose homeostasis (Bayascas et al. 2006). All together, these genetic models confirmed both the implication of PDK1 in mediating metabolic responses to insulin as well as in promoting cell viability.

4.2 PDK1 and T-Cell Development

The role of PDK1 in T-cell development has been also thoroughly investigated by using these series of tissue-specific conditional mice models. Deletion of PDK1 in the thymus completely blocked T-cell differentiation (Hinton et al. 2004), whereas reducing the expression of PDK1 by using hypomorphic alleles was shown to be permissive for T-cell differentiation, but blocked proliferative expansion of alpha/beta (α/β) and not gamma delta (γ/δ) T lymphocytes (Hinton et al. 2004; Kelly et al. 2006). Moreover, T-cell-specific conditional knock-in mice expressing the PIF pocket PDK1 Leu155Glu mutation were shown to undergo normal differentiation, but they were deficient for proliferative expansion, thus delimitating the importance of the PKB branch for the initial T-cell differentiation and that of the PIF pocket-dependent branch for the proliferative expansion (Kelly et al. 2007). More recently, a specific role of PDK1 in regulating T-cell migration has been suggested, which required binding of PDK1 to PtdIns(3,4,5)P_3, optimal PKB activation and maximal Foxo phosphorylation (Waugh et al. 2009). A role for PDK1 in modulating the pathogen-mediated activation of dendritic cells has been also demonstrated (Zaru et al. 2008).

4.3 PDK1, Growth and Cancer

Generation of PDK1 hypomorphic alleles was meant to be also a good genetic strategy to overcome the embryonic lethal periods caused by a complete PDK1 deficiency. That idea was proven right when PDK1 hypomorphic mice expressing reduced levels of PDK1 were obtained and were shown to be viable. These mice exhibited no overt phenotype other than being smaller because of a reduction in cell size rather than cell number, in spite of the fact that no detectable defects in the PDK1 downstream signalling pathways were observed (Lawlor et al. 2002). Interestingly, the PDK1 Lys465Glu knock-in mice described before were shown to be similarly smaller when compared to the PDK1 hypomorphic mice, a phenotype that might be due to a deficiency in PKB activation (Bayascas et al. 2008).

The PDK1 hypomorphic mice became a very fruitful model to explore different aspects of the physiological roles of PDK1, for example in regulating the intestinal electrolyte transport by activating the Na+/H+ exchanger (Sandu et al. 2006); reduction of PDK1 expression leads also to impairment of intestinal absorption and renal re-absorption of amino acids (Rexhepaj et al. 2006), calcium influx and erythrocyte cell death (Foller et al. 2008), as well as gastric acid secretion (Rotte et al. 2008). The epistatic relationships between PTEN and PDK1 in migration and malignant transformation of lymphocytes (Finlay et al. 2009) and in regulating nervous system development (Chalhoub et al. 2009) have been also recently addressed. PTEN, the lipid phosphatase that antagonise PI 3-kinase activity by degrading PtdIns(3,4,5)P_3, functions in cells as a tumour suppressor that is frequently mutated in human cancer. PTEN haploinsufficient mice develop a variety of tumours and have become a widely accepted mouse model prone to develop cancer resulting from elevated PtdIns(3,4,5)P_3 levels. Strikingly, when the PDK1 hypomorphic mice were crossed with the PTEN +/− mice, those PTEN heterozygous mice expressing reduced levels of PDK1 were markedly protected from developing a wide range of tumours, indicating that PDK1 is a key effector in mediating tumorigenesis resulting from loss of PTEN and further validating PDK1 as a valuable anticancer target (Bayascas et al. 2005), as discussed next.

5 PDK1 as a Druggable Target

PDK1 governs, in physiological conditions, a critical signalling node whose deregulation has dramatic consequences in pathologies such us diabetes and cancer. Impairment of the insulin responses leads to glucose intolerance and insulin resistance that normally precede the onset of type II diabetes (Biddinger and Kahn 2006). By contrast, hyperactivation of the same signalling pathway by extracellular growth factors leads to deregulation of growth, survival, proliferation, apoptosis and ultimately to transformation (Vanhaesebroeck et al. 2001). Thus, it seems reasonable to propose that selective activation of PDK1 might be a good strategy for the treatment of diabetes. Likewise, developing specific PDK1 inhibitors appears to be a good approach to treat cancer. It should be taken into account that positive modulators of PDK1 for the treatment of diabetes might be detrimental for cancer and vice versa. Also, the approach to develop therapeutically beneficial compounds targeting PDK1 is completely different in cancer or diabetes research.

Deregulation of the PI 3-kinase pathway is common in human cancer, and many human tumours of quite different origin possess elevated levels of PtdIns(3,4,5)P_3, which are meant to arise from mutations in different apical elements of this signalling pathway, encompassing tyrosine kinase receptors, PI 3-kinase or PTEN (Li et al. 1997; Vanhaesebroeck et al. 2001; Cantley 2002; Yuan and Cantley 2008). Accordingly, mice model in which one of the two copies of the PTEN tumour suppressor gene was disrupted developed a variety of tumours with high incidence and is nowadays considered as one of the best models to study PtdIns(3,4,5)P_3-driven tumorigenesis (Suzuki et al. 1998; Di et al. 1998; Podsypanina et al. 1999). Although PDK1 was shown to be itself oncogenic when expressed in mammary epithelial cells (Zeng et al. 2002; Xie et al. 2003), there are, however, not many reports showing PDK1 alterations in human disease. For example, increased PDK1 expression has been reported in invasive breast cancers, suggesting its importance in the metastatic process (Xie et al. 2006). Overexpression of PDK1 has been reported in 45% of patients with acute myeloid leukaemia, which is closely associated with hyperphosphorylation of PKC isoforms (Pearn et al. 2007). A role of PDK1 in ovarian carcinoma progression has been also proposed (Ahmed et al. 2008). PDK1 overexpression and amplification of the PDK1 gene are common occurrences in breast cancer harbouring upstream lesions on the PI 3-kinase pathway, thus potentiating the oncogenic effect of having elevated PtdIns (3,4,5)P_3 levels (Maurer et al. 2009; Vasudevan et al. 2009). The finding that hypomorphic alleles expressing reduced levels of the PDK1 protein greatly rescued the PTEN heterozygous mice suffering from cancer was considered as the first in vivo evidence of PDK1 mediating neoplasia and convincingly validated PDK1 as a promising anticancer target (Bayascas et al. 2005). The PDK1 hypomorphic mice express 10–20% of PDK1 protein when compared to control littermates, a situation that might be equivalent to the administration of a drug which would reduce the endogenous PDK1 activity by 80–90%. In contrast to the PDK1 PH domain knock-in mice, which suffered from significant insulin resistance presumably due to defective PKB activation, the PDK1 hypomorphic mice did not exhibit either signalling lesions or deleterious phenotypes, which strongly suggests that undesirable side effects of such kind of inhibitory compound would not be expected. Moreover, a role of the PDK1 signalling pathway in mediating resistance of breast cancer cells to tamoxifen has been recently reported, which suggests that PDK1 inhibitors are likely to have additional utility in sensitizing breast tumours to this broadly used anticancer drug (Iorns et al. 2009). In recent years, a number of ATP-competitive PDK1 inhibitors have been suggested to be therapeutically beneficial in inhibiting cancer progression. While more work is needed to design improved, more effective and selective PDK1 inhibitors, validation of these compounds as drugs that could be employed in clinical trials to treat cancers becomes an attractive prospect that might not be a long way off (Peifer and Alessi 2008).

Although most of the PDK1 drug discovery activity has been focussed on developing inhibitors targeting the PDK1 ATP-binding site, biochemical and structural data on the mechanism of action of PDK1 indicate that both the PH domain and the PIF pocket might as well be exploited as druggable targets. While genetic inhibition of the PDK1 PH domain–phosphoinositide interaction resulted in mice prone to diabetes (Bayascas et al. 2008), the potential benefit that the subsequent deficient PKB activation observed in these mice might have in protecting them from PtdIns(3,4,5)P_3-driven tumorigenesis needs to be evaluated. Unfortunately, there is not a single compound inhibiting the binding of the PDK1 PH domain to PtdIns(3,4,5)P_3 described so far. Various lipid-based phosphatidylinositol analogues have been reported to inhibit PKB activation by targeting its PH domain. Although some of them are currently in clinical trials, their specificity against other PH-domain-containing proteins has been questioned (Lindsley et al. 2007).

By contrast, two low-molecular-weight compounds targeting the PDK1 PIF pocket, which were postulated to have the potential of allosterically modulate the activity of this kinase (Biondi et al. 2000), have been discovered (Engel et al. 2006) and further developed (Stroba et al. 2009). These compounds have the ability to specifically activate PDK1 by mimicking the conformational transition which normally occurs upon interaction of PDK1 with the phosphorylated hydrophobic motif of substrates (Hindie et al. 2009), without directly affecting the intrinsic catalytic activity of other related AGC kinases. However, activation of PDK1 substrates that require hydrophobic motif phosphorylation for docking, such as S6K or SGK, was severely impaired by this molecules, further demonstrating that the inhibitor and the substrate docking site both compete for the same PIF pocket binding domain in PDK1 (Engel et al. 2006). Equivalent compounds specifically targeting AGC kinases others than PDK1 are predicted to inhibit their intrinsic activity by interfering with the intramolecular interaction between their own PIF pocket and hydrophobic motifs, which is needed to achieve the active conformation. Persistent activation of S6K by nutrients activates negative feedback loops, which ultimately lead to the attenuation of the insulin signal and could contribute to the diet-induced insulin resistance. Accordingly, S6K1-deficient mice were shown to be protected from age- and diet-induced obesity (Um et al. 2004). Therefore, developing these allosteric modulators into drugs that could enhance PDK1 activity, preventing at the same time S6K activation, would potentially be of great interest for the treatment of diabetes.

6 Concluding Remarks

Since it was first identified as the PKB/akt kinase, PDK1 has emerged as a major transducer of PI 3-kinase actions, by regulating a number of AGC kinase family members controlling cellular responses to many extracellular signals. PDK1 is quite unique in many aspects. First, it has represented the fist example of a master regulatory kinase activating as much as 23 different downstream kinases. Second, because PDK1 is itself constitutively active, the signalling strategy employed by PDK1 to specifically activate different substrates in response to agonists is greatly dependent on docking site interactions and conformational transitions. Third, PDK1 is one of the few kinases represented in higher eukaryotic genomes as a single-copy gene, which makes it an amenable subject for genetic analysis and has certainly facilitated the generation of a fruitful collection of transgenic mice. Finally, the pivotal role that PDK1 plays in controlling such an important signal transduction node makes it a prominent candidate for the rational design of inhibitors that could be beneficial to treat human diseases such as diabetes and cancer.

Acknowledgements I am grateful to all the scientists and supporting staff members of the MRC Protein Phosphorylation Unit at the University of Dundee in Scotland, and especially to Professor Dario Alessi, for their help, advice, and all the knowledge and training I received during the 5 years I spent in such a world leading centre. I also thank the Spanish Government Ministerio de Educación y Ciencia (Ramon y Cajal Programme-2006) and Ministerio de Sanidad y Consumo (Project FIS PI070701) for current financial support.

References

Ahmed N, Riley C, Quinn MA (2008) An immunohistochemical perspective of PPAR beta and one of its putative targets PDK1 in normal ovaries, benign and malignant ovarian tumours. Br J Cancer 98(8):1415–1424

Alessi DR, Deak M, Casamayor A, Caudwell FB, Morrice N, Norman DG, Gaffney P, Reese CB, MacDougall CN, Harbison D, Ashworth A, Bownes M (1997a) 3-Phosphoinositide-dependent protein kinase-1 (PDK1): structural and functional homology with the Drosophila DSTPK61 kinase. Curr Biol 7(10):776–789

Alessi DR, James SR, Downes CP, Holmes AB, Gaffney PR, Reese CB, Cohen P (1997b) Characterization of a 3-phosphoinositide-dependent protein kinase which phosphorylates and activates protein kinase Balpha. Curr Biol 7(4):261–269

Alessi DR, Kozlowski MT, Weng QP, Morrice N, Avruch J (1998) 3-Phosphoinositide-dependent protein kinase 1 (PDK1) phosphorylates and activates the p70 S6 kinase in vivo and in vitro. Curr Biol 8(2):69–81

Alessi DR, Pearce LR, Garcia-Martinez JM (2009) New insights into mTOR signaling: mTORC2 and beyond. Sci Signal 2(67):e27

Andjelkovic M, Alessi DR, Meier R, Fernandez A, Lamb NJ, Frech M, Cron P, Cohen P, Lucocq JM, Hemmings BA (1997) Role of translocation in the activation and function of protein kinase B. J Biol Chem 272(50):31515–31524

Anjum R, Blenis J (2008) The RSK family of kinases: emerging roles in cellular signalling. Nat Rev Mol Cell Biol 9(10):747–758

Balendran A, Casamayor A, Deak M, Paterson A, Gaffney P, Currie R, Downes CP, and Alessi DR (1999) PDK1 acquires PDK2 activity in the presence of a synthetic peptide derived from the carboxyl terminus of PRK2. Curr Biol 9(8):393–404

Balendran A, Biondi RM, Cheung PC, Casamayor A, Deak M, and Alessi DR (2000a) A 3-phosphoinositide-dependent protein kinase-1 (PDK1) docking site is required for the phosphorylation of protein kinase Czeta (PKCzeta) and PKC-related kinase 2 by PDK1. J Biol Chem 275(27):20806–20813

Balendran A, Hare GR, Kieloch A, Williams MR, and Alessi DR (2000b) Further evidence that 3-phosphoinositide-dependent protein kinase-1 (PDK1) is required for the stability and phosphorylation of protein kinase C (PKC) isoforms. FEBS Lett 484(3):217–223

Bayascas JR, Alessi DR (2005) Regulation of Akt/PKB Ser473 phosphorylation. Mol Cell 18(2):143–145

Bayascas JR, Leslie NR, Parsons R, Fleming S, Alessi DR (2005) Hypomorphic mutation of PDK1 suppresses tumorigenesis in PTEN(+/−) mice. Curr Biol 15(20):1839–1846

Bayascas JR, Sakamoto K, Armit L, Arthur JS, Alessi DR (2006) Evaluation of approaches to generation of tissue-specific knock-in mice. J Biol Chem 281(39):28772–28781

Bayascas JR, Wullschleger S, Sakamoto K, Garcia-Martinez JM, Clacher C, Komander D, Van Aalten DM, Boini KM, Lang F, Lipina C, Logie L, Sutherland C, Chudek JA, van Diepen JA, Voshol PJ, Lucocq JM, Alessi DR (2008) Mutation of the PDK1 PH domain inhibits protein kinase B/Akt, leading to small size and insulin resistance. Mol Cell Biol 28(10):3258–3272

Biddinger SB, Kahn CR (2006) From mice to men: insights into the insulin resistance syndromes. Annu Rev Physiol 68:123–158

Biondi RM (2004) Phosphoinositide-dependent protein kinase 1, a sensor of protein conformation. Trends Biochem Sci 29(3):136–142

Biondi RM, Cheung PC, Casamayor A, Deak M, Currie RA, Alessi DR (2000) Identification of a pocket in the PDK1 kinase domain that interacts with PIF and the C-terminal residues of PKA. EMBO J 19(5):979–988

Biondi RM, Kieloch A, Currie RA, Deak M, and Alessi DR (2001) The PIF-binding pocket in PDK1 is essential for activation of S6K and SGK, but not PKB. EMBO J 20(16):4380–4390.

Biondi RM, Komander D, Thomas CC, Lizcano JM, Deak M, Alessi DR, Van Aalten DM (2002) High resolution crystal structure of the human PDK1 catalytic domain defines the regulatory phosphopeptide docking site. EMBO J 21(16):4219–4228

Calleja V, Alcor D, Laguerre M, Park J, Vojnovic B, Hemmings BA, Downward J, Parker PJ, Larijani B (2007) Intramolecular and intermolecular interactions of protein kinase B define its activation in vivo. PLoS Biol 5(4):e95

Calleja V, Laguerre M, Parker PJ, Larijani B (2009) Role of a novel PH-kinase domain interface in PKB/Akt regulation: structural mechanism for allosteric inhibition. PLoS Biol 7(1):e17

Cantley LC (2002) The phosphoinositide 3-kinase pathway. Science 296(5573):1655–1657

Casamayor A, Morrice NA, Alessi DR (1999) Phosphorylation of Ser-241 is essential for the activity of 3-phosphoinositide-dependent protein kinase-1: identification of five sites of phosphorylation in vivo. Biochem J 342(Pt 2):287–292

Chalhoub N, Zhu G, Zhu X, Baker SJ (2009) Cell type specificity of PI3K signaling in Pdk1- and Pten-deficient brains. Genes Dev 23(14):1619–1624

Chou MM, Hou W, Johnson J, Graham LK, Lee MH, Chen CS, Newton AC, Schaffhausen BS, Toker A (1998) Regulation of protein kinase C zeta by PI 3-kinase and PDK-1. Curr Biol 8 (19):1069–1077

Chung J, Kuo CJ, Crabtree GR, Blenis J (1992) Rapamycin-FKBP specifically blocks growth-dependent activation of and signaling by the 70 kd S6 protein kinases. Cell 69(7):1227–1236

Collins BJ, Deak M, Arthur JS, Armit LJ, Alessi DR (2003) In vivo role of the PIF-binding docking site of PDK1 defined by knock-in mutation. EMBO J 22(16):4202–4211

Collins BJ, Deak M, Murray-Tait V, Storey KG, Alessi DR (2005) In vivo role of the phosphate groove of PDK1 defined by knockin mutation. J Cell Sci 118(Pt 21):5023–5034

Currie RA, Walker KS, Gray A, Deak M, Casamayor A, Downes CP, Cohen P, Alessi DR, Lucocq J (1999) Role of phosphatidylinositol 3, 4, 5-trisphosphate in regulating the activity and localization of 3-phosphoinositide-dependent protein kinase-1. Biochem J 337(Pt 3): 575–583

Dalby KN, Morrice N, Caudwell FB, Avruch J, Cohen P (1998) Identification of regulatory phosphorylation sites in mitogen-activated protein kinase (MAPK)-activated protein kinase-1a/p90rsk that are inducible by MAPK. J Biol Chem 273(3):1496–1505

Dann SG, Selvaraj A, Thomas G (2007) mTOR Complex1-S6K1 signaling: at the crossroads of obesity, diabetes and cancer. Trends Mol Med 13(6):252–259

Dettori R, Sonzogni S, Meyer L, Lopez-Garcia LA, Morrice NA, Zeuzem S, Engel M, Piiper A, Neimanis S, Frodin M, Biondi RM (2009) Regulation of the interaction between protein kinase C-related protein kinase 2 (PRK2) and its upstream kinase, 3-phosphoinositide-dependent protein kinase 1 (PDK1). J Biol Chem 284(44):30318–30327

Di CA, Pesce B, Cordon-Cardo C, Pandolfi PP (1998) Pten is essential for embryonic development and tumour suppression. Nat Genet 19(4):348–355

Dummler B, Hemmings BA (2007) Physiological roles of PKB/Akt isoforms in development and disease. Biochem Soc Trans 35(Pt 2):231–235

Dutil EM, Toker A, Newton AC (1998) Regulation of conventional protein kinase C isozymes by phosphoinositide-dependent kinase 1 (PDK-1). Curr Biol 8(25):1366–1375

Engel M, Hindie V, Lopez-Garcia LA, Stroba A, Schaeffer F, Adrian I, Imig J, Idrissova L, Nastainczyk W, Zeuzem S, Alzari PM, Hartmann RW, Piiper A, Biondi RM (2006) Allosteric activation of the protein kinase PDK1 with low molecular weight compounds. EMBO J 25 (23):5469–5480

Finlay DK, Sinclair LV, Feijoo C, Waugh CM, Hagenbeek TJ, Spits H, Cantrell DA (2009) Phosphoinositide-dependent kinase 1 controls migration and malignant transformation but not cell growth and proliferation in PTEN-null lymphocytes. J Exp Med 206(11): 2441–2454

Foller M, Mahmud H, Koka S, Lang F (2008) Reduced Ca2+ entry and suicidal death of erythrocytes in PDK1 hypomorphic mice. Pflugers Arch 455(5):939–949

Frias MA, Thoreen CC, Jaffe JD, Schroder W, Sculley T, Carr SA, Sabatini DM (2006) mSin1 is necessary for Akt/PKB phosphorylation, and its isoforms define three distinct mTORC2s. Curr Biol 16(18):1865–1870

Frodin M, Jensen CJ, Merienne K, Gammeltoft S (2000) A phosphoserine-regulated docking site in the protein kinase RSK2 that recruits and activates PDK1. EMBO J 19(12):2924–2934

Frodin M, Antal TL, Dummler BA, Jensen CJ, Deak M, Gammeltoft S, Biondi RM (2002) A phosphoserine/threonine-binding pocket in AGC kinases and PDK1 mediates activation by hydrophobic motif phosphorylation. EMBO J 21(20):5396–5407

Garcia-Martinez JM, Alessi DR (2008) mTOR complex 2 (mTORC2) controls hydrophobic motif phosphorylation and activation of serum- and glucocorticoid-induced protein kinase 1 (SGK1). Biochem J 416(3):375–385

Garcia-Martinez JM, Moran J, Clarke RG, Gray A, Cosulich SC, Chresta CM, Alessi DR (2009) Ku-0063794 is a specific inhibitor of the mammalian target of rapamycin (mTOR). Biochem J 421(1):29–42

Guertin DA, Sabatini DM (2007) Defining the role of mTOR in cancer. Cancer Cell 12(1):9–22

Guertin DA, Stevens DM, Thoreen CC, Burds AA, Kalaany NY, Moffat J, Brown M, Fitzgerald KJ, Sabatini DM (2006) Ablation in mice of the mTORC components raptor, rictor, or mLST8 reveals that mTORC2 is required for signaling to Akt-FOXO and PKCalpha, but not S6K1. Dev Cell 11(6):859–871

Hara K, Maruki Y, Long X, Yoshino K, Oshiro N, Hidayat S, Tokunaga C, Avruch J, Yonezawa K (2002) Raptor, a binding partner of target of rapamycin (TOR), mediates TOR action. Cell 110 (2):177–189

Hashimoto N, Kido Y, Uchida T, Asahara S, Shigeyama Y, Matsuda T, Takeda A, Tsuchihashi D, Nishizawa A, Ogawa W, Fujimoto Y, Okamura H, Arden KC, Herrera PL, Noda T, Kasuga M (2006) Ablation of PDK1 in pancreatic beta cells induces diabetes as a result of loss of beta cell mass. Nat Genet 38(5):589–593

Hauge C, Frodin M (2006) RSK and MSK in MAP kinase signalling. J Cell Sci 119(Pt 15): 3021–3023

Hauge C, Antal TL, Hirschberg D, Doehn U, Thorup K, Idrissova L, Hansen K, Jensen ON, Jorgensen TJ, Biondi RM, Frodin M (2007) Mechanism for activation of the growth factor-activated AGC kinases by turn motif phosphorylation. EMBO J 26(9):2251–2261

Hindie V, Stroba A, Zhang H, Lopez-Garcia LA, Idrissova L, Zeuzem S, Hirschberg D, Schaeffer F, Jorgensen TJ, Engel M, Alzari PM, Biondi RM (2009) Structure and allosteric effects of low-molecular-weight activators on the protein kinase PDK1. Nat Chem Biol 5(10):758–764

Hinton HJ, Alessi DR, Cantrell DA (2004) The serine kinase phosphoinositide-dependent kinase 1 (PDK1) regulates T cell development. Nat Immunol 5(5):539–545

Iorns E, Lord CJ, Ashworth A (2009) Parallel RNAi and compound screens identify the PDK1 pathway as a target for tamoxifen sensitization. Biochem J 417(1):361–370

Jacinto E, Facchinetti V, Liu D, Soto N, Wei S, Jung SY, Huang Q, Qin J, Su B (2006) SIN1/MIP1 maintains rictor-mTOR complex integrity and regulates Akt phosphorylation and substrate specificity. Cell 127(1):125–137

Jensen CJ, Buch MB, Krag TO, Hemmings BA, Gammeltoft S, Frodin M (1999) 90-kDa ribosomal S6 kinase is phosphorylated and activated by 3-phosphoinositide-dependent protein kinase-1. J Biol Chem 274(38):27168–27176

Kelly AP, Hinton HJ, Clarke RG, Cantrell DA (2006) Phosphoinositide-dependent kinase 1 (PDK1) haplo-insufficiency inhibits production of alpha/beta (alpha/beta) but not gamma delta (gamma/delta) T lymphocytes. FEBS Lett 580(8):2135–2140

Kelly AP, Finlay DK, Hinton HJ, Clarke RG, Fiorini E, Radtke F, Cantrell DA (2007) Notch-induced T cell development requires phosphoinositide-dependent kinase 1. EMBO J 26 (14):3441–3450

Kim DH, Sarbassov DD, Ali SM, King JE, Latek RR, Erdjument-Bromage H, Tempst P, Sabatini DM (2002) mTOR interacts with raptor to form a nutrient-sensitive complex that signals to the cell growth machinery. Cell 110(2):163–175

Kim DH, Sarbassov DD, Ali SM, Latek RR, Guntur KV, Erdjument-Bromage H, Tempst P, Sabatini DM (2003) GbetaL, a positive regulator of the rapamycin-sensitive pathway required for the nutrient-sensitive interaction between raptor and mTOR. Mol Cell 11(4):895–904

Kobayashi T, Cohen P (1999) Activation of serum- and glucocorticoid-regulated protein kinase by agonists that activate phosphatidylinositide 3-kinase is mediated by 3-phosphoinositide-dependent protein kinase-1 (PDK1) and PDK2. Biochem J 339(Pt 2):319–328

Komander D, Fairservice A, Deak M, Kular GS, Prescott AR, Peter DC, Safrany ST, Alessi DR, van Aalten DM (2004) Structural insights into the regulation of PDK1 by phosphoinositides and inositol phosphates. EMBO J 23(20):3918–3928

Komander D, Kular G, Deak M, Alessi DR, Van Aalten DM (2005) Role of T-loop phosphorylation in PDK1 activation, stability, and substrate binding. J Biol Chem 280(19):18797–18802

Lang F, Bohmer C, Palmada M, Seebohm G, Strutz-Seebohm N, Vallon V (2006) (Patho) physiological significance of the serum- and glucocorticoid-inducible kinase isoforms. Physiol Rev 86(4):1151–1178

Lawlor MA, Mora A, Ashby PR, Williams MR, Murray-Tait V, Malone L, Prescott AR, Lucocq JM, Alessi DR (2002) Essential role of PDK1 in regulating cell size and development in mice. EMBO J 21(14):3728–3738

Le Good JA, Ziegler WH, Parekh DB, Alessi DR, Cohen P, Parker PJ (1998) Protein kinase C isotypes controlled by phosphoinositide 3-kinase through the protein kinase PDK1. Science 281(5385):2042–2045

Li J, Yen C, Liaw D, Podsypanina K, Bose S, Wang SI, Puc J, Miliaresis C, Rodgers L, McCombie R, Bigner SH, Giovanella BC, Ittmann M, Tycko B, Hibshoosh H, Wigler MH, Parsons R (1997) PTEN, a putative protein tyrosine phosphatase gene mutated in human brain, breast, and prostate cancer. Science 275(5308):1943–1947

Lindsley CW, Barnett SF, Yaroschak M, Bilodeau MT, Layton ME (2007) Recent progress in the development of ATP-competitive and allosteric Akt kinase inhibitors. Curr Top Med Chem 7 (14):1349–1363

Lizcano JM, Alessi DR (2002) The insulin signalling pathway. Curr Biol 12(7):R236–R238

Manning BD, Cantley LC (2007) AKT/PKB signaling: navigating downstream. Cell 129(7):1261–1274

Maurer M, Su T, Saal LH, Koujak S, Hopkins BD, Barkley CR, Wu J, Nandula S, Dutta B, Xie Y, Chin YR, Kim DI, Ferris JS, Gruvberger-Saal SK, Laakso M, Wang X, Memeo L, Rojtman A, Matos T, Yu JS, Cordon-Cardo C, Isola J, Terry MB, Toker A, Mills GB, Zhao JJ, Murty VV, Hibshoosh H, Parsons R (2009) 3-Phosphoinositide-dependent kinase 1 potentiates upstream lesions on the phosphatidylinositol 3-kinase pathway in breast carcinoma. Cancer Res 69(15):6299–6306

McManus EJ, Collins BJ, Ashby PR, Prescott AR, Murray-Tait V, Armit LJ, Arthur JS, Alessi DR (2004) The in vivo role of PtdIns(3, 4, 5)P3 binding to PDK1 PH domain defined by knockin mutation. EMBO J 23(10):2071–2082

Mellor H, Parker PJ (1998) The extended protein kinase C superfamily. Biochem J 332(Pt 2): 281–292

Milburn CC, Deak M, Kelly SM, Price NC, Alessi DR, Van Aalten DM (2003) Binding of phosphatidylinositol 3, 4, 5-trisphosphate to the pleckstrin homology domain of protein kinase B induces a conformational change. Biochem J 375(Pt 3):531–538

Mora A, Davies AM, Bertrand L, Sharif I, Budas GR, Jovanovic S, Mouton V, Kahn CR, Lucocq JM, Gray GA, Jovanovic A, Alessi DR (2003) Deficiency of PDK1 in cardiac muscle results in heart failure and increased sensitivity to hypoxia. EMBO J 22(18):4666–4676

Mora A, Komander D, Van Aalten DM, Alessi DR (2004) PDK1, the master regulator of AGC kinase signal transduction. Semin Cell Dev Biol 15(2):161–170

Mora A, Lipina C, Tronche F, Sutherland C, Alessi DR (2005) Deficiency of PDK1 in liver results in glucose intolerance, impairment of insulin-regulated gene expression and liver failure. Biochem J 385(Pt 3):639–648

Newton AC (2001) Protein kinase C: structural and spatial regulation by phosphorylation, cofactors, and macromolecular interactions. Chem Rev 101(8):2353–2364

Newton AC (2003) Regulation of the ABC kinases by phosphorylation: protein kinase C as a paradigm. Biochem J 370(Pt 2):361–371

Newton AC (2010) Protein kinase C: poised to signal. Am J Physiol Endocrinol Metab 298(3): E395–E402

Okamoto Y, Ogawa W, Nishizawa A, Inoue H, Teshigawara K, Kinoshita S, Matsuki Y, Watanabe E, Hiramatsu R, Sakaue H, Noda T, Kasuga M (2007) Restoration of glucokinase expression in the liver normalizes postprandial glucose disposal in mice with hepatic deficiency of PDK1. Diabetes 56(4):1000–1009

Parekh D, Ziegler W, Yonezawa K, Hara K, Parker PJ (1999) Mammalian TOR controls one of two kinase pathways acting upon nPKCdelta and nPKCepsilon. J Biol Chem 274(49):34758–34764

Park J, Leong ML, Buse P, Maiyar AC, Firestone GL, Hemmings BA (1999) Serum and glucocorticoid-inducible kinase (SGK) is a target of the PI 3-kinase-stimulated signaling pathway. EMBO J 18(11):3024–3033

Pearce LR, Huang X, Boudeau J, Pawlowski R, Wullschleger S, Deak M, Ibrahim AF, Gourlay R, Magnuson MA, Alessi DR (2007) Identification of Protor as a novel Rictor-binding component of mTOR complex-2. Biochem J 405(3):513–522

Pearce LR, Komander D, Alessi DR (2010) The nuts and bolts of AGC protein kinases. Nat Rev Mol Cell Biol 11(1):9–22

Pearn L, Fisher J, Burnett AK, Darley RL (2007) The role of PKC and PDK1 in monocyte lineage specification by Ras. Blood 109(10):4461–4469

Peifer C, Alessi DR (2008) Small-molecule inhibitors of PDK1. ChemMedChem 3(12):1810–1838

Podsypanina K, Ellenson LH, Nemes A, Gu J, Tamura M, Yamada KM, Cordon-Cardo C, Catoretti G, Fisher PE, Parsons R (1999) Mutation of Pten/Mmac1 in mice causes neoplasia in multiple organ systems. Proc Natl Acad Sci U S A 96(4):1563–1568

Price DJ, Grove JR, Calvo V, Avruch J, Bierer BE (1992) Rapamycin-induced inhibition of the 70-kilodalton S6 protein kinase. Science 257(5072):973–977

Pullen N, Dennis PB, Andjelkovic M, Dufner A, Kozma SC, Hemmings BA, Thomas G (1998) Phosphorylation and activation of p70s6k by PDK1. Science 279(5351):707–710

Rexhepaj R, Grahammer F, Volkl H, Remy C, Wagner CA, Sandulache D, Artunc F, Henke G, Nammi S, Capasso G, Alessi DR, Lang F (2006) Reduced intestinal and renal amino acid transport in PDK1 hypomorphic mice. FASEB J 20(13):2214–2222

Richards SA, Fu J, Romanelli A, Shimamura A, Blenis J (1999) Ribosomal S6 kinase 1 (RSK1) activation requires signals dependent on and independent of the MAP kinase ERK. Curr Biol 9 (15):810–820

Riojas RA, Kikani CK, Wang C, Mao X, Zhou L, Langlais PR, Hu D, Roberts JL, Dong LQ, Liu F (2006) Fine tuning PDK1 activity by phosphorylation at Ser163. J Biol Chem 281(31): 21588–21593

Rotte A, Bhandaru M, Ackermann TF, Boini KM, Lang F (2008) Role of PDK1 in regulation of gastric acid secretion. Cell Physiol Biochem 22(5–6):725–734

Sabatini DM, Erdjument-Bromage H, Lui M, Tempst P, Snyder SH (1994) RAFT1: a mammalian protein that binds to FKBP12 in a rapamycin-dependent fashion and is homologous to yeast TORs. Cell 78(1):35–43

Sandu C, Artunc F, Palmada M, Rexhepaj R, Grahammer F, Hussain A, Yun C, Alessi DR, Lang F (2006) Impaired intestinal NHE3 activity in the PDK1 hypomorphic mouse. Am J Physiol Gastrointest Liver Physiol 291(5):G868–G876

Sarbassov DD, Ali SM, Kim DH, Guertin DA, Latek RR, Erdjument-Bromage H, Tempst P, Sabatini DM (2004) Rictor, a novel binding partner of mTOR, defines a rapamycin-insensitive and raptor-independent pathway that regulates the cytoskeleton. Curr Biol 14(14):1296–1302

Sarbassov DD, Guertin DA, Ali SM, Sabatini DM (2005) Phosphorylation and regulation of Akt/PKB by the rictor-mTOR complex. Science 307(5712):1098–1101

Stephens L, Anderson K, Stokoe D, Erdjument-Bromage H, Painter GF, Holmes AB, Gaffney PR, Reese CB, McCormick F, Tempst P, Coadwell J, Hawkins PT (1998) Protein kinase B kinases that mediate phosphatidylinositol 3,4,5-trisphosphate-dependent activation of protein kinase B. Science 279(5351):710–714

Stokoe D, Stephens LR, Copeland T, Gaffney PR, Reese CB, Painter GF, Holmes AB, McCormick F, Hawkins PT (1997) Dual role of phosphatidylinositol-3,4,5-trisphosphate in the activation of protein kinase B. Science 277(5325):567–570

Stroba A, Schaeffer F, Hindie V, Lopez-Garcia L, Adrian I, Frohner W, Hartmann RW, Biondi RM, Engel M (2009) 3,5-Diphenylpent-2-enoic acids as allosteric activators of the protein kinase PDK1: structure-activity relationships and thermodynamic characterization of binding as paradigms for PIF-binding pocket-targeting compounds. J Med Chem 52(15):4683–4693

Suzuki A, de la Pompa JL, Stambolic V, Elia AJ, Sasaki T, del BB I, Ho A, Wakeham A, Itie A, Khoo W, Fukumoto M, Mak TW (1998) High cancer susceptibility and embryonic lethality associated with mutation of the PTEN tumor suppressor gene in mice. Curr Biol 8(21):1169–1178

Thomas CC, Deak M, Alessi DR, Van Aalten DM (2002) High-resolution structure of the pleckstrin homology domain of protein kinase b/akt bound to phosphatidylinositol (3,4,5)-trisphosphate. Curr Biol 12(14):1256–1262

Um SH, Frigerio F, Watanabe M, Picard F, Joaquin M, Sticker M, Fumagalli S, Allegrini PR, Kozma SC, Auwerx J, Thomas G (2004) Absence of S6K1 protects against age- and diet-induced obesity while enhancing insulin sensitivity. Nature 431(7005):200–205

Vanhaesebroeck B, Leevers SJ, Ahmadi K, Timms J, Katso R, Driscoll PC, Woscholski R, Parker PJ, Waterfield MD (2001) Synthesis and function of 3-phosphorylated inositol lipids. Annu Rev Biochem 70:535–602

Vasudevan KM, Barbie DA, Davies MA, Rabinovsky R, McNear CJ, Kim JJ, Hennessy BT, Tseng H, Pochanard P, Kim SY, Dunn IF, Schinzel AC, Sandy P, Hoersch S, Sheng Q, Gupta PB, Boehm JS, Reiling JH, Silver S, Lu Y, Stemke-Hale K, Dutta B, Joy C, Sahin AA, Gonzalez-Angulo AM, Lluch A, Rameh LE, Jacks T, Root DE, Lander ES, Mills GB, Hahn WC, Sellers WR, Garraway LA (2009) AKT-independent signaling downstream of oncogenic PIK3CA mutations in human cancer. Cancer Cell 16(1):21–32

Waugh C, Sinclair L, Finlay D, Bayascas JR, Cantrell D (2009) Phosphoinositide (3,4,5)-triphosphate binding to phosphoinositide-dependent kinase 1 regulates a protein kinase B/Akt signaling threshold that dictates T-cell migration, not proliferation. Mol Cell Biol 29(21):5952–5962

Whiteman EL, Cho H, Birnbaum MJ (2002) Role of Akt/protein kinase B in metabolism. Trends Endocrinol Metab 13(10):444–451

Wick MJ, Ramos FJ, Chen H, Quon MJ, Dong LQ, Liu F (2003) Mouse 3-phosphoinositide-dependent protein kinase-1 undergoes dimerization and trans-phosphorylation in the activation loop. J Biol Chem 278(44):42913–42919

Williams MR, Arthur JS, Balendran A, van der KJ, Poli V, Cohen P, Alessi DR (2000) The role of 3-phosphoinositide-dependent protein kinase 1 in activating AGC kinases defined in embryonic stem cells. Curr Biol 10(8):439–448

Xie Z, Zeng X, Waldman T, Glazer RI (2003) Transformation of mammary epithelial cells by 3-phosphoinositide- dependent protein kinase-1 activates beta-catenin and c-Myc, and down-regulates caveolin-1. Cancer Res 63(17):5370–5375

Xie Z, Yuan H, Yin Y, Zeng X, Bai R, Glazer RI (2006) 3-phosphoinositide-dependent protein kinase-1 (PDK1) promotes invasion and activation of matrix metalloproteinases. BMC Cancer 6:77

Yang J, Cron P, Good VM, Thompson V, Hemmings BA, Barford D (2002a) Crystal structure of an activated Akt/protein kinase B ternary complex with GSK3-peptide and AMP-PNP. Nat Struct Biol 9(12):940–944

Yang J, Cron P, Thompson V, Good VM, Hess D, Hemmings BA, Barford D (2002b) Molecular mechanism for the regulation of protein kinase B/Akt by hydrophobic motif phosphorylation. Mol Cell 9(6):1227–1240

Yang Q, Inoki K, Ikenoue T, Guan KL (2006) Identification of Sin1 as an essential TORC2 component required for complex formation and kinase activity. Genes Dev 20(20):2820–2832

Yang KJ, Shin S, Piao L, Shin E, Li Y, Park KA, Byun HS, Won M, Hong J, Kweon GR, Hur GM, Seok JH, Chun T, Brazil DP, Hemmings BA, Park J (2008) Regulation of 3-phosphoinositide-dependent protein kinase-1 (PDK1) by Src involves tyrosine phosphorylation of PDK1 and Src homology 2 domain binding. J Biol Chem 283(3):1480–1491

Yuan TL, Cantley LC (2008) PI3K pathway alterations in cancer: variations on a theme. Oncogene 27(41):5497–5510

Zaru R, Mollahan P, Watts C (2008) 3-phosphoinositide-dependent kinase 1 deficiency perturbs Toll-like receptor signaling events and actin cytoskeleton dynamics in dendritic cells. J Biol Chem 283(2):929–939

Zeng X, Xu H, Glazer RI (2002) Transformation of mammary epithelial cells by 3-phosphoinositide-dependent protein kinase-1 (PDK1) is associated with the induction of protein kinase Calpha. Cancer Res 62(12):3538–3543

Protein Kinase B (PKB/Akt), a Key Mediator of the PI3K Signaling Pathway

Elisabeth Fayard, Gongda Xue, Arnaud Parcellier, Lana Bozulic, and Brian A. Hemmings

Contents

Abstract Protein kinase B (PKB/Akt) is a serine/threonine protein kinase that created serious interest when it was revealed as a mediator of the PI3K pathway. It comprises three isoforms that play both unique and redundant roles. Upon binding to phosphatidylinositol-(3,4,5)-trisphosphate (PIP3) generated by PI3K, PKB is phosphorylated by PDK1 at T308. To achieve full kinase activity, PKB needs to be phosphorylated at a second key residue, S473, by members of the PI3K-related kinase family mTORC2 or DNA-PK, depending on the stimulus and the context. Besides, a number of phosphatases and interacting partners have been shown to further modulate its subcellular localization, phosphorylation, and kinase activity. This review aims at illustrating the remarkable complexity in the

E. Fayard, G. Xue, A. Parcellier, L. Bozulic, and B.A. Hemmings (✉)
Friedrich Miescher Institute for Biomedical Research, Maulbeerstrasse 66, 4058 Basel, Switzerland
e-mail: brian.hemmings@fmi.ch

C. Rommel et al. (eds.). *Phosphoinositide 3-kinase in Health and Disease*, Volume 1
Current Topics in Microbiology and Immunology 346, DOI 10.1007/82_2010_58

regulation of PKB signaling downstream of PI3K. Such regulation could be attributed to the specific roles of the PKB isoforms, their expression pattern, subcellular localization, targets, phosphorylation by upstream kinases in a stimulus- and context-dependent manner and by phosphatases, and interaction with binding partners. This allows this key kinase to fulfill physiological functions in numerous processes, including embryonic development, thymocyte development, adipocyte differentiation, glucose homeostasis, and to avoid pathological loss of control such as tumor formation.

1 Classification, Structure, and Substrates of PKB

Conserved from primitive metazoans to humans, the serine/threonine protein kinase B (PKB, also known as Akt) belongs to the AGC group of protein kinases (named after PKA, PKG, and PKC) (Hanada et al. 2004) and has emerged as a critical signaling molecule. PKB was first cloned by three independent groups (Bellacosa et al. 1991; Coffer and Woodgett 1991; Jones et al. 1991) more than a decade following the original identification of its viral homolog, the v-Akt proto-oncogene, which is expressed by a transforming retrovirus (AKT-8) isolated from a spontaneous thymic lymphoma of an AKR mouse (Staal et al. 1977; Staal 1987).

PKB comprises three mammalian isoforms: PKBα (Akt1), PKBβ (Akt2), and PKBγ (Akt3). PKBα is ubiquitously detected, whereas PKBβ is mainly expressed in insulin-sensitive tissues and PKBγ is mostly found in brain and testis. Each of these isoforms is encoded by a separate gene but shares more than 80% amino acid sequence identity and a comparable structural organization that includes three functional domains: an amino-terminal (N-terminal) pleckstrin homology (PH) domain, a central catalytic domain, and a carboxyl-terminal (C-terminal) regulatory domain containing the hydrophobic motif (FPQFSY, where F is phenylalanine, P is proline, Q is glutamine, S is serine, and Y is tyrosine) (Fayard et al. 2005; Franke 2008). This architecture is conserved among species from fly, worm, mouse, to human. Both the central catalytic domain and the hydrophobic motif are highly preserved among AGC members, including PKC, p70 S6 kinase (S6K), ribosomal p90 S6 kinase (RSK), and serum and glucocorticoid induced kinase (SGK) (Hanada et al. 2004). Phosphorylation of both T308 in the activation loop of the catalytic domain and S473 in the hydrophobic motif of the regulatory domain is required for full activation of PKB (Alessi et al. 1996a, 1997). The crystal structures of the catalytic domain in its inactive and active states were solved, and this provided an explanation of how these two phosphorylation sites contribute to enzymatic activation of the kinase (Yang et al. 2002a, b). The phosphorylation of T308 induces a catalytically active conformation, which is stabilized upon the phosphorylation of S473. This stabilization is due to intramolecular interactions between the hydrophobic motif and its acceptor structure within the catalytic domain, named the hydrophobic groove.

The key function of PKB in signaling became obvious when it was revealed as a mediator of the phosphoinositide 3-kinase (PI3K) pathway. Activated upon PI3K signaling, PKB phosphorylates a wide range of substrates influencing diverse cellular and physiological processes, including cell cycle progression, cell growth, cell differentiation, cell survival/suppression of apoptosis, metabolism, angiogenesis, and motility (Fayard et al. 2005; Franke 2008). Most of the PKB substrates contain the minimal consensus sequence RXRXX[S/T]B, where R is arginine, X is any amino acid, S/T is either a serine or a threonine residue, and B is a bulky hydrophobic residue (Alessi et al. 1996b; Cross et al. 1995). Up to now, over 50 proteins have been identified as putative PKB substrates (Manning and Cantley 2007). For example, a number of them mediate the pro-survival effect of PKB. They include the pro-apoptotic Bcl-2-antagonist of death (BAD) (Datta et al. 1997), the NF-κB regulator IκB kinase (IKK) (Ozes et al. 1999), the E3 ubiquitin ligase mouse double minute 2 (Mdm2) (Mayo and Donner 2001; Zhou et al. 2001), and the Forkhead family of transcription factors (FoxOs) (Brunet et al. 1999). Mdm2 and FoxOs are also involved in facilitating the G1/S transition of the cell cycle via the transcriptional regulation of the cyclin-dependent kinase inhibitors $p21^{Cip1}$ (Feng et al. 2004b) and $p27^{Kip1}$ (Medema et al. 2000). In addition, $p21^{Cip1}$ was reported to be directly phosphorylated and subsequently inhibited by PKB (Rossig et al. 2001). Via phosphorylation and inhibition of glycogen synthase kinase-3 (GSK3) (Cross et al. 1995), PKB increases cyclin D1 protein stability (Diehl et al. 1998), thereby promoting cell cycle progression. By inactivating GSK3, PKB also activates glycogen synthase and, consequently, glycogen synthesis. Another substrate repressed by PKB and involved in metabolism is PGC1α, a coactivator of transcription factors that affect gluconeogenesis, such as Foxo1, HNF4α, and PPARα (Li et al. 2007). The phosphorylation by PKB of TSC2 leads to the induction of cell growth via the activation of the mTORC1 complex (Inoki et al. 2002; Manning et al. 2002; Potter et al. 2002), which mediates ribosome biogenesis and translation initiation through S6K and eukaryotic initiation factor 4E (eIF4E)-binding protein 1 (4E-BP1), respectively (Wullschleger et al. 2006). By phosphorylating the endothelial nitric oxide synthase (eNOS), PKB mediates its activation, which leads to an increase of nitric oxide (NO) production (Dimmeler et al. 1999; Michell et al. 1999). This results in the regulation of angiogenesis.

2 Regulation of PKB Activity

2.1 Regulation of PKB Activity by Phosphorylation Events

As already mentioned, PKB acts as a major signal transducer downstream of the PI3K pathway, which is itself activated upon autophosphorylation of receptor tyrosine kinases induced by ligands (such as insulin or other growth factors) (Burgering and Coffer 1995; Franke et al. 1995; Kohn et al. 1995) and stimulation of G protein-coupled receptors (Lopez-Ilasaca et al. 1997). The PI3K enzymes phosphorylate as lipid kinases the 3′-hydroxyl group of the inositol ring of phosphoinositides and its derivatives. They are divided into three major classes (I, II, and III) on the basis of their molecular structure and substrate specificity (Domin and Waterfield 1997). Class I PI3Ks convert at the plasma membrane phosphatidylinositol-(4,5)-bisphosphate (PI(4,5)P2) into the second messenger phosphatidylinositol-(3,4,5)-trisphosphate (PIP3). PIP3 contains a binding site for intracellular enzymes carrying a PH domain. PIP3-mediated membrane recruitment brings PKB in close proximity to another PH domain-containing serine/threonine kinase, phosphoinositide-dependent kinase 1 (PDK1). Co-localization of the proteins leads to the phosphorylation of the catalytic domain of PKB by PDK1 at T308 – one of the residues key for PKB activation (Alessi et al. 1997). In addition to recruiting PKB at the plasma membrane, binding to PIP3 also induces a large conformational change within its PH domain that enhances the rate at which it is phosphorylated by PDK1 (Calleja et al. 2003; Milburn et al. 2003). Recently, PKB in its inactive state was shown to be pre-complexed with PDK1 in the cytosol and switched to its active state upon membrane recruitment (Calleja et al. 2007). Artificial targeting of PKB to the plasma membrane (either by way of myristoylation/palmitoylation or fusion of a gag sequence to the enzyme's N-terminus) results in its constitutive activation (Andjelkovic et al. 1997). To achieve full kinase activity, PKB needs to be phosphorylated at a second key residue located in the hydrophobic motif within the regulatory domain: S473 (Alessi et al. 1996a). Several candidate kinases were proposed to be responsible for the phosphorylation of the S473 residue (Fayard et al. 2005), whose regulation appears to be more complex than that of T308. Indeed, S473 phosphorylation was recently shown to be regulated in a stimulus- and context-dependent manner by members of the PI3K-related kinase family (PIKKs, also referred to as class IV PI3Ks).

The current consensus proposes that, under conditions of growth factor stimulation, S473 is phosphorylated supposedly at the plasma membrane by mTOR, when complexed with rapamycin-insensitive companion of mTOR (rictor), mammalian LST8/G-protein β-subunit like protein (mLST8/GβL), mammalian stress-activated protein kinase interacting protein 1 (mSin1), and protor (the whole complex is referred to as mTORC2) (Frias et al. 2006; Guertin et al. 2006; Jacinto et al. 2006; Pearce et al. 2007; Sarbassov et al. 2005). The growth-related signaling mediated by PKB seems to depend on mTORC2 in a tissue-, cell type-, and stage-specific manner. For instance, although activation of PKB does not require mTORC2 in skeletal muscle (Bentzinger et al. 2008), it is the main PKB S473 kinase in several human cancer cell lines (Sarbassov et al. 2005) and in a mouse adipose cell line (Hresko and Mueckler 2005) under growth-stimulated conditions. mTORC2 is also the dominant kinase that phosphorylates S473 in PKB during embryogenesis as observed in a whole-body rictor knockout (Shiota et al. 2006). Interestingly, loss of mTORC2 function differentially affects the phosphorylation of predicted PKB substrates. Indeed, PKB signaling to forkhead box O3 (FoxO3), but not to TSC2 or GSK3, required mTORC2 upon insulin stimulation of primary mouse fibroblasts (Guertin et al. 2006; Jacinto et al. 2006). In a situation of stress, such as following

DNA damage or in the presence of CpG DNA, DNA dependent protein kinase (DNA-PK) acts as the major PKB S473 kinase (Boehme et al. 2008; Bozulic et al. 2008; Dragoi et al. 2005; Feng et al. 2004a). DNA-PK, in which the Ku70/Ku80 complex serves as a DNA double-strand break-targeting subunit for the DNA-PK catalytic subunit (DNA-PKcs), is activated by the interaction with DNA double-strand breaks and is involved in genome/transcriptome surveillance (Rivera-Calzada et al. 2005, 2007). Upon DNA damage leading to the formation of DNA double-strand breaks, DNA-PK allowed the activation of PKB in the nucleus, which then induced cell survival likely via at least in part its effect on the transcription of p53-regulated genes, such as $p21^{Cip1}$ (Bozulic et al. 2008). This was reported to be via the inactivation of GSK3 by PKB in the nucleus and the subsequent decreased phosphorylation of Mdm2, leading to the accumulation of p53 and upregulation of $p21^{Cip1}$, which induces cell-cycle arrest and indirectly promotes cell survival (Boehme et al. 2008; Bozulic et al. 2008). Moreover, phosphorylated levels of the pro-apoptotic transcription factor FoxO4 (but not FoxO3), leading to reduced FoxO4 activity, were increased following DNA damage in a DNA-PK-dependent manner (Surucu et al. 2008). In aggregate, although PIP3-dependency for phosphorylation of PKB S473 by mTORC2 and DNA-PK still needs to be solved, the precise regulation of PKB by specific upstream S473 kinases upon different stimuli and in different subcellular locations contributes to the specificity of PKB signaling (Fig. 1). Differently activated forms of PKB may, hence, regulate their numerous targets in a specific manner, highlighting the complexity of the PKB pathway.

Opposing the action of T308 and S473 kinases, several phosphatases have been identified that negatively regulate PKB activity. For instance, protein phosphatase 2A (PP2A) directly does so by dephosphorylating both T308 and S473 (Andjelkovic et al. 1996; Ugi et al. 2004), while PH domain leucine-rich repeat protein phosphatases, PHLPP1 and PHLPP2, specifically dephosphorylate S473 (Brognard et al. 2007; Gao et al. 2005). Of note, each PHLPP inactivates a given PKB isoform, highlighting a mechanism that selectively terminates PKB downstream pathways. Indeed, PHLPP1 specifically modulates the phosphorylation of Mdm2 and GSK3 by PKBβ, whereas PHLPP2 affects PKBγ-mediated phosphorylation of $p27^{Kip1}$ (importantly, $p27^{Kip1}$ is a substrate of PKB in human but not in mouse) (Brognard et al. 2007). The tumor suppressor phosphatase and tensin homology deleted on chromosome ten (PTEN) (Stambolic et al. 1998) and the SH2 domain-containing inositol polyphosphate 5-phosphatase (SHIP) (Huber et al. 1999) indirectly inhibit PKB activity by converting PIP3 to PI(4,5)P2, and PI(3,4)P2, respectively.

2.2 *Regulation of PKB Activity by Binding Partners*

Besides continuous interest in identifying specific upstream kinases and phosphatases that regulate PKB activity, a number of proteins binding to PKB have been shown to further modulate their subcellular localization, phosphorylation, and

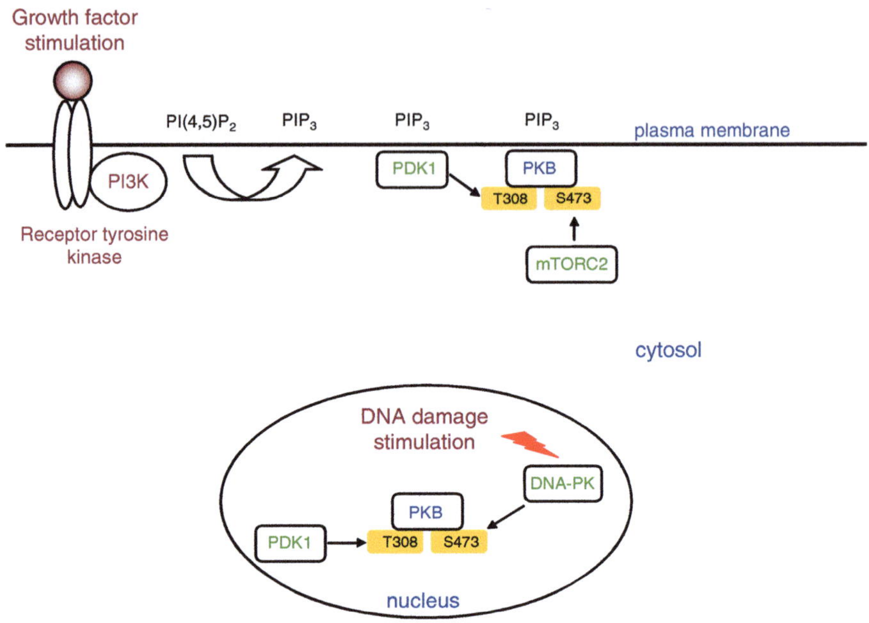

Fig. 1 Regulation of PKB activity upon specific stimuli. Upon growth factor stimulation, receptor tyrosine kinases are activated, which results in PI3K activation. Active PI3K converts at the plasma membrane PI(4,5)P2 into the second messenger PIP3. PIP3-mediated membrane recruitment brings PKB in close proximity to PDK1. Co-localization of the proteins leads to the phosphorylation of T308 on PKB. Full activation is achieved when S473 is phosphorylated by mTORC2. The mechanism by which growth factor stimulation activates mTORC2 is, however, unknown. Upon DNA damage stimulation (γ-irradiation, doxorubicin), the generation of DNA double-strand breaks activates DNA-PK, leading to the phosphorylation of S473 on PKB. We propose that PDK1 phosphorylates T308 in the nucleus

kinase activity (Table 1), thereby illustrating a remarkable complexity in PKB regulation. Based on their regulatory effects, these binding partners are mentioned in three parts.

2.2.1 PKB Activators

α-actinin 4 (ACTN4), an actin-binding protein, has been revealed as a binding partner of PKBα in a retrovirus-based protein complementation assay screen (Ding et al. 2006). Silencing ACTN4 prevented membrane translocation and activation of PKB, which resulted in the inhibition of cell proliferation via enhanced expression of $p27^{Kip1}$. Similarly, the adaptor protein containing PH domain, phosphotyrosine-binding domain, and leucine zipper motif (APPL1), an adaptor protein that scaffolds inactive PKBβ and p110α in the cytoplasm, facilitated membrane translocation and activation of PKBβ upon mitogenic stimulation (Mitsuuchi et al. 1999). The proto-oncogene TCL1 (T-cell leukemia/lymphoma 1), instead of facilitating membrane

Table 1 PKB binding partners with defined regulatory functions on PKB

PKB binding partners	Effect of binding
ACTN4	ACTN4 physically interacts with PKB. Silencing of ACTN4 inhibits translocation of PKB to the plasma membrane, thus reducing PKB phosphorylation.
APE	APE binds to the C-terminal domain of PKB and enhances its basal phosphorylation. Knocking down APE reduces PKB phosphorylation at T308 and S473.
APPL1	APPL1 functions as an adaptor that tethers inactive PKBb and p110a in the cytoplasm and expedites them to the plasma membrane upon mitogenic stimulation. Overexpression of APPL1 enhances PKB activation by IGF1 and promotes PKB-mediated suppression of androgen receptor transactivation.
Ft1	Ft1 directly associates with the C-terminal part of PKB and enhances its phosphorylation by promoting PKB binding to PDK1.
Hsp90/Cdc37	PKB physically interacts with the chaperone complex Hsp90/cdc37 through its kinase domain to maintain its stability. Disruption of this interaction rapidly induces PKB ubiquitination, which results in its degradation by the proteasome. Loss of cdc37 reduces PKB activity in human colon cancer cells.
PIKE-A	PIKE-A stimulates PKB kinase activity via direct interaction with the C-terminal domain of PKB, which results in facilitated human cancer cell invasion and prevention of apoptosis. Cdk5-mediated interaction of PIKE-A and PKB is crucial for glioblastoma cell invasion.
RasGAP	RasGAP binds to the PH domain of PKB, enhances S473 phosphorylation through an ILK-dependent pathway, and protects cells from apoptosis via PKB activation. Downregulation of RasGAP inhibits serum-induced PKB activity.
TCL1	TCL1 physically interacts with PKB by binding to its PH domain and enhances cell proliferation and survival. Disruption of this interaction significantly inhibits PKB activity. Interaction with TCL1 can also promote PKB translocation to the nucleus.
TRB3	TRB3 inhibits insulin and growth factor-induced PKB phosphorylation by direct binding to the kinase domain of PKB in the activation loop. Knocking down TRB3 expression in the liver of diabetic mice restores insulin responsiveness.
Keratin-10	Keratin-10 sequesters PKB to the cytoskeleton and inhibits its translocation to the plasma membrane via physical binding to PKB at the N-terminal part.

localization of PKB, enhanced its activity via direct interaction with its PH domain, thus enhancing cell proliferation and survival (Laine et al. 2000). Disrupting the interaction between TCL1 and PKB significantly inhibited the activity of PKB and its downstream responses (Hiromura et al. 2004). Interestingly, this interaction also promoted nuclear translocation of PKB (Pekarsky et al. 2000), the consequence of which requires further evaluation. Another PH-domain-interacting protein, Ras GTPase activating protein (RasGAP), enhanced the phosphorylation of PKB at S473 and its activity through an ILK-dependent pathway, allowing cells to be protected from apoptosis (Yue et al. 2004). Downregulation of RasGAP inhibited serum-induced stimulation of PKB. Akt-phosphorylation enhancer (APE) was identified as a protein associated specifically with the C-terminal domain of PKB, resulting in enhanced PKB phosphorylation and activation without growth factor stimulation (Anai et al. 2005). Knocking down APE significantly reduced PKB phosphorylation and cell proliferation. Also discovered as binding to the C-terminal part of PKB, fused toes protein homolog 1 (Ft1) was shown to enhance its phosphorylation by promoting the interaction with PDK1 (Remy and Michnick 2004). Similarly, the GTPase phosphatidylinositol 3-kinase enhancer A (PIKE-A) also stimulated PKB kinase activity via direct interaction with its C-terminus. This resulted in cellular invasion and anti-apoptotic effects (Ahn et al. 2004a, b), in line with the recent study showing the role of the Cdk5-mediated interaction of PIKE-A with PKB in glioblastoma cell invasion (Liu et al. 2008). Inhibition of heat shock protein 90 (Hsp90) resulted in reduced activation of the PKB pathway. This observation allowed the highlighting of the association of the chaperone complex Hsp90/cdc37 with intracellular PKB through its kinase domain (Basso et al. 2002). Disrupting the interaction of PKB with Hsp90/cdc37 rapidly induced its ubiquitination and subsequent degradation by the proteasome, suggesting that Hsp90/cdc37 prevents PKB from degradation by the ubiquitination-proteasome pathway. This correlates with the fact that cdc37 deficiency reduced PKB phosphorylation and activity (Gray et al. 2007; Smith et al. 2009).

2.2.2 PKB Inhibitors

Some of the proteins that bind directly to PKB have been revealed as PKB inhibitors. One of them, tribbles homolog 3 (TRB3), bound the kinase domain of PKB in the activation loop and inhibited its phosphorylation at both T308 and S473 upon growth factors and insulin stimulation (Du et al. 2003). Knocking down TRB3 expression in the liver of diabetic mice restored insulin responsiveness (Koo et al. 2004). In addition, a component of the intermediate filament cytoskeleton, keratin-10, sequestered PKB to the cytoskeleton and inhibited its translocation to the plasma membrane by binding to its N-terminus (Paramio et al. 2001; Santos et al. 2002).

2.2.3 PKB-Interacting Proteins with Undefined Functions

The effect of other PKB-binding proteins is, however, still poorly understood. For example, growth factor receptor-binding protein-10 (Grb10), JNK-interacting protein 1 (JIP-1), periplakin, inosine-50 monophosphate dehydrogenase (IMPDH), and myosin II were observed to interact with the PH domain of PKB but did not reveal any clear outcome on PKB activity (Frost and Lang 2007; Ingley and Hemmings 2000; Jahn et al. 2002; Kim et al. 2002, 2003; Pan et al. 2006; Smith et al. 2007; Song and Lee 2005; Tanaka et al. 1999; van den Heuvel et al. 2002;

Wang et al. 2007; Wick et al. 2003). Moreover, while prohibitin 2 was reported to directly bind to the C-terminal regulatory domain of PKB (Heron-Milhavet et al. 2008; Sun et al. 2004), the resulting effect on PKB activity requires further investigation.

2.3 *Regulation of PKB by CTMP*

Carboxyl-terminal modulator protein (CTMP) was initially identified as a cytosolic interactor of PKB that prevented its activation at the plasma membrane in response to various stimuli and that exhibited tumor suppressor-like functions (Chae et al. 2005; Maira et al. 2001). This notion was strengthened by the observation that primary glioblastoma exhibited reduced levels of CTMP transcripts due to promoter hypermethylation (Knobbe et al. 2004). Nevertheless, the physiological localization and function of CTMP were still poorly understood until we identified CTMP as a mitochondrial protein capable of sensitizing cells to apoptosis (Parcellier et al. 2009). Bioinformatics analysis of the amino acid sequence using subcellular localization prediction tools highlighted an N-terminal mitochondrial localization signal (MLS) conserved in CTMP orthologs. Once synthesized, CTMP translocated to the mitochondria and underwent maturation through cleavage of its MLS by mitochondrial peptidases. We also showed that CTMP was released into the cytosol upon apoptotic stimuli in a Bcl-2-dependent manner and promoted apoptosis with a concomitant delay/inhibition of PKB phosphorylation on S473 (Parcellier et al. 2009). Overexpression and loss-of-function studies revealed that CTMP needs to be processed in the mitochondria and released into the cytosol to regulate apoptosis and suggested that this pro-apoptotic effect of CTMP was mediated at least partially by the inhibition of PKB activity. In line with this work, CTMP was shown to play a crucial role in ischemia-induced neuronal death by inhibiting PKB (Miyawaki et al. 2009).

The initial data supporting a negative role of CTMP on PKB activity via direct interaction were obtained using a CTMP construct in which a flag-tag was attached to the N-terminal end of CTMP (Maira et al. 2001). This N-terminal tag containing positively charged amino acids likely interfered with the translocation of CTMP into the mitochondria. The protein inserted instead into the plasma membrane, facilitating the direct interaction between CTMP and PKB. Following its activation at the plasma membrane, PKB was shown to translocate to different subcellular compartments, including the mitochondria outer membrane, in a cell type- and stimulus-dependent manner (Ahmad et al. 2006; Andjelkovic et al. 1997; Bijur and Jope 2003; Kunkel et al. 2005; Sasaki et al. 2003). The biological significance of the translocation of active PKB to the mitochondria is, however, ill-defined. Altogether, these observations allow proposing a dynamic model by which CTMP modulates PKB activity in a specific subcellular compartment (the mitochondria or the cytosol) depending on the nature of the stimulus (survival or apoptosis). Under apoptotic conditions, no direct interaction between PKB and CTMP could be observed, which argues for an indirect mechanism by which

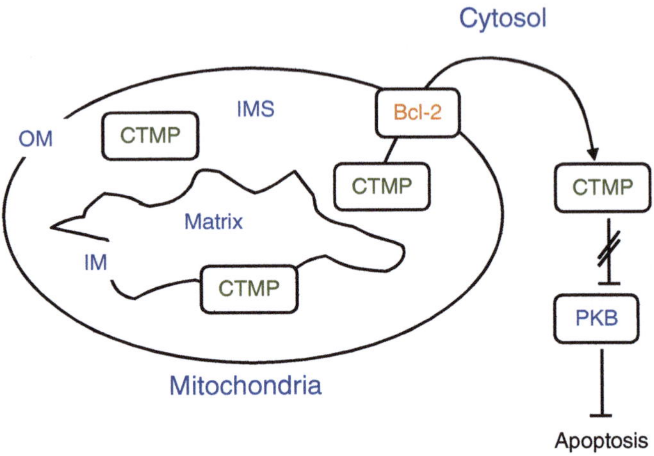

Fig. 2 Subcellular localization and function of CTMP. CTMP is a mitochondrial protein. It is found both in association with the inner mitochondrial membrane (IM) and free in the inter-mitochondrial space (IMS). Upon cell death stimuli, CTMP is released to the cytosol in a Bcl-2-dependent manner, where it promotes apoptosis likely by delaying or inhibiting PKB activation. *OM* outer mitochondrial membrane

CTMP delays PKB phosphorylation and, thus, its activation (Fig. 2). Reports showing that the initiation of apoptosis triggered by the release of cytochrome c is coordinated by "mitochondrial-shaping" proteins strengthens the notion that the same set of signaling proteins regulates antagonist functions within the mitochondria (Cereghetti and Scorrano 2006; Gottlieb 2006; Pellegrini and Scorrano 2007). *In vitro* and *in vivo* observations allow us to suggest a role for CTMP in the mitochondria fission process (unpublished data). Thus, CTMP potentially mediates its pro-apoptotic effect by modulating the activity of some of the regulators of mitochondrial dynamics.

3 The Role of PKB in Physiological and Pathological Conditions

The shift of studies from a cellular context to a whole organism setting has revealed a wide range of *in vivo* functions of PKB. Analysis of genetically modified mice, in which transgenic expression or targeted deletion of the PKBs have been achieved, highlighted both unique and redundant roles of the different isoforms in physiological and pathological conditions. These studies uncovered the involvement of PKB in the regulation of numerous processes, including embryonic development, thymocyte development, adipocyte differentiation, glucose homeostasis, and tumor formation (Fig. 3).

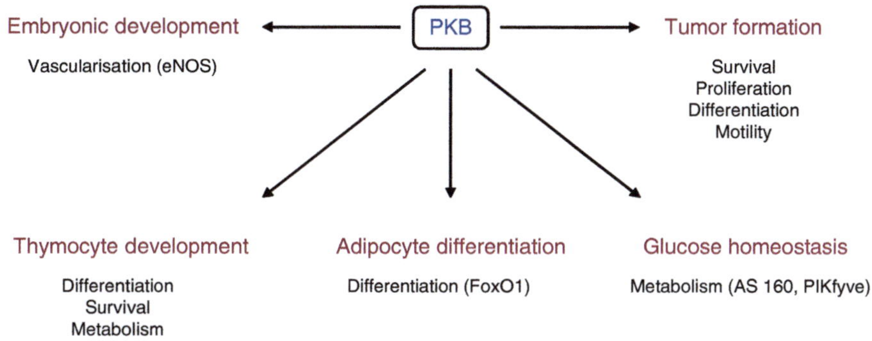

Fig. 3 The role of PKB in physiological and pathological conditions. In mouse models, PKB has been shown to play a crucial role in a number of physiological and pathological situations, including embryonic development, thymocyte development, adipocyte differentiation, glucose homeostasis, and tumor formation. The processes and the substrates that mediate the effect of PKB and that have been highlighted *in vivo* are specified

3.1 Effects of Knocking-Out Individual PKB Isoforms in Mice

In order to assess the influence of individual PKB isoforms *in vivo*, each of the PKBs has been disrupted in the mouse germ line via homologous recombination. The specific phenotypes of PKBα-, PKBβ-, and PKBγ-deficient mice could be due to the relative tissue expression of the isoforms but also suggest distinct roles for each isoform in regulating different biological events. For example, ablation of PKBα in mice led to placental hypotrophy, partial neonatal mortality, and reduced animal size from the embryonic stages (Chen et al. 2001; Cho et al. 2001b; Yang et al. 2003). Targeted deletion of PKBβ caused hyperglycemia and impaired insulin action in skeletal muscle, fat, and liver. This was accompanied by a mild growth retardation, which was dependent on the genetic background, and age-dependent loss of adipose tissue (Cho et al. 2001a; Garofalo et al. 2003). Mice lacking PKBγ exhibited a specific deficiency in postnatal development of the brain, which was characterized by a 20–25% size reduction due at least partially to decreased cell size and cell number (Easton et al. 2005; Tschopp et al. 2005). Interestingly, combined loss of PKBβ and PKBγ resulted in impaired glucose homeostasis and brain size to the same extent as mice lacking individual isoforms (Dummler et al. 2006), indicating that these isoforms play a unique function and do not compensate for each other.

3.2 PKB in Embryonic Development

Mice lacking individual PKB isoforms are mostly viable. However, combined deficiency of PKBα/PKBγ (Yang et al. 2005) or PKBα/PKBβ (Peng et al. 2003)

caused lethality at the embryonic and neonatal stages, respectively, with the latter probably owing to respiratory failure. This suggests, besides unique effects of each isoform, an extensive functional redundancy among them *in vivo*. PKBα appears, however, to have a dominant role in embryonic development and postnatal survival that is further supported by the fact that mice deficient for both PKBβ and PKBγ and, most astonishing, mice retaining only one allele of PKBα (PKBα$^{+/-}$PKBβ$^{-/-}$PKBγ$^{-/-}$) are viable despite a dramatic reduction of total PKB levels in many tissues (Dummler et al. 2006). The absence of PKBα led to hypotrophy of the placenta that showed impaired vascularization combined with reduced phosphorylation of eNOS (Yang et al. 2003), this likely contributing to embryonic and neonatal lethality.

During embryonic stages prior to lethality, combined loss of PKBα and one of the other PKB isoforms led to additional and more severe phenotypes compared to the corresponding single mutants. For instance, PKBα$^{-/-}$PKBβ$^{-/-}$ embryos exhibited extreme intrauterine growth deficiency, impaired adipogenesis, delayed bone development, as well as prominent atrophy of the skin (due to proliferation defects of basal keratinocytes) and skeletal muscle (caused by a marked decrease in individual muscle cell size) (Peng et al. 2003). Furthermore, PKBα$^{-/-}$PKBγ$^{-/-}$ embryos showed severe developmental defects in the cardiovascular and nervous systems as well as impaired vascularization (Yang et al. 2005). In aggregate, this suggests that a crucial threshold of PKB activity is required in some physiological processes.

3.3 PKB in Thymocyte Development

The arm of the immune system that assures antigen-specificity and immunological memory involves the action of T lymphocytes. These cells express a unique T cell receptor (TCR) which specifically recognizes a given antigen when presented on the surface of antigen-presenting cells. The formation of a functional T cell population with a broad TCR repertoire depends on a precisely controlled developmental process which takes place in the thymus. T lymphocytes developing within the thymus are referred to as thymocytes. Cells at early stages of thymocyte development are characterized by the lack of CD4 and CD8 expression (they are designated double-negative (DN) thymocytes) and by the formation of a pre-TCR. Whereas thymocytes that fail to express a pre-TCR die by apoptosis, signaling via this receptor rescues cells (referred to as pre-thymocytes) from programmed cell death, initiates cell proliferation, and allows for further development (Fehling et al. 1995; Hoffman et al. 1996). The events following successful pre-TCR signaling are referred to as β-selection, which is one of the checkpoints key to thymocyte development (Mallick et al. 1993). β-selected pre-thymocytes then progress to a double-positive (DP) phenotype (characterized by the concomitant expression of CD4 and CD8 and the formation of a final TCR) which will attain the function and phenotype of mature T cells.

In mice, the absence of PKBα only or both PKBα and PKBβ resulted in cell-autonomous accumulation of thymocytes at the pre-thymocyte stage, whereas a lack of all three PKB isoforms caused a block at this stage (Fayard et al. 2007; Juntilla et al. 2007; Mao et al. 2007). Hence, PKBβ and PKBγ isoforms may in part compensate the loss of PKBα, which is the main PKB isoform expressed in early developing thymocytes (Fayard et al. 2007; Juntilla et al. 2007; Mao et al. 2007). PKB was shown to act as a signal transducer downstream of the pre-TCR. Indeed, activation of pre-TCR signaling induced PKB phosphorylation (Mao et al. 2007). Furthermore, the deletion of PKBα affected, in thymocytes, the expression of a number of genes known to be involved in pre-TCR and/or TCR signaling (Fayard et al. 2007). This was associated with the lack of normal downregulation of CD25 surface expression (Fayard et al. 2007; Mao et al. 2007), a phenotypic change dependent on the pre-TCR (Aifantis et al. 1997). In addition, introduction of a constitutively active form of PKBα was sufficient to rescue the development of immature thymocytes that lacked the expression of a pre-TCR (Ciofani and Zuniga-Pflucker 2005; Mao et al. 2007; Patra et al. 2006).

Using PKB-deficient mouse models, the role of PKB in pre-thymocyte proliferation has been controversial (Juntilla et al. 2007; Mao et al. 2007). *In vivo* experiments using a mutated form of PDK1 ($PDK1^{L155E}$, where L is leucine and E is glutamate), allowing the selective activation of PKB but not other PDK1 substrates (Biondi et al. 2001; Collins et al. 2003), have recently shed light on this issue. The expression of $PDK1^{L155E}$ corrected in part the otherwise complete block of PDK1-deficient cells at the pre-thymocyte stage (Hinton et al. 2004; Kelly et al. 2007). Thus, activation of PKB by PDK1 was largely sufficient for pre-thymocytes to differentiate to later stages, but it did not instruct cell proliferation (Kelly et al. 2007).

The absence of PKB isoforms resulted in an increased apoptosis of pre-thymocytes (Juntilla et al. 2007; Mao et al. 2007). The tendency of PKB-deficient pre-thymocytes to undergo spontaneous apoptosis was unlikely to be caused by a dysregulation of anti-apoptotic proteins such as Bcl-2 members (Ciofani and Zuniga-Pflucker 2005; Juntilla et al. 2007; Mandal et al. 2005). It was rather effected by an altered metabolism due to reduced glucose uptake which correlated with decreased Glut1 expression (Ciofani and Zuniga-Pflucker 2005; Juntilla et al. 2007). Progression through β-selection is also accompanied by a sustained presence of the nutrient receptors, such as the transferrin receptor CD71 and a subunit of the L-amino acid transporter CD98 (Kelly et al. 2007). Their cell surface expression is limiting for pre-thymocyte growth, is dependent on signals from pre-TCR, and has been shown to be mediated by PKB (Kelly et al. 2007). Therefore PKB is required to meet the high metabolic demands associated with β-selected thymocyte proliferation.

3.4 PKB in Adipocyte Differentiation

Physiological evidence for an implication of PKB in adipocyte differentiation has been inferred from the analysis of PKB-deficient mice. While subcutaneous fat was modestly reduced in the absence of PKBα (Yang et al. 2003), the lack of both PKBα and PKBβ isoforms resulted in the lack of differentiated brown adipose tissue in neonates (Peng et al. 2003). In addition, mouse embryonic fibroblasts (MEFs) derived from PKBα$^{-/-}$ or from PKBα$^{-/-}$PKBβ$^{-/-}$ embryos were unable to differentiate into adipocytes in a standard *in vitro* assay (Baudry et al. 2006; Peng et al. 2003). Conversely, a constitutively active form of PKBα was able to induce spontaneous differentiation of PKBα$^{-/-}$ MEFs (Baudry et al. 2006) and 3T3-L1 pre-adipocytes (Kohn et al. 1996; Magun et al. 1996). Interestingly, in the presence of adipogenic treatment, differentiation of MEFs lacking PKBα only or both PKBα and PKBβ into adipocytes was restored by ectopic expression of PKBα but not PKBβ (Baudry et al. 2006; Yun et al. 2008). These results, and the fact that the deletion of PKBβ in MEFs had only a modest effect on adipocyte differentiation (Peng et al. 2003), suggest that the regulation of adipocyte differentiation is a specific function of PKBα.

In PKBα$^{-/-}$ and PKBα$^{-/-}$PKBβ$^{-/-}$ MEFs, as well as in 3T3-L1 pre-adipocytes impaired for PKBα expression by RNA interference, phosphorylation of FoxO1 was severely impaired, leading to its subsequent activation (Baudry et al. 2006; Peng et al. 2003; Xu and Liao 2004). PKBα was found to be the major PKB isoform involved in regulating the phosphorylation and nuclear export of FoxO1 in MEFs (Yun et al. 2008). This PKB deficiency-dependent activated FoxO1 as well as a constitutively active form of FoxO1 in which all three putative PKB phosphorylation sites were mutated both coincided with a failure of induced peroxisome proliferator-activated receptor γ (PPARγ) expression levels and a block in adipocyte differentiation (Baudry et al. 2006; Nakae et al. 2003; Peng et al. 2003; Xu and Liao 2004). PPARγ is a key regulator of the transcriptional program involved in adipocyte differentiation and is induced prior to the transcriptional activation of most adipocyte-specific genes. In aggregate, this suggests that the induction of PPARγ expression and adipocyte differentiation are linked with PKBα-mediated phosphorylation and inactivation of FoxO1. It was also shown that the expression of krüppel-like factor 15 (Klf15), a transcription factor playing an important role in adipogenesis through the regulation of PPARγ gene expression (Mori et al. 2005), was affected by the absence of PKBα during differentiation of MEFs into adipocytes (Baudry et al. 2006). The mechanism by which PKBα regulates Klf15 gene expression is, however, unknown. In addition, a period of cell cycle progression referred to as mitotic clonal expansion when p27^{Kip1} expression is reduced is a prerequisite for adipocyte differentiation. Of note, the expression of p27^{Kip1} and clonal expansion, which are mediated by FoxO1, was also regulated in a PKBα-specific manner in MEFs undergoing adipocyte differentiation (Yun et al. 2008).

3.5 PKB in Glucose Homeostasis

PKBβ-deficient mice exhibited pre-diabetes-like syndrome with elevated fed and fasting plasma glucose levels (Cho et al. 2001a; Garofalo et al. 2003). In addition to hyperglycemia, these mice showed a significantly reduced glucose uptake in muscle and adipose tissue upon insulin stimulation that characterizes peripheral insulin resistance. Insulin action was also impaired in liver, where hepatic glucose production and output were not normally suppressed. A compensatory increase of pancreatic islet mass and subsequent hyperinsulinemia may have resulted from decreased responsiveness to the hormone in peripheral tissues. A subset of these mice, however, developed a type 2 form of diabetes which was accompanied by pancreatic β-cell failure with increased apoptotic islets, loss of insulin secretion, and severe hyperglycemia (Garofalo et al. 2003).

In contrast, normal glucose homeostasis was maintained in both PKBα$^{-/-}$ and PKBγ$^{-/-}$ mice, and analysis of PKBβ$^{-/-}$PKBγ$^{-/-}$ animals showed that the absence of PKBγ did not intensify the phenotype of PKBβ$^{-/-}$ mice (Cho et al. 2001b; Dummler et al. 2006; Easton et al. 2005; Tschopp et al. 2005). A role for PKBα was, however, recently uncovered in glucose homeostasis and in the genesis of diabetes. Haplodeficiency of PKBα in PKBβ-null mice led to overt type 2 diabetes, which was manifested by hyperglycemia due to β-cell dysfunction combined with impaired glucose homeostasis (Chen et al. 2009). Insulin injection was not sufficient to alleviate hyperglycemia, suggesting that defective glucose homeostasis was independent of insulin levels. Likely reflecting lipoatrophy, decreased plasma leptin concentration was reported in PKBα$^{+/-}$PKBβ$^{-/-}$ and, to some extent, in PKBβ$^{-/-}$ mice and was proposed to be the major contributing factor to the impaired glucose homeostasis and the predominant cause of diabetes in PKB-deficient mice (Chen et al. 2009; Garofalo et al. 2003). Indeed, restoring leptin levels was sufficient to obtain normal blood glucose and insulin levels in both PKBβ$^{-/-}$ and PKBα$^{+/-}$PKBβ$^{-/-}$ mice (Chen et al. 2009).

While PKBα seems to be the main PKB isoform required for adipocyte differentiation (see above), PKBβ is the preferred isoform in mediating insulin-induced glucose uptake in adipocytes and is likely to play a predominant role in Glut4 translocation from intracellular storage vesicles to the cell surface (Bae et al. 2003; Jiang et al. 2003; Katome et al. 2003). For instance, while glucose uptake was reduced in adipocytes isolated from PKBβ-deficient mice upon insulin stimulation, ectopic expression of PKBβ but not comparable levels of ectopic PKBα expression, could completely restore insulin-mediated glucose uptake and Glut4 translocation in adipocytes derived from PKBβ$^{-/-}$ MEFs (Bae et al. 2003). Nevertheless, although PKBβ is the major isoform required in this response, PKBα could also play a role. Knockdown of both PKBα and PKBβ in 3T3-L1 adipocytes produced a more severe reduction in glucose uptake than knockdown of PKBβ only (Jiang et al. 2003; Katome et al. 2003). In aggregate, these results suggest that a cell autonomous role of PKB in adipocyte glucose uptake could account, at least in part, for the phenotype observed in PKBβ$^{-/-}$ and PKBα$^{+/-}$PKBβ$^{-/-}$ mice. Even though PKB was shown to induce translocation of Glut4 to the plasma membrane in adipocytes (Cong et al. 1997; Kohn et al. 1996; Tanti et al. 1997), the mechanism involved is unclear. A number of potential PKB substrates, including Akt substrate of 160 kDa (AS160) (Sano et al. 2003) and PI5-kinase (PIKfyve) (Berwick et al. 2004) were, however, proposed to mediate this effect of PKB.

Assessing the phosphorylation status of these proteins in PKBβ-deficient insulin-responsive tissues would validate or invalidate their role downstream of PKB in Glut4 translocation and could clarify the effect of deleting PKB in glucose intolerance and insulin resistance.

The role of PKB deficiency in diabetes was underscored by the discovery of a family with an inherited loss of function PKBβ mutation (R274H, where H is histidine) (George et al. 2004). This mutation results in a dominant negative PKBβ which inhibits the action of other PKB isoforms. Individuals carrying such a mutated form of PKBβ suffer from severe insulin resistance and diabetes combined with lipoatrophy.

3.6 PKB in Tumor Formation

PKB is commonly hyperactivated in human cancers. Its aberrant activation occurs by a variety of mechanisms including mutations in components of the PI3K pathway, such as PTEN and the PI3K subunits, or amplification of PKB genes (Franke 2008). Mutations within PKBβ and PKBγ genes were highlighted in different cancer types (Davies et al. 2008; Greenman et al. 2007). A rare mutation in the PH domain of PKBα ($PKB\alpha^{E17K}$, where E is glutamic acid and K is lysine), resulting in increased plasma membrane recruitment, has also been described in human breast, colorectal, ovarian, lung, and melanoma clinical cancer specimens (Bleeker et al. 2008; Carpten et al. 2007; Davies et al. 2008; Do et al. 2008; Malanga et al. 2008). PKB likely contributes to cancer progression by promoting cell proliferation, cell survival, metabolic capacity of cells, and angiogenesis. Its hyperactivation correlates with low prognosis and is found more frequently in poorly differentiated tumors that are invasive, fast growing, and resistant to treatment (Franke 2008).

PKBα was first identified in a spontaneously formed thymic lymphoma of an AKR mouse as the cellular homolog of the v-Akt proto-oncogene (Staal et al. 1977; Staal 1987). v-Akt, expressed by the transforming retrovirus AKT-8, encodes a fusion protein between the viral protein gag and PKBα. The gag domain possesses a myristoylation signal that mediates targeting to the plasma membrane and, thereby, confers constitutive kinase activity to v-Akt. Upon inoculation of susceptible mice, v-Akt caused thymic lymphomas (Staal and Hartley 1988). Similarly, the transgenic expression of a constitutively active form of PKBα in thymocytes and T cells resulted in lymphoma formation (Malstrom et al. 2001; Patra et al. 2006; Rathmell et al. 2003). DP thymocytes carrying the constitutively active PKBα transgene were more resistant to spontaneous apoptosis, which correlated with heightened endogenous levels of the anti-apoptotic molecule Bcl-XL (Jones et al. 2000; Ma et al. 1995), suggesting that PKB-mediated dysregulation of Bcl-XL protein levels could participate in thymic lymphoma formation.

To further assess the effect of PKB hyperactivation in tumorigenesis, other transgenic mice have been generated that express a constitutively active form of

PKB (Yang et al. 2004). For instance, prostate-restricted activation of PKB was sufficient to induce cell proliferation, giving rise to prostatic intraepithelial neoplasia (Majumder et al. 2003). Importantly, in Pten$^{+/-}$ mice where PKB is hyperactivated, deletion of PKBα was sufficient to inhibit prostate neoplasia, endometrial carcinoma, thyroid and adrenal medulla tumors, intestinal polyps, and to markedly reduce lymphoid hyperplasia (Chen et al. 2006). This indicates that tumors induced by PTEN inactivation (Di Cristofano et al. 1998; Podsypanina et al. 1999; Suzuki et al. 1998) are, to a large extent, dependent on PKBα.

In several cases, constitutive activation of PKB alone in mice appeared to be insufficient for tumorigenesis. For instance, constitutively active PKBα in mammary glands, providing critical cell survival signals, contributed to tumor progression but only when coexpressed with polyomavirus middle T oncoprotein that has been decoupled from the PI3K pathway by mutation (Hutchinson et al. 2001). Transgenic expression of activated PKBα in mammary epithelium, correlating with enhanced proliferation, dramatically accelerated the progression of tumor induced by an ErbB2 transgene (Hutchinson et al. 2004). Moreover, constitutively active PKBα in neural progenitor cells required coexpression of v-Ki-ras2 Kirsten rat sarcomaviral oncogene homolog (K-Ras) to induce glioblastomas in mice (Holland et al. 2000). Furthermore, expression of any two of the three oncogenes c-myc, K-ras, and PKBα was necessary to form ovarian carcinoma in a p53-deficient mouse model (Orsulic et al. 2002).

When constitutively active, all three PKB isoforms possess *in vitro* transformation ability (Mende et al. 2001). However, opposing functions of PKBα and PKBβ have been reported in cells. For instance, constitutive activation of PKBα reduced migration, invasion, and metastatic progression of breast cancer cells due to increased differentiation (Hutchinson et al. 2004; Yoeli-Lerner et al. 2005). In contrast, overexpression of PKBβ in breast and ovarian cancer cells has been associated with cell motility, invasion, and metastatic potential (Arboleda et al. 2003). Interestingly, in nontransformed cells, PKBα was shown to be essential for cell proliferation while PKBβ promoted cell cycle exit (Heron-Milhavet et al. 2006).

4 Conclusion

Downstream of PI3K, the regulation of PKB signaling requires fine-tuning to fulfill physiological functions and avoid pathological loss of control. While hyperactivation of this pathway gives rise to tumor formation, its downregulation leads to metabolic disorders. Therefore, precisely restoring normal levels of PKB activity, rather than applying uncontrolled inhibition or activation, as well as specifically modulating the individual isoforms could be of significant therapeutic value. Furthermore, cellular processes and molecules downstream of PKB seem to differ depending on the physiological and pathological conditions. Therefore, targeting the players downstream of PKB that are specific to the pathology of interest could avoid potential side-effects. In view of the numerous substrates modulated by PKB, the precise regulation of PKB signaling in the normal *in vivo* situation could be attributed to specific roles of the different PKBs *in vivo*, their expression pattern, subcellular localization, regulation by upstream kinases in a stimulus- and context-dependent manner and by phosphatases, and interaction with binding partners. Altogether, this highlights the complexity of PKB signaling regulation, which is further emphasized by the fact that many of the downstream molecules modulated by PKB are also subject to regulation by other signaling cascade.

Acknowledgments EF is supported by European Molecular Biology Organization (EMBO) Long Term Fellowship (ALTF 506-2005) and Marie-Curie Fellowship (MEIF-CT-2006-025075). AP is supported by the Swiss Cancer League (OCS-01167-09-2001). The Friedrich Miescher Institute for Biomedical research is part of the Novartis Research Foundation.

References

Ahmad N, Wang Y, Haider KH, Wang B, Pasha Z, Uzun O, Ashraf M (2006) Cardiac protection by mitoKATP channels is dependent on Akt translocation from cytosol to mitochondria during late preconditioning. Am J Physiol Heart Circ Physiol 290:H2402–H2408

Ahn JY, Hu Y, Kroll TG, Allard P, Ye K (2004a) PIKE-A is amplified in human cancers and prevents apoptosis by up-regulating Akt. Proc Natl Acad Sci USA 101:6993–6998

Ahn JY, Rong R, Kroll TG, Van Meir EG, Snyder SH, Ye K (2004b) PIKE (phosphatidylinositol 3-kinase enhancer)-A GTPase stimulates Akt activity and mediates cellular invasion. J Biol Chem 279:16441–16451

Aifantis I, Buer J, von Boehmer H, Azogui O (1997) Essential role of the pre-T cell receptor in allelic exclusion of the T cell receptor beta locus. Immunity 7:601–607

Alessi DR, Andjelkovic M, Caudwell B, Cron P, Morrice N, Cohen P, Hemmings BA (1996a) Mechanism of activation of protein kinase B by insulin and IGF-1. Embo J 15:6541–6551

Alessi DR, Caudwell FB, Andjelkovic M, Hemmings BA, Cohen P (1996b) Molecular basis for the substrate specificity of protein kinase B; comparison with MAPKAP kinase-1 and p70 S6 kinase. FEBS Lett 399:333–338

Alessi DR, James SR, Downes CP, Holmes AB, Gaffney PR, Reese CB, Cohen P (1997) Characterization of a 3-phosphoinositide-dependent protein kinase which phosphorylates and activates protein kinase Balpha. Curr Biol 7:261–269

Anai M, Shojima N, Katagiri H, Ogihara T, Sakoda H, Onishi Y, Ono H, Fujishiro M, Fukushima Y, Horike N et al (2005) A novel protein kinase B (PKB)/AKT-binding protein enhances PKB kinase activity and regulates DNA synthesis. J Biol Chem 280:18525–18535

Andjelkovic M, Jakubowicz T, Cron P, Ming XF, Han JW, Hemmings BA (1996) Activation and phosphorylation of a pleckstrin homology domain containing protein kinase (RAC-PK/PKB) promoted by serum and protein phosphatase inhibitors. Proc Natl Acad Sci USA 93:5699–5704

Andjelkovic M, Alessi DR, Meier R, Fernandez A, Lamb NJ, Frech M, Cron P, Cohen P, Lucocq JM, Hemmings BA (1997) Role of translocation in the activation and function of protein kinase B. J Biol Chem 272:31515–31524

Arboleda MJ, Lyons JF, Kabbinavar FF, Bray MR, Snow BE, Ayala R, Danino M, Karlan BY, Slamon DJ (2003) Overexpression of AKT2/protein kinase Bbeta leads to up-regulation of beta1 integrins, increased invasion, and metastasis of human breast and ovarian cancer cells. Cancer Res 63:196–206

Bae SS, Cho H, Mu J, Birnbaum MJ (2003) Isoform-specific regulation of insulin-dependent glucose uptake by Akt/protein kinase B. J Biol Chem 278:49530–49536

Basso AD, Solit DB, Chiosis G, Giri B, Tsichlis P, Rosen N (2002) Akt forms an intracellular complex with heat shock protein 90 (Hsp90) and Cdc37 and is destabilized by inhibitors of Hsp90 function. J Biol Chem 277:39858–39866

Baudry A, Yang ZZ, Hemmings BA (2006) PKBalpha is required for adipose differentiation of mouse embryonic fibroblasts. J Cell Sci 119:889–897

Bellacosa A, Testa JR, Staal SP, Tsichlis PN (1991) A retroviral oncogene, akt, encoding a serine-threonine kinase containing an SH2-like region. Science 254:274–277

Bentzinger CF, Romanino K, Cloetta D, Lin S, Mascarenhas JB, Oliveri F, Xia J, Casanova E, Costa CF, Brink M et al (2008) Skeletal muscle-specific ablation of raptor, but not of rictor, causes metabolic changes and results in muscle dystrophy. Cell Metab 8:411–424

Berwick DC, Dell GC, Welsh GI, Heesom KJ, Hers I, Fletcher LM, Cooke FT, Tavare JM (2004) Protein kinase B phosphorylation of PIKfyve regulates the trafficking of GLUT4 vesicles. J Cell Sci 117:5985–5993

Bijur GN, Jope RS (2003) Rapid accumulation of Akt in mitochondria following phosphatidylinositol 3-kinase activation. J Neurochem 87:1427–1435

Biondi RM, Kieloch A, Currie RA, Deak M, Alessi DR (2001) The PIF-binding pocket in PDK1 is essential for activation of S6K and SGK, but not PKB. EMBO J 20:4380–4390

Bleeker FE, Felicioni L, Buttitta F, Lamba S, Cardone L, Rodolfo M, Scarpa A, Leenstra S, Frattini M, Barbareschi M et al (2008) AKT1(E17K) in human solid tumours. Oncogene 27:5648–5650

Boehme KA, Kulikov R, Blattner C (2008) p53 stabilization in response to DNA damage requires Akt/PKB and DNA-PK. Proc Natl Acad Sci USA 105:7785–7790

Bozulic L, Surucu B, Hynx D, Hemmings BA (2008) PKBalpha/Akt1 acts downstream of DNA-PK in the DNA double-strand break response and promotes survival. Mol Cell 30:203–213

Brognard J, Sierecki E, Gao T, Newton AC (2007) PHLPP and a second isoform, PHLPP2, differentially attenuate the amplitude of Akt signaling by regulating distinct Akt isoforms. Mol Cell 25:917–931

Brunet A, Bonni A, Zigmond MJ, Lin MZ, Juo P, Hu LS, Anderson MJ, Arden KC, Blenis J, Greenberg ME (1999) Akt promotes cell survival by phosphorylating and inhibiting a Forkhead transcription factor. Cell 96:857–868

Burgering BM, Coffer PJ (1995) Protein kinase B (c-Akt) in phosphatidylinositol-3-OH kinase signal transduction. Nature 376:599–602

Calleja V, Ameer-Beg SM, Vojnovic B, Woscholski R, Downward J, Larijani B (2003) Monitoring conformational changes of proteins in cells by fluorescence lifetime imaging microscopy. Biochem J 372:33–40

Calleja V, Alcor D, Laguerre M, Park J, Vojnovic B, Hemmings BA, Downward J, Parker PJ, Larijani B (2007) Intramolecular and Intermolecular Interactions of Protein Kinase B Define Its Activation *In vivo*. PLoS Biol 5:e95

Carpten JD, Faber AL, Horn C, Donoho GP, Briggs SL, Robbins CM, Hostetter G, Boguslawski S, Moses TY, Savage S et al (2007) A transforming mutation in the pleckstrin homology domain of AKT1 in cancer. Nature 448:439–444

Cereghetti GM, Scorrano L (2006) The many shapes of mitochondrial death. Oncogene 25: 4717–4724

Chae KS, Martin-Caraballo M, Anderson M, Dryer SE (2005) Akt activation is necessary for growth factor-induced trafficking of functional K(Ca) channels in developing parasympathetic neurons. J Neurophysiol 93:1174–1182

Chen WS, Xu PZ, Gottlob K, Chen ML, Sokol K, Shiyanova T, Roninson I, Weng W, Suzuki R, Tobe K et al (2001) Growth retardation and increased apoptosis in mice with homozygous disruption of the Akt1 gene. Genes Dev 15:2203–2208

Chen ML, Xu PZ, Peng XD, Chen WS, Guzman G, Yang X, Di Cristofano A, Pandolfi PP, Hay N (2006) The deficiency of Akt1 is sufficient to suppress tumor development in Pten+/− mice. Genes Dev 20:1569–1574

Chen WS, Peng XD, Wang Y, Xu PZ, Chen ML, Luo Y, Jeon SM, Coleman K, Haschek WM, Bass J et al (2009) Leptin deficiency and beta-cell dysfunction underlie type 2 diabetes in compound Akt knockout mice. Mol Cell Biol 29:3151–3162

Cho H, Mu J, Kim JK, Thorvaldsen JL, Chu Q, Crenshaw EB 3rd, Kaestner KH, Bartolomei MS, Shulman GI, Birnbaum MJ (2001a) Insulin resistance and a diabetes mellitus-like syndrome in mice lacking the protein kinase Akt2 (PKB beta). Science 292:1728–1731

Cho H, Thorvaldsen JL, Chu Q, Feng F, Birnbaum MJ (2001b) Akt1/PKBalpha is required for normal growth but dispensable for maintenance of glucose homeostasis in mice. J Biol Chem 276:38349–38352

Ciofani M, Zuniga-Pflucker JC (2005) Notch promotes survival of pre-T cells at the beta-selection checkpoint by regulating cellular metabolism. Nat Immunol 6:881–888

Coffer PJ, Woodgett JR (1991) Molecular cloning and characterisation of a novel putative protein-serine kinase related to the cAMP-dependent and protein kinase C families. Eur J Biochem 201:475–481

Collins BJ, Deak M, Arthur JS, Armit LJ, Alessi DR (2003) *In vivo* role of the PIF-binding docking site of PDK1 defined by knock-in mutation. EMBO J 22:4202–4211

Cong LN, Chen H, Li Y, Zhou L, McGibbon MA, Taylor SI, Quon MJ (1997) Physiological role of Akt in insulin-stimulated translocation of GLUT4 in transfected rat adipose cells. Mol Endocrinol 11:1881–1890

Cross DA, Alessi DR, Cohen P, Andjelkovich M, Hemmings BA (1995) Inhibition of glycogen synthase kinase-3 by insulin mediated by protein kinase B. Nature 378:785–789

Datta SR, Dudek H, Tao X, Masters S, Fu H, Gotoh Y, Greenberg ME (1997) Akt phosphorylation of BAD couples survival signals to the cell-intrinsic death machinery. Cell 91:231–241

Davies MA, Stemke-Hale K, Tellez C, Calderone TL, Deng W, Prieto VG, Lazar AJ, Gershenwald JE, Mills GB (2008) A novel AKT3 mutation in melanoma tumours and cell lines. Br J Cancer 99:1265–1268

Di Cristofano A, Pesce B, Cordon-Cardo C, Pandolfi PP (1998) Pten is essential for embryonic development and tumour suppression. Nat Genet 19:348–355

Diehl JA, Cheng M, Roussel MF, Sherr CJ (1998) Glycogen synthase kinase-3beta regulates cyclin D1 proteolysis and subcellular localization. Genes Dev 12:3499–3511

Dimmeler S, Fleming I, Fisslthaler B, Hermann C, Busse R, Zeiher AM (1999) Activation of nitric oxide synthase in endothelial cells by Akt-dependent phosphorylation. Nature 399:601–605

Ding Z, Liang J, Lu Y, Yu Q, Songyang Z, Lin SY, Mills GB (2006) A retrovirus-based protein complementation assay screen reveals functional AKT1-binding partners. Proc Natl Acad Sci USA 103:15014–15019

Do H, Solomon B, Mitchell PL, Fox SB, Dobrovic A (2008) Detection of the transforming AKT1 mutation E17K in non-small cell lung cancer by high resolution melting. BMC Res Notes 1:14

Domin J, Waterfield MD (1997) Using structure to define the function of phosphoinositide 3-kinase family members. FEBS Lett 410:91–95

Dragoi AM, Fu X, Ivanov S, Zhang P, Sheng L, Wu D, Li GC, Chu WM (2005) DNA-PKcs, but not TLR9, is required for activation of Akt by CpG-DNA. EMBO J 24:779–789

Du K, Herzig S, Kulkarni RN, Montminy M (2003) TRB3: a tribbles homolog that inhibits Akt/PKB activation by insulin in liver. Science 300:1574–1577

Dummler B, Tschopp O, Hynx D, Yang ZZ, Dirnhofer S, Hemmings BA (2006) Life with a single isoform of Akt: mice lacking Akt2 and Akt3 are viable but display impaired glucose homeostasis and growth deficiencies. Mol Cell Biol 26:8042–8051

Easton RM, Cho H, Roovers K, Shineman DW, Mizrahi M, Forman MS, Lee VM, Szabolcs M, de Jong R, Oltersdorf T et al (2005) Role for Akt3/protein kinase Bgamma in attainment of normal brain size. Mol Cell Biol 25:1869–1878

Fayard E, Tintignac LA, Baudry A, Hemmings BA (2005) Protein kinase B/Akt at a glance. J Cell Sci 118:5675–5678

Fayard E, Gill J, Paolino M, Hynx D, Hollander GA, Hemmings BA (2007) Deletion of PKBalpha/Akt1 affects thymic development. PLoS ONE 2:e992

Fehling HJ, Krotkova A, Saint-Ruf C, von Boehmer H (1995) Crucial role of the pre-T-cell receptor alpha gene in development of alpha beta but not gamma delta T cells. Nature 375:795–798

Feng J, Park J, Cron P, Hess D, Hemmings BA (2004a) Identification of a PKB/Akt hydrophobic motif Ser-473 kinase as DNA-dependent protein kinase. J Biol Chem 279:41189–41196

Feng J, Tamaskovic R, Yang Z, Brazil DP, Merlo A, Hess D, Hemmings BA (2004b) Stabilization of Mdm2 via decreased ubiquitination is mediated by protein kinase B/Akt-dependent phosphorylation. J Biol Chem 279:35510–35517

Franke TF (2008) PI3K/Akt: getting it right matters. Oncogene 27:6473–6488

Franke TF, Yang SI, Chan TO, Datta K, Kazlauskas A, Morrison DK, Kaplan DR, Tsichlis PN (1995) The protein kinase encoded by the Akt proto-oncogene is a target of the PDGF-activated phosphatidylinositol 3-kinase. Cell 81:727–736

Frias MA, Thoreen CC, Jaffe JD, Schroder W, Sculley T, Carr SA, Sabatini DM (2006) mSin1 is necessary for Akt/PKB phosphorylation, and its isoforms define three distinct mTORC2s. Curr Biol 16:1865–1870

Frost RA, Lang CH (2007) Protein kinase B/Akt: a nexus of growth factor and cytokine signaling in determining muscle mass. J Appl Physiol 103:378–387

Gao T, Furnari F, Newton AC (2005) PHLPP: a phosphatase that directly dephosphorylates Akt, promotes apoptosis, and suppresses tumor growth. Mol Cell 18:13–24

Garofalo RS, Orena SJ, Rafidi K, Torchia AJ, Stock JL, Hildebrandt AL, Coskran T, Black SC, Brees DJ, Wicks JR et al (2003) Severe diabetes, age-dependent loss of adipose tissue, and mild growth deficiency in mice lacking Akt2/PKB beta. J Clin Invest 112:197–208

George S, Rochford JJ, Wolfrum C, Gray SL, Schinner S, Wilson JC, Soos MA, Murgatroyd PR, Williams RM, Acerini CL et al (2004) A family with severe insulin resistance and diabetes due to a mutation in AKT2. Science 304:1325–1328

Gottlieb E (2006) OPA1 and PARL keep a lid on apoptosis. Cell 126:27–29

Gray PJ Jr, Stevenson MA, Calderwood SK (2007) Targeting Cdc37 inhibits multiple signaling pathways and induces growth arrest in prostate cancer cells. Cancer Res 67:11942–11950

Greenman C, Stephens P, Smith R, Dalgliesh GL, Hunter C, Bignell G, Davies H, Teague J, Butler A, Stevens C et al (2007) Patterns of somatic mutation in human cancer genomes. Nature 446:153–158

Guertin DA, Stevens DM, Thoreen CC, Burds AA, Kalaany NY, Moffat J, Brown M, Fitzgerald KJ, Sabatini DM (2006) Ablation in mice of the mTORC components raptor, rictor, or mLST8 reveals that mTORC2 is required for signaling to Akt-FOXO and PKCalpha, but not S6K1. Dev Cell 11:859–871

Hanada M, Feng J, Hemmings BA (2004) Structure, regulation and function of PKB/AKT–a major therapeutic target. Biochim Biophys Acta 1697:3–16

Heron-Milhavet L, Franckhauser C, Rana V, Berthenet C, Fisher D, Hemmings BA, Fernandez A, Lamb NJ (2006) Only Akt1 is required for proliferation, while Akt2 promotes cell cycle exit through p21 binding. Mol Cell Biol 26:8267–8280

Heron-Milhavet L, Mamaeva D, Rochat A, Lamb NJ, Fernandez A (2008) Akt2 is implicated in skeletal muscle differentiation and specifically binds Prohibitin2/REA. J Cell Physiol 214: 158–165

Hinton HJ, Alessi DR, Cantrell DA (2004) The serine kinase phosphoinositide-dependent kinase 1 (PDK1) regulates T cell development. Nat Immunol 5:539–545

Hiromura M, Okada F, Obata T, Auguin D, Shibata T, Roumestand C, Noguchi M (2004) Inhibition of Akt kinase activity by a peptide spanning the betaA strand of the proto-oncogene TCL1. J Biol Chem 279:53407–53418

Hoffman ES, Passoni L, Crompton T, Leu TM, Schatz DG, Koff A, Owen MJ, Hayday AC (1996) Productive T-cell receptor beta-chain gene rearrangement: coincident regulation of cell cycle and clonality during development *in vivo*. Genes Dev 10:948–962

Holland EC, Celestino J, Dai C, Schaefer L, Sawaya RE, Fuller GN (2000) Combined activation of Ras and Akt in neural progenitors induces glioblastoma formation in mice. Nat Genet 25:55–57

Hresko RC, Mueckler M (2005) mTOR.RICTOR is the Ser473 kinase for Akt/protein kinase B in 3T3-L1 adipocytes. J Biol Chem 280:40406–40416

Huber M, Helgason CD, Damen JE, Scheid M, Duronio V, Liu L, Ware MD, Humphries RK, Krystal G (1999) The role of SHIP in growth factor induced signalling. Prog Biophys Mol Biol 71:423–434

Hutchinson J, Jin J, Cardiff RD, Woodgett JR, Muller WJ (2001) Activation of Akt (protein kinase B) in mammary epithelium provides a critical cell survival signal required for tumor progression. Mol Cell Biol 21:2203–2212

Hutchinson JN, Jin J, Cardiff RD, Woodgett JR, Muller WJ (2004) Activation of Akt-1 (PKB-alpha) can accelerate ErbB-2-mediated mammary tumorigenesis but suppresses tumor invasion. Cancer Res 64:3171–3178

Ingley E, Hemmings BA (2000) PKB/Akt interacts with inosine-5' monophosphate dehydrogenase through its pleckstrin homology domain. FEBS Lett 478:253–259

Inoki K, Li Y, Zhu T, Wu J, Guan KL (2002) TSC2 is phosphorylated and inhibited by Akt and suppresses mTOR signalling. Nat Cell Biol 4:648–657

Jacinto E, Facchinetti V, Liu D, Soto N, Wei S, Jung SY, Huang Q, Qin J, Su B (2006) SIN1/MIP1 maintains rictor-mTOR complex integrity and regulates Akt phosphorylation and substrate specificity. Cell 127:125–137

Jahn T, Seipel P, Urschel S, Peschel C, Duyster J (2002) Role for the adaptor protein Grb10 in the activation of Akt. Mol Cell Biol 22:979–991

Jiang ZY, Zhou QL, Coleman KA, Chouinard M, Boese Q, Czech MP (2003) Insulin signaling through Akt/protein kinase B analyzed by small interfering RNA-mediated gene silencing. Proc Natl Acad Sci USA 100:7569–7574

Jones PF, Jakubowicz T, Pitossi FJ, Maurer F, Hemmings BA (1991) Molecular cloning and identification of a serine/threonine protein kinase of the second-messenger subfamily. Proc Natl Acad Sci USA 88:4171–4175

Jones RG, Parsons M, Bonnard M, Chan VS, Yeh WC, Woodgett JR, Ohashi PS (2000) Protein kinase B regulates T lymphocyte survival, nuclear factor kappaB activation, and Bcl-X(L) levels *in vivo*. J Exp Med 191:1721–1734

Juntilla MM, Wofford JA, Birnbaum MJ, Rathmell JC, Koretzky GA (2007) Akt1 and Akt2 are required for alphabeta thymocyte survival and differentiation. Proc Natl Acad Sci USA 104:12105–12110

Katome T, Obata T, Matsushima R, Masuyama N, Cantley LC, Gotoh Y, Kishi K, Shiota H, Ebina Y (2003) Use of RNA interference-mediated gene silencing and adenoviral overexpression to elucidate the roles of AKT/protein kinase B isoforms in insulin actions. J Biol Chem 278: 28312–28323

Kelly AP, Finlay DK, Hinton HJ, Clarke RG, Fiorini E, Radtke F, Cantrell DA (2007) Notch-induced T cell development requires phosphoinositide-dependent kinase 1. EMBO J 26:3441–3450

Kim AH, Yano H, Cho H, Meyer D, Monks B, Margolis B, Birnbaum MJ, Chao MV (2002) Akt1 regulates a JNK scaffold during excitotoxic apoptosis. Neuron 35:697–709

Kim AH, Sasaki T, Chao MV (2003) JNK-interacting protein 1 promotes Akt1 activation. J Biol Chem 278:29830–29836

Knobbe CB, Reifenberger J, Blaschke B, Reifenberger G (2004) Hypermethylation and transcriptional downregulation of the carboxyl-terminal modulator protein gene in glioblastomas. J Natl Cancer Inst 96:483–486

Kohn AD, Kovacina KS, Roth RA (1995) Insulin stimulates the kinase activity of RAC-PK, a pleckstrin homology domain containing ser/thr kinase. Embo J 14:4288–4295

Kohn AD, Summers SA, Birnbaum MJ, Roth RA (1996) Expression of a constitutively active Akt Ser/Thr kinase in 3T3-L1 adipocytes stimulates glucose uptake and glucose transporter 4 translocation. J Biol Chem 271:31372–31378

Koo SH, Satoh H, Herzig S, Lee CH, Hedrick S, Kulkarni R, Evans RM, Olefsky J, Montminy M (2004) PGC-1 promotes insulin resistance in liver through PPAR-alpha-dependent induction of TRB-3. Nat Med 10:530–534

Kunkel MT, Ni Q, Tsien RY, Zhang J, Newton AC (2005) Spatio-temporal dynamics of protein kinase B/Akt signaling revealed by a genetically encoded fluorescent reporter. J Biol Chem 280:5581–5587

Laine J, Kunstle G, Obata T, Sha M, Noguchi M (2000) The protooncogene TCL1 is an Akt kinase coactivator. Mol Cell 6:395–407

Li X, Monks B, Ge Q, Birnbaum MJ (2007) Akt/PKB regulates hepatic metabolism by directly inhibiting PGC-1alpha transcription coactivator. Nature 447:1012–1016

Liu R, Tian B, Gearing M, Hunter S, Ye K, Mao Z (2008) Cdk5-mediated regulation of the PIKE-A-Akt pathway and glioblastoma cell invasion. Proc Natl Acad Sci USA 105:7570–7575

Lopez-Ilasaca M, Crespo P, Pellici PG, Gutkind JS, Wetzker R (1997) Linkage of G protein-coupled receptors to the MAPK signaling pathway through PI 3-kinase gamma. Science 275:394–397

Ma A, Pena JC, Chang B, Margosian E, Davidson L, Alt FW, Thompson CB (1995) Bclx regulates the survival of double-positive thymocytes. Proc Natl Acad Sci USA 92:4763–4767

Magun R, Burgering BM, Coffer PJ, Pardasani D, Lin Y, Chabot J, Sorisky A (1996) Expression of a constitutively activated form of protein kinase B (c-Akt) in 3T3-L1 preadipose cells causes spontaneous differentiation. Endocrinology 137:3590–3593

Maira SM, Galetic I, Brazil DP, Kaech S, Ingley E, Thelen M, Hemmings BA (2001) Carboxyl-terminal modulator protein (CTMP), a negative regulator of PKB/Akt and v-Akt at the plasma membrane. Science 294:374–380

Majumder PK, Yeh JJ, George DJ, Febbo PG, Kum J, Xue Q, Bikoff R, Ma H, Kantoff PW, Golub TR et al (2003) Prostate intraepithelial neoplasia induced by prostate restricted Akt activation: the MPAKT model. Proc Natl Acad Sci USA 100:7841–7846

Malanga D, Scrima M, De Marco C, Fabiani F, De Rosa N, De Gisi S, Malara N, Savino R, Rocco G, Chiappetta G et al (2008) Activating E17K mutation in the gene encoding the protein kinase AKT1 in a subset of squamous cell carcinoma of the lung. Cell Cycle 7:665–669

Mallick CA, Dudley EC, Viney JL, Owen MJ, Hayday AC (1993) Rearrangement and diversity of T cell receptor beta chain genes in thymocytes: a critical role for the beta chain in development. Cell 73:513–519

Malstrom S, Tili E, Kappes D, Ceci JD, Tsichlis PN (2001) Tumor induction by an Lck-MyrAkt transgene is delayed by mechanisms controlling the size of the thymus. Proc Natl Acad Sci USA 98:14967–14972

Mandal M, Borowski C, Palomero T, Ferrando AA, Oberdoerffer P, Meng F, Ruiz-Vela A, Ciofani M, Zuniga-Pflucker JC, Screpanti I et al (2005) The BCL2A1 gene as a pre-T cell receptor-induced regulator of thymocyte survival. J Exp Med 201:603–614

Manning BD, Cantley LC (2007) AKT/PKB signaling: navigating downstream. Cell 129: 1261–1274

Manning BD, Tee AR, Logsdon MN, Blenis J, Cantley LC (2002) Identification of the tuberous sclerosis complex-2 tumor suppressor gene product tuberin as a target of the phosphoinositide 3-kinase/akt pathway. Mol Cell 10:151–162

Mao C, Tili EG, Dose M, Haks MC, Bear SE, Maroulakou I, Horie K, Gaitanaris GA, Fidanza V, Ludwig T et al (2007) Unequal contribution of Akt isoforms in the double-negative to double-positive thymocyte transition. J Immunol 178:5443–5453

Mayo LD, Donner DB (2001) A phosphatidylinositol 3-kinase/Akt pathway promotes translocation of Mdm2 from the cytoplasm to the nucleus. Proc Natl Acad Sci USA 98:11598–11603

Medema RH, Kops GJ, Bos JL, Burgering BM (2000) AFX-like Forkhead transcription factors mediate cell-cycle regulation by Ras and PKB through p27kip1. Nature 404:782–787

Mende I, Malstrom S, Tsichlis PN, Vogt PK, Aoki M (2001) Oncogenic transformation induced by membrane-targeted Akt2 and Akt3. Oncogene 20:4419–4423

Michell BJ, Griffiths JE, Mitchelhill KI, Rodriguez-Crespo I, Tiganis T, Bozinovski S, de Montellano PR, Kemp BE, Pearson RB (1999) The Akt kinase signals directly to endothelial nitric oxide synthase. Curr Biol 9:845–848

Milburn CC, Deak M, Kelly SM, Price NC, Alessi DR, Van Aalten DM (2003) Binding of phosphatidylinositol 3,4,5-trisphosphate to the pleckstrin homology domain of protein kinase B induces a conformational change. Biochem J 375:531–538

Mitsuuchi Y, Johnson SW, Sonoda G, Tanno S, Golemis EA, Testa JR (1999) Identification of a chromosome 3p14.3-21.1 gene, APPL, encoding an adaptor molecule that interacts with the oncoprotein-serine/threonine kinase AKT2. Oncogene 18:4891–4898

Miyawaki T, Ofengeim D, Noh KM, Latuszek-Barrantes A, Hemmings BA, Follenzi A, Zukin RS (2009) The endogenous inhibitor of Akt, CTMP, is critical to ischemia-induced neuronal death. Nat Neurosci 12:618–626

Mori T, Sakaue H, Iguchi H, Gomi H, Okada Y, Takashima Y, Nakamura K, Nakamura T, Yamauchi T, Kubota N et al (2005) Role of Kruppel-like factor 15 (KLF15) in transcriptional regulation of adipogenesis. J Biol Chem 280:12867–12875

Nakae J, Kitamura T, Kitamura Y, Biggs WH 3rd, Arden KC, Accili D (2003) The forkhead transcription factor Foxo1 regulates adipocyte differentiation. Dev Cell 4:119–129

Orsulic S, Li Y, Soslow RA, Vitale-Cross LA, Gutkind JS, Varmus HE (2002) Induction of ovarian cancer by defined multiple genetic changes in a mouse model system. Cancer Cell 1:53–62

Ozes ON, Mayo LD, Gustin JA, Pfeffer SR, Pfeffer LM, Donner DB (1999) NF-kappaB activation by tumour necrosis factor requires the Akt serine-threonine kinase. Nature 401:82–85

Pan J, Pei DS, Yin XH, Hui L, Zhang GY (2006) Involvement of oxidative stress in the rapid Akt1 regulating a JNK scaffold during ischemia in rat hippocampus. Neurosci Lett 392:47–51

Paramio JM, Segrelles C, Ruiz S, Jorcano JL (2001) Inhibition of protein kinase B (PKB) and PKCzeta mediates keratin K10-induced cell cycle arrest. Mol Cell Biol 21:7449–7459

Parcellier A, Tintignac LA, Zhuravleva E, Cron P, Schenk S, Bozulic L, Hemmings BA (2009) Carboxy-Terminal Modulator Protein (CTMP) is a mitochondrial protein that sensitizes cells to apoptosis. Cell Signal 21:639–650

Patra AK, Drewes T, Engelmann S, Chuvpilo S, Kishi H, Hunig T, Serfling E, Bommhardt UH (2006) PKB rescues calcineurin/NFAT-induced arrest of Rag expression and pre-T cell differentiation. J Immunol 177:4567–4576

Pearce LR, Huang X, Boudeau J, Pawlowski R, Wullschleger S, Deak M, Ibrahim AF, Gourlay R, Magnuson MA, Alessi DR (2007) Identification of Protor as a novel Rictor-binding component of mTOR complex-2. Biochem J 405:513–522

Pekarsky Y, Koval A, Hallas C, Bichi R, Tresini M, Malstrom S, Russo G, Tsichlis P, Croce CM (2000) Tcl1 enhances Akt kinase activity and mediates its nuclear translocation. Proc Natl Acad Sci USA 97:3028–3033

Pellegrini L, Scorrano L (2007) A cut short to death: Parl and Opa1 in the regulation of mitochondrial morphology and apoptosis. Cell Death Differ 14:1275–1284

Peng XD, Xu PZ, Chen ML, Hahn-Windgassen A, Skeen J, Jacobs J, Sundararajan D, Chen WS, Crawford SE, Coleman KG et al (2003) Dwarfism, impaired skin development, skeletal muscle atrophy, delayed bone development, and impeded adipogenesis in mice lacking Akt1 and Akt2. Genes Dev 17:1352–1365

Podsypanina K, Ellenson LH, Nemes A, Gu J, Tamura M, Yamada KM, Cordon-Cardo C, Catoretti G, Fisher PE, Parsons R (1999) Mutation of Pten/Mmac1 in mice causes neoplasia in multiple organ systems. Proc Natl Acad Sci USA 96:1563–1568

Potter CJ, Pedraza LG, Xu T (2002) Akt regulates growth by directly phosphorylating Tsc2. Nat Cell Biol 4:658–665

Rathmell JC, Elstrom RL, Cinalli RM, Thompson CB (2003) Activated Akt promotes increased resting T cell size, CD28-independent T cell growth, and development of autoimmunity and lymphoma. Eur J Immunol 33:2223–2232

Remy I, Michnick SW (2004) Regulation of apoptosis by the Ft1 protein, a new modulator of protein kinase B/Akt. Mol Cell Biol 24:1493–1504

Rivera-Calzada A, Maman JD, Spagnolo L, Pearl LH, Llorca O (2005) Three-dimensional structure and regulation of the DNA-dependent protein kinase catalytic subunit (DNA-PKcs). Structure 13:243–255

Rivera-Calzada A, Spagnolo L, Pearl LH, Llorca O (2007) Structural model of full-length human Ku70-Ku80 heterodimer and its recognition of DNA and DNA-PKcs. EMBO Rep 8:56–62

Rossig L, Jadidi AS, Urbich C, Badorff C, Zeiher AM, Dimmeler S (2001) Akt-dependent phosphorylation of p21(Cip1) regulates PCNA binding and proliferation of endothelial cells. Mol Cell Biol 21:5644–5657

Sano H, Kane S, Sano E, Miinea CP, Asara JM, Lane WS, Garner CW, Lienhard GE (2003) Insulin-stimulated phosphorylation of a Rab GTPase-activating protein regulates GLUT4 translocation. J Biol Chem 278:14599–14602

Santos M, Paramio JM, Bravo A, Ramirez A, Jorcano JL (2002) The expression of keratin k10 in the basal layer of the epidermis inhibits cell proliferation and prevents skin tumorigenesis. J Biol Chem 277:19122–19130

Sarbassov DD, Guertin DA, Ali SM, Sabatini DM (2005) Phosphorylation and regulation of Akt/PKB by the rictor-mTOR complex. Science 307:1098–1101

Sasaki K, Sato M, Umezawa Y (2003) Fluorescent indicators for Akt/protein kinase B and dynamics of Akt activity visualized in living cells. J Biol Chem 278:30945–30951

Shiota C, Woo JT, Lindner J, Shelton KD, Magnuson MA (2006) Multiallelic disruption of the rictor gene in mice reveals that mTOR complex 2 is essential for fetal growth and viability. Dev Cell 11:583–589

Smith FM, Holt LJ, Garfield AS, Charalambous M, Koumanov F, Perry M, Bazzani R, Sheardown SA, Hegarty BD, Lyons RJ et al (2007) Mice with a disruption of the imprinted Grb10 gene exhibit altered body composition, glucose homeostasis, and insulin signaling during postnatal life. Mol Cell Biol 27:5871–5886

Smith JR, Clarke PA, de Billy E, Workman P (2009) Silencing the cochaperone CDC37 destabilizes kinase clients and sensitizes cancer cells to HSP90 inhibitors. Oncogene 28:157–169

Song JJ, Lee YJ (2005) Dissociation of Akt1 from its negative regulator JIP1 is mediated through the ASK1-MEK-JNK signal transduction pathway during metabolic oxidative stress: a negative feedback loop. J Cell Biol 170:61–72

Staal SP (1987) Molecular cloning of the akt oncogene and its human homologues AKT1 and AKT2: amplification of AKT1 in a primary human gastric adenocarcinoma. Proc Natl Acad Sci USA 84:5034–5037

Staal SP, Hartley JW (1988) Thymic lymphoma induction by the AKT8 murine retrovirus. J Exp Med 167:1259–1264

Staal SP, Hartley JW, Rowe WP (1977) Isolation of transforming murine leukemia viruses from mice with a high incidence of spontaneous lymphoma. Proc Natl Acad Sci USA 74:3065–3067

Stambolic V, Suzuki A, de la Pompa JL, Brothers GM, Mirtsos C, Sasaki T, Ruland J, Penninger JM, Siderovski DP, Mak TW (1998) Negative regulation of PKB/Akt-dependent cell survival by the tumor suppressor PTEN. Cell 95:29–39

Sun L, Liu L, Yang XJ, Wu Z (2004) Akt binds prohibitin 2 and relieves its repression of MyoD and muscle differentiation. J Cell Sci 117:3021–3029

Surucu B, Bozulic L, Hynx D, Parcellier A, Hemmings BA (2008) *In vivo* Analysis of Protein Kinase B (PKB)/Akt Regulation in DNA-PKcs-null Mice Reveals a Role for PKB/Akt in DNA Damage Response and Tumorigenesis. J Biol Chem 283:30025–30033

Suzuki A, de la Pompa JL, Stambolic V, Elia AJ, Sasaki T, del Barco Barrantes I, Ho A, Wakeham A, Itie A, Khoo W et al (1998) High cancer susceptibility and embryonic lethality associated with mutation of the PTEN tumor suppressor gene in mice. Curr Biol 8:1169–1178

Tanaka M, Konishi H, Touhara K, Sakane F, Hirata M, Ono Y, Kikkawa U (1999) Identification of myosin II as a binding protein to the PH domain of protein kinase B. Biochem Biophys Res Commun 255:169–174

Tanti JF, Grillo S, Gremeaux T, Coffer PJ, Van Obberghen E, Le Marchand-Brustel Y (1997) Potential role of protein kinase B in glucose transporter 4 translocation in adipocytes. Endocrinology 138:2005–2010

Tschopp O, Yang ZZ, Brodbeck D, Dummler BA, Hemmings-Mieszczak M, Watanabe T, Michaelis T, Frahm J, Hemmings BA (2005) Essential role of protein kinase B gamma

(PKB gamma/Akt3) in postnatal brain development but not in glucose homeostasis. Development 132:2943–2954
Ugi S, Imamura T, Maegawa H, Egawa K, Yoshizaki T, Shi K, Obata T, Ebina Y, Kashiwagi A, Olefsky JM (2004) Protein phosphatase 2A negatively regulates insulin's metabolic signaling pathway by inhibiting Akt (protein kinase B) activity in 3T3-L1 adipocytes. Mol Cell Biol 24:8778–8789
van den Heuvel AP, de Vries-Smits AM, van Weeren PC, Dijkers PF, de Bruyn KM, Riedl JA, Burgering BM (2002) Binding of protein kinase B to the plakin family member periplakin. J Cell Sci 115:3957–3966
Wang L, Balas B, Christ-Roberts CY, Kim RY, Ramos FJ, Kikani CK, Li C, Deng C, Reyna S, Musi N et al (2007) Peripheral disruption of the Grb10 gene enhances insulin signaling and sensitivity *in vivo*. Mol Cell Biol 27:6497–6505
Wick KR, Werner ED, Langlais P, Ramos FJ, Dong LQ, Shoelson SE, Liu F (2003) Grb10 inhibits insulin-stimulated insulin receptor substrate (IRS)-phosphatidylinositol 3-kinase/Akt signaling pathway by disrupting the association of IRS-1/IRS-2 with the insulin receptor. J Biol Chem 278:8460–8467
Wullschleger S, Loewith R, Hall MN (2006) TOR signaling in growth and metabolism. Cell 124:471–484
Xu J, Liao K (2004) Protein kinase B/AKT 1 plays a pivotal role in insulin-like growth factor-1 receptor signaling induced 3T3-L1 adipocyte differentiation. J Biol Chem 279:35914–35922
Yang J, Cron P, Good VM, Thompson V, Hemmings BA, Barford D (2002a) Crystal structure of an activated Akt/protein kinase B ternary complex with GSK3-peptide and AMP-PNP. Nat Struct Biol 9:940–944
Yang J, Cron P, Thompson V, Good VM, Hess D, Hemmings BA, Barford D (2002b) Molecular mechanism for the regulation of protein kinase B/Akt by hydrophobic motif phosphorylation. Mol Cell 9:1227–1240
Yang ZZ, Tschopp O, Hemmings-Mieszczak M, Feng J, Brodbeck D, Perentes E, Hemmings BA (2003) Protein kinase B alpha/Akt1 regulates placental development and fetal growth. J Biol Chem 278:32124–32131
Yang ZZ, Tschopp O, Baudry A, Dummler B, Hynx D, Hemmings BA (2004) Physiological functions of protein kinase B/Akt. Biochem Soc Trans 32:350–354
Yang ZZ, Tschopp O, Di-Poi N, Bruder E, Baudry A, Dummler B, Wahli W, Hemmings BA (2005) Dosage-dependent effects of Akt1/protein kinase Balpha (PKBalpha) and Akt3/PKBgamma on thymus, skin, and cardiovascular and nervous system development in mice. Mol Cell Biol 25:10407–10418
Yoeli-Lerner M, Yiu GK, Rabinovitz I, Erhardt P, Jauliac S, Toker A (2005) Akt blocks breast cancer cell motility and invasion through the transcription factor NFAT. Mol Cell 20:539–550
Yue Y, Lypowy J, Hedhli N, Abdellatif M (2004) Ras GTPase-activating protein binds to Akt and is required for its activation. J Biol Chem 279:12883–12889
Yun SJ, Kim EK, Tucker DF, Kim CD, Birnbaum MJ, Bae SS (2008) Isoform-specific regulation of adipocyte differentiation by Akt/protein kinase Balpha. Biochem Biophys Res Commun 371:138–143
Zhou BP, Liao Y, Xia W, Zou Y, Spohn B, Hung MC (2001) HER-2/neu induces p53 ubiquitination via Akt-mediated MDM2 phosphorylation. Nat Cell Biol 3:973–982

PI3Ks in Lymphocyte Signaling and Development

Klaus Okkenhaug and David A. Fruman

Contents

Abstract Lymphocyte development and function are regulated by tyrosine kinase and G-protein coupled receptors. Each of these classes of receptors activates phosphoinositide 3-kinase (PI3K). In this chapter, we summarize current understanding of how PI3K contributes to key aspects of the adaptive immune system.

K. Okkenhaug (✉)
Laboratory of Lymphocyte Signalling and Development, The Babraham Institute, Cambridge, UK
klaus.okkenhaug@bbsrc.ac.uk

D.A. Fruman (✉)
Department of Molecular Biology and Biochemistry, Institute for Immunology, University of California-Irvine, Irvine, CA 92697-3900, USA
dfruman@uci.edu

C. Rommel et al. (eds.). *Phosphoinositide 3-kinase in Health and Disease*, Volume 1
Current Topics in Microbiology and Immunology 346, DOI 10.1007/82_2010_45

Abbreviations

APC	Antigen presenting cell
BAFF	B-cell activation factor of the tumor necrosis factor family
BCAP	B cell adaptor protein
BcR	B-cell antigen receptor
DC	Dendritic cell
dko	Double knockout
DN	Double negative
DP	Double positive
FO	Follicular
Foxo	Forkhead box subgroup O
GEF	Guanine nucleotide exchange factor
GPCR	G-protein coupled receptor
IFN-γ	Interferon-gamma
IL-10	Interleukin-10
IL-2	Interleukin-2
IL-4	Interleukin-4
IL-7	Interleukin-7
KLF2	Krüppel-like factor-2
LPS	Lipopolysaccharide
MHC	Major histocompatibility complex
mTOR	Mammalian target of rapamycin
mTORC1	mTOR complex-1
MZ	Marginal zone
NK	Natural killer
PI3K	Phosphoinositide 3-kinase
PIP_2	Phosphatidylinositol-4,5-bisphosphate
PIP_3	Phosphatidylinositol-3,4,5-trisphosphate
PKCβ	Protein kinase C-beta
PKCθ	Protein kinase C-theta
PLCγ	Phospholipase C-gamma
pMHC	Peptide-MHC
RAG	Recombination-activating gene
S1P	Sphingosine-1-phosphate
SH2	Src homology-2
SLP-65	SH2 domain containing protein of 65 kDa
TcR	T-cell antigen receptor
Th1	T helper type-1
Th2	T helper type-2
TLR	Toll-like receptor
Tregs	Regulatory T cells

1 Introduction

The adaptive immune response protects the host from recurring infections by viruses, bacteria, and parasites. The enormous diversity and exquisite specificity of the antigen receptors expressed by B and T lymphocytes ensure that any pathogen can be recognized, while clonal expansion and the generation of long-lived memory cells ensure effective clearance and, in many cases, life-long immunity from reinfection by the same pathogen. In addition, natural killer (NK) lymphocytes provide more immediate protection and can, in some cases, serve as a back-up mechanism to kill pathogens that evade immune recognition by T cells by suppressing major histocompatibility complex (MHC) expression. The diversity of the antigen receptor repertoire is generated by recombination of gene components in B-cell and T-cell precursors. After gene recombination, each clone is selected so that clones bearing nonfunctional or autoreactive antigen receptors are eliminated. In the case of T cells, clones bearing antigen receptors capable of recognizing MHC structures are positively selected. This system originally evolved in teleost fishes and is required for protection against a wide variety of common pathogens, many of which have coevolved mechanisms to evade the adaptive immune response. The extreme susceptibility of patients with severe combined or acquired immune deficiency to common infections illustrates how dependent we are on the adaptive immune responses for survival. However, the evolution of the adaptive immune system has also come at a cost, particularly in long-lived animals such as humans. Self–non-self discrimination is an imperfect process, and several common autoimmune diseases are caused by autoreactive T cells and/or B cells. In addition, the gut, vaginal tract, respiratory tract, and skin are host to a rich microbial flora of mostly harmless or beneficial microbes. Inappropriate activation of the immune system against commensal bacteria causes unwarranted inflammation and disease. Moreover, the immune system can respond inappropriately to innocuous substances such as pollen and cause allergic reactions. Finally, the rejection of transplanted organs is mediated by the adaptive immune response. Therefore, therapeutic strategies are aimed at dampening unwanted immune responses without rendering the patient unprotected from infections.

Currently, the most widely used immunosuppressive drugs are corticosteroids, cyclosporine, and cytotoxic or cytostatic drugs such as methotrexate or rapamycin. In addition, various recombinant protein-based therapies have been developed; the most widely used of these have been the anti-tumor necrosis factor-α drugs used to treat rheumatoid arthritis and other autoimmune diseases (Feldmann and Maini 2008). Abatacept (CTLA4-Ig) and Rituximab (anti-CD20) are more recent examples of agents that target T cells and B cells specifically and help treat autoimmunity (Bluestone et al. 2006; Edwards et al. 2004). Yet, there remains significant clinical need for additional drugs that either suppress or modulate adaptive immune responses, as not all patients respond favorably to currently available drugs. The phosphoinositide 3-kinase (PI3K) enzymes, which are the focus of this series, have received considerable interest in this context

(Rommel et al. 2007; Ruckle et al. 2006). PI3Ks play central roles in signaling not only by antigen receptors but also by various costimulatory, cytokine, and chemokine receptors that control lymphocyte biology (Okkenhaug et al. 2007; Fruman and Bismuth 2009). In general, the PI3K isoforms p110α and p110β transmit signals from tyrosine kinases and G-protein coupled receptors (GPCRs), respectively (Roberts and Ilić 2010). However, in lymphocytes, tyrosine kinase-associated receptors signal primarily through p110δ, whereas GPCRs signal primarily through p110γ, although there are notable exceptions, which we highlight in this review. Since p110δ and p110γ are expressed at low or undetectable levels in most organs, inhibitors against p110δ and p110γ should, in theory, be selective for the immune system and nontoxic. In this chapter, we will consider the roles played by PI3K during development and activation of lymphocytes and also consider both the advantages and liabilities associated with pharmacological targeting of these PI3K isoforms.

Our understanding of the roles of PI3Ks in lymphocyte biology has been advanced both by the development of small molecule inhibitors that target the PI3Ks with high selectivity, and by gene mutation through homologous recombination in mice (for a list of gene targeting studies, see Table 1). Initial studies demonstrated that B-cell and T-cell proliferation could be blocked by the broad-spectrum PI3K inhibitors wortmannin and LY294002 (Crabbe et al. 2007). However, these compounds have many off-target effects, and a specific role for PI3K could therefore not always be established unequivocally. The demonstration that p85α knockout mice showed impaired B-cell development and humoral immune responses provided clear evidence that PI3Ks are intimately involved in signal transduction by the B-cell antigen receptor (BcR) (Fruman et al. 1999; Suzuki et al. 1999). These phenotypes were also evident in mice where p110δ had been inactivated by deletion or by point mutations in the genes (Clayton et al. 2002; Okkenhaug et al. 2002; Jou et al. 2002). Thus, a picture emerged, in which p85α would recruit p110δ to antigen and coreceptor complexes at the plasma membrane and that the PI3K activity thus generated was essential for B-cell proliferation. The situation for T cells was more complicated. While p85α was dispensable for normal T-cell proliferation and cytokine production, p110δ did contribute to these responses – especially when T cells were stimulated with peptide-MHC rather than antibodies (Okkenhaug et al. 2002, 2006). This apparent discrepancy was in part resolved subsequently when it was demonstrated that p85α and p85β play mutually redundant roles in transmitting signals from the T-cell antigen receptor (TcR) (Deane et al. 2007). Nonetheless, in terms of development and proliferative responses by purified cells, the T cells appeared to be less severely affected by PI3K deficiency than B cells. However, while the development of various mature subsets and their proliferative responses are the most immediately obvious measurements to make, both T cells and B cells live complicated lives, and recently a more detailed appreciation of how PI3Ks can both enhance and suppress certain B cell and T cell responses has evolved. The challenge of understanding the apparently contradictory roles of PI3K in adaptive immune function is a significant one. Understanding these roles is going to have a

Table 1 Role of PI3K isoforms in lymphocyte development or function

PI3K isoform(s) targeted	Genetic or pharmacological	Cell type or animal model	References
p110δ	Null allele	T and B cells	Clayton et al. (2002), Jou et al. (2002)
		B cells	Llorian et al. (2007), Janas et al. (2008)
		NK cells	Kim et al. (2007b), Zebedin et al. (2008), Saudemont et al. (2009b)
p110δ	Kinase-dead D910A knock-in	T and B cells	Okkenhaug et al. (2002), Reif et al. (2004)
		T cells	Okkenhaug et al. (2006), Garcon et al. (2008), Jarmin et al. (2008), Mirenda et al. (2007), Nashed et al. (2007), Patton et al. (2006), Liu et al. (2009), Soond et al. (2010), Sinclair et al. (2008)
		B cells	Al-Alwan et al. (2007), Bilancio et al. (2006)
		NK or NKT	Guo et al. (2008), Saudemont et al. (2009b), Kishimoto et al. (2007)
p110δ	Selective inhibitor (IC87114)	B cells	Bilancio et al. (2006), Zhang et al. (2008)
		In vivo (asthma)	Lee et al. (2006)
p110γ	Null or kinase-dead	T and B cells	Reif et al. (2004), Nombela-Arrieta et al. (2004), Sasaki et al. (2000), Hirsch et al. (2000), Li et al. (2000)
		T cells	Garcon et al. (2008), Alcazar et al. (2007), Martin et al. (2008), Thomas et al. (2008)
		NK	Tassi et al. (2007), Guo et al. (2008), Saudemont et al. (2009b)
p110γ	Selective inhibitors	In vivo (arthritis, lupus)	Barber et al. (2005), Camps et al. (2005)
p110δ/p110γ	Double knockout or knockout/knock-in	T cells	Webb et al. (2005), Swat et al. (2006), Janas et al. (2010), Ji et al. (2007)
		NK cells	Tassi et al. (2007)
		In vivo (arthritis)	Randis et al. (2008)
p85α	Null	B cells	Suzuki et al. (1999, 2003), Donahue et al. (2004), Donahue and Fruman (2007), Hess et al. (2004)
		T cells	Shiroki et al. (2007)
p85α/p55α/p50α	Null	B cells	Fruman et al. (1999)
p85α/p55α/p50α	T cell-specific conditional	T cells	Deane et al. (2007)
p85α/p55α/p50α	B cell-specific conditional	B cells	Oak et al. (2009)
p85β	Null	T and B cells	Deane et al. (2004)
		T cells	Alcazar et al. (2009b)
p85α/p55α/p50α/p85β	p85 dko = double knockout (conditional p85α/p55α/p50α, null p85β)	T cells	Deane et al. (2007), Oak et al. (2006)
		B cells	Oak et al. (2009)

significant impact on the development of small molecules against the PI3Ks expressed in lymphocytes.

2 PI3K in B Cells

2.1 B-Cell Development

B cells develop from lymphoid progenitors in the fetal liver and bone marrow (Hardy et al. 2007). At an early stage, a lineage split occurs resulting in two general subsets: B-1 B cells that reside in body cavities, and B-2 B cells that reside in secondary lymphoid organs. The B-2 subset is further subdivided into marginal zone (MZ) B cells, which occupy a unique anatomical and functional niche in the MZ of the spleen; and follicular (FO) B cells, which recirculate through blood, lymph, and secondary lymphoid tissues. Mice lacking p85α or p110δ have markedly fewer B-1 and MZ B cells as well as reduced numbers of FO cells (Fruman et al. 1999; Suzuki et al. 1999, 2003; Clayton et al. 2002; Okkenhaug et al. 2002; Jou et al. 2002; Donahue et al. 2004).

Developing B cells undergo an ordered series of gene rearrangement steps (Herzog et al. 2009). First, heavy chain rearrangement occurs at the pro-B-cell stage. If successful, light chain rearrangement proceeds at the pre-B-cell stage. Cells that make productive heavy and light chain rearrangements express surface IgM and are classified as immature B cells. These cells are then screened for self-reactivity in the bone marrow and spleen, and autoreactive cells are either anergized, deleted, or induced to re-express recombination-activating gene (RAG) proteins and undergo receptor editing. In the absence of self-antigen, immature B cells suppress RAG expression via tonic BcR signaling and advance to the mature B-cell state. Mice lacking p85α or p110δ have reduced numbers of cells making it through the pro-B to pre-B transition as well as defects in tonic signaling at the immature stage (Fruman et al. 1999; Suzuki et al. 1999; Clayton et al. 2002; Okkenhaug et al. 2002; Llorian et al. 2007; Verkoczy et al. 2007). In addition, PI3K inhibitors can lead to dedifferentiation of immature B cells in vitro (Tze et al. 2005).

Activation of naive B cells by foreign antigen, in the presence of T-cell help or Toll-like receptor (TLR) costimulation, results in clonal expansion and differentiation. Some activated B cells develop rapidly into antibody-secreting plasma cells to provide an early source of antibody. Other B-cell clones continue differentiation within the germinal centers within lymphoid follicles, where the antibody constant region is changed by isotype switching, and the variable region is modified by somatic hypermutation. B cells surviving the germinal center reaction then develop either into plasma cells secreting high-affinity antibodies, or memory B cells capable of mounting rapid secondary responses. B cells lacking p85α or p110δ mediate reduced T-independent antibody responses in response to polymeric antigens in vivo (Fruman et al. 1999; Suzuki et al. 1999; Clayton et al. 2002; Okkenhaug et al. 2002; Jou et al. 2002).

2.2 *BcR Signaling, Antigen Presentation, and Metabolism*

Each of the B-cell fate decisions outlined above is controlled by signaling through the BcR (Fig. 1). Considering the various defects in PI3K-deficient B cells, it is therefore not surprising that PI3K activation is a fundamental aspect of BcR signaling. The mechanism through which the BcR is coupled to PI3K signaling has recently been resolved. Two proteins with YxxM motifs (providing the optimal binding site for the Src homology-2 (SH2) domains of p85) are known to be important for B-cell signaling: CD19 and B-cell adaptor protein (BCAP). While deleting either of these alone is insufficient to ablate PI3K signaling in B cells, combined deficiency of CD19 and BCAP reduces PI3K signaling below a detectable threshold and strongly impacts B-cell development at the stage where the B-cell progenitors first express a pre-BcR (composed of an immunoglobulin heavy chain and a surrogate light chain) (Aiba et al. 2008). As CD19 and BCAP are both substrates for BcR-dependent Syk activity, we can conclude that we now have a detailed model of how the BcR connects with PI3K activity. However, it should be noted that the guanine exchange factor Vav and its substrate Rac have also been implicated in the activation of PI3K downstream of the BcR (Vigorito et al. 2004; Inabe et al. 2002; Walmsley et al. 2003). In recognition of this complexity, we have proposed a "signalosome" model of BcR signaling in which PI3K, its lipid products, and other components function as an integrated molecular machine in which all parts are functionally interconnected (Fruman et al. 2000).

A primary role of PI3K in BcR signaling is to promote signalosome assembly and membrane localization, leading to phospholipase C-gamma (PLCγ) activation and phosphatidylinositol-4,5-bisphosphate (PIP_2) hydrolysis (Fruman and Bismuth 2009; Donahue et al. 2004). Thus, signaling events controlled by PLCγ such as elevation of cytoplasmic calcium and activation of protein kinase C-beta (PKCβ) are also impaired in PI3K-deficient B cells. B cells lacking PI3K or CD19 exhibit defective accumulation of protein aggregates, termed microsignalosomes, at or near the plasma membrane (Depoil et al. 2008; Weber et al. 2008). Consequently, PI3K-deficient B cells proliferate poorly in response to antigen receptor triggering. This response can be rescued by the provision of signals via CD40 or TLR4, which may explain why humoral immune responses can be realized in PI3K-deficient mice, albeit at reduced levels. Engagement of CD40 or TLRs provides a mechanism for activating the NFκB pathway, whose induction by BcR signaling is attenuated in PI3K-deficient cells. Interestingly, a B-cell-intrinsic mechanism to limit activation involves engagement of the inhibitory receptor FcγRIIb1, which recruits the inositol lipid phosphatase SHIP1 and reduces phosphatidylinositol-3,4,5-trisphosphate (PIP_3) levels. Mice lacking FcγRIIb1 develop a lupus-like syndrome, and some human lupus patients show decreased expression or function of FcγRIIb1 which

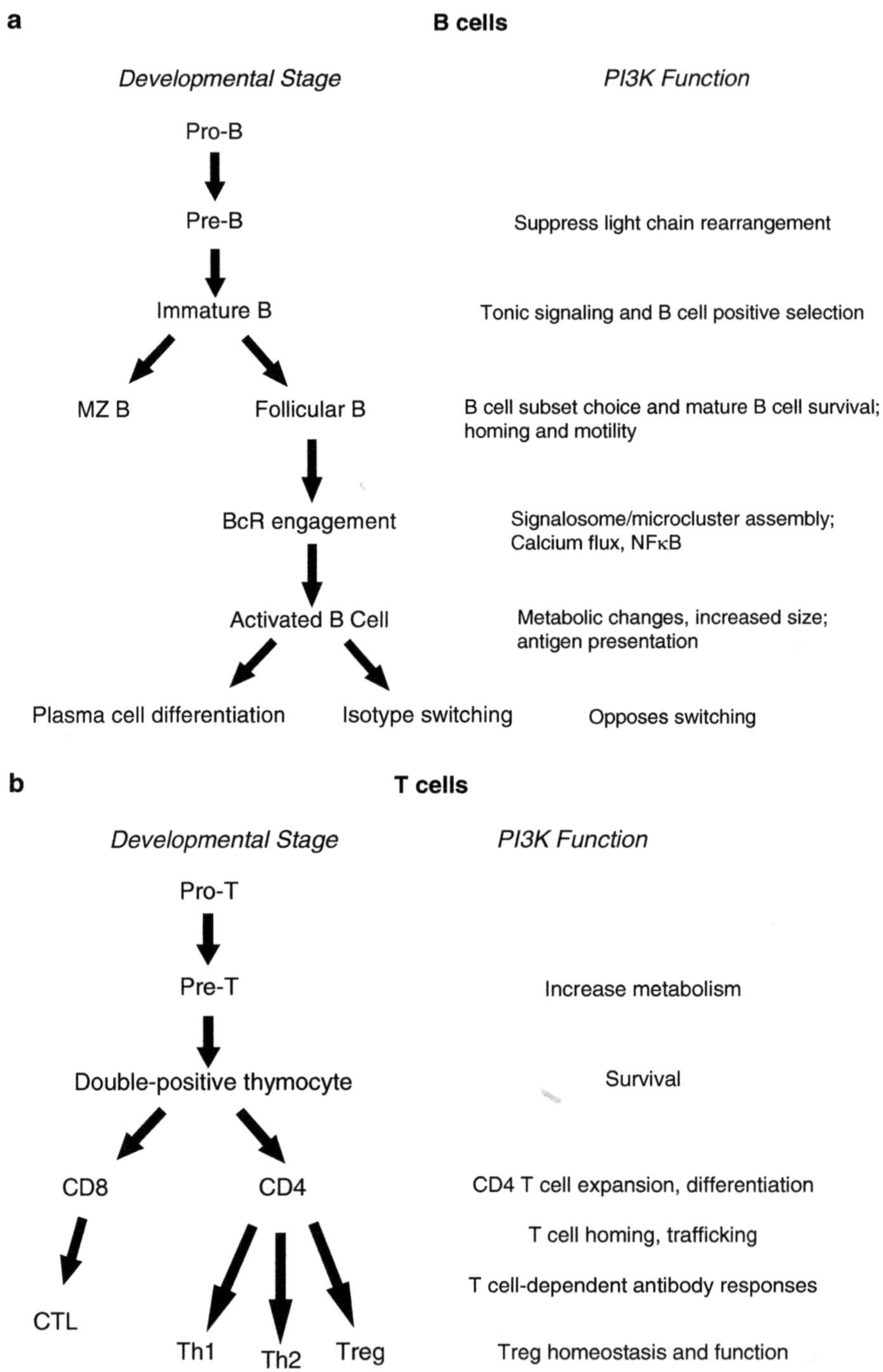

Fig. 1 Simplified flowcharts of lymphocyte development and function, showing key points of action of PI3K in B cells (**a**) and T cells (**b**)

indirectly suggests a role for SHIP in preventing autoimmunity (Floto et al. 2005; Brauweiler et al. 2000; Tarasenko et al. 2007a). Of interest, BcR-mediated PIP_3 production is suppressed in anergic B cells, providing another example of how intrinsic attenuation of PI3K activation limits B-cell responsiveness (Browne et al. 2009).

BcR clustering promotes antigen internalization, processing, and presentation of the antigen to T cells. Several studies using anti-Ig antibodies as model antigens suggested a role for PI3K signaling in BcR aggregation, internalization, and early stages of antigen presentation since internalization was inhibited by coligation of FcγRIIB (Phee et al. 2001; Wagle et al. 1999). A more recent study suggested that p110δ is required for efficient antigen presentation, but through regulation of a later step in antigen processing (Al-Alwan et al. 2007). Although defective antigen presentation may contribute to the reduced T-dependent antibody responses seen in p110δ-deficient mice (Clayton et al. 2002; Okkenhaug et al. 2002; Jou et al. 2002), other functions for p110δ in B and T cells are also likely to determine this phenotype.

In addition to providing "core" signals required for proliferation and differentiation, PI3Ks contribute to the fitness of B cells by increasing cellular growth and metabolism. These functions are mediated in part by the Akt serine/threonine kinases, which are activated by BcR and/or CD19 engagement in a PI3K-dependent manner. As in most cells, Akt contributes to B-cell size increase through promoting activation of mammalian target of rapamycin (mTOR) complex-1 (mTORC1) (Donahue and Fruman 2007). However, this mechanism is stimulus dependent, with lipopolysaccharide (LPS) inducing mTORC1 activity in a PI3K-independent manner. MZ cells exhibit high basal mTORC1 activity that might contribute to the larger size and more rapid activation of these cells compared to FO cells. Akt is also thought to mediate changes in glucose metabolism in activated B cells. BcR engagement induces a glycolytic switch and this can be mimicked by expression of an inducible, constitutively active Akt construct (Doughty et al. 2006). Activation of the PI3K/Akt signaling axis also contributes to survival in response to interleukin-4 (IL-4), B-cell activation factor of the tumor necrosis factor family (BAFF), and other cytokines (Bilancio et al. 2006; Henley et al. 2008). Thus PI3Ks contribute to diverse signaling events that control B-cell development and antibody production.

2.3 *Immunoglobulin Gene Rearrangement and Isotype Switching*

A surprising observation was that SH2 domain containing protein of 65 kDa (SLP-65) (otherwise known as BLNK), an adaptor protein important for signalosome assembly, appears to suppress PI3K signaling in pre-B cells (Herzog et al. 2008). The mechanism by which SLP-65 suppresses PI3K signaling is not known.

However, a picture of why it may be important to suppress PI3K signaling at various stages of development has recently emerged.

The only group of transcription factors whose activity is entirely dependent on PI3K signaling in diverse cell types is the Forkhead Box Subgroup O (Foxo) proteins (Coffer and Burgering 2004). There are four Foxo isoforms; among these, Foxo1 and Foxo3a appear to be the most important in B cells and T cells (Dengler et al. 2008; Amin and Schlissel 2008; Kerdiles et al. 2009; Lin et al. 2004; Ouyang et al. 2009). In its unphosphorylated form, Foxo translocates to the nucleus, where it drives the transcription of a number of genes whose products regulate apoptosis and cell cycle progression such as Bim, p27, and FasL (Coffer and Burgering 2004). More recently, it was found that Foxo promotes expression of the *Rag-1* and *Rag-2* genes (Dengler et al. 2008; Amin and Schlissel 2008). When phosphorylated by Akt, Foxo is found in the cytoplasm associated with 14-3-3 proteins and unable to regulate gene transcription. Thus, active PI3K/Akt signaling inhibits Foxo-dependent transcription. With respect to the *Rag* genes, it has been proposed that suppression of PI3K signaling is important for recombination of the immunoglobulin light chain locus to occur. Furthermore, it seems that subsequent reactivation of PI3K is also important to suppress Rag expression and to prevent inappropriate expression of a second light chain (Llorian et al. 2007; Herzog et al. 2008; Amin and Schlissel 2008). Concurrently, tonic signaling by surface IgM on immature B cells acts through PI3K to positively select cells for further differentiation into mature B-cell subsets and to promote survival of mature B cells (Verkoczy et al. 2007; Tze et al. 2005; Srinivasan et al. 2009).

Immunoglobulin isotype switching is also regulated by the suppression of Foxo function, and hence indirectly by PI3K. Class-switch recombination was enhanced in the presence of p110δ inhibitors and increased further by constitutively active Foxo proteins, but suppressed in Pten-deficient B cells where PI3K signaling was elevated (Omori et al. 2006). These results seemed to contradict other studies with p85α or p110δ deficient mice, which showed reduced switching to other isotypes after immunization. However, further analysis revealed increased propensity for both p85α- and p110δ-deficient B cells to produce immunoglobulins of the IgE subclass, both in vitro and in vivo (Zhang et al. 2008; Doi et al. 2008). A similar effect was observed in mice that were treated with the p110δ-selective inhibitor IC87114 (Zhang et al. 2008). An increased percentage of cells showed sequential switching from IgG1 to IgE, consistent with unrestrained class-switch recombination in the absence of p110δ activity. The excessive IgE production in $p85\alpha^{-/-}$, $p110\delta^{D910A}$, and IC87114-treated mice suggests that therapeutic use of p110δ inhibitors could lead to increased allergic responses. This should be compensated by diminished mast cell function, but only as long as the inhibitor is administered (Ali et al. 2004, 2008). Thus, while, in general, PI3Ks are thought to integrate signals from the BcR, costimulatory, and cytokine receptors to promote B-cell activation and differentiation, PI3Ks also negatively regulate the expression of Rag proteins and genes involved in immunoglobulin class-switch recombination and may thus paradoxically also limit aspects of B-cell development and differentiation.

2.4 *B-Cell Chemotaxis and Trafficking*

B cell trafficking is strongly influenced by PI3K signaling. Treatment of B cells with wortmannin reduces chemokine-driven migration in vitro, and alters homing after adoptive transfer to mice (Reif et al. 2004; Nombela-Arrieta et al. 2004; Matheu et al. 2007). B-cell chemotaxis is partially dependent on the class IA isoform p110δ, an exception to the rule that class IB (p110γ) functions downstream of GPCRs in lymphocytes. Wortmannin-treated B cells and p85α-deficient B cells also display reduced basal motility in lymph nodes (Matheu et al. 2007). Proper recirculation of B cells requires the lymph node homing receptor L-selectin (CD62L), whose expression is controlled by Foxo1 (Dengler et al. 2008). BcR stimulation downregulates L-selectin expression, and this is partially prevented in p85α-deficient B cells (Hess et al. 2004).

Figure 1a outlines key actions of PI3K in B-cell development and function, as highlighted in this section.

3 PI3K in T Cells

3.1 *T-Cell Development and Differentiation*

T cells develop in the thymus from $CD4^{-}CD8^{-}TcR^{-}$ precursors that arrive from the bone marrow. These precursors are referred to as double negative (DN) cells (because of the lack of expression of CD4 or CD8) and are subdivided into DN1–4 based on the differential expression of CD44 and CD25. At the DN3 stage, T cells express RAG enzymes that rearrange the TcR β-chain genes so that a TcRβ chain can be expressed on the cell surface. T cells that express a productively rearranged TcRβ chain are selected to become $CD4^{+}CD8^{+}$ double positive (DP) cells that rearrange and express TcRα chains. Concurrent with this developmental stage, TcRαβ+ DP T cells are positively selected to become CD4 or CD8 single positive mature thymocytes if they bind self-peptide presented by MHC class I or II molecules with intermediate affinity. DP cells are negatively selected though apoptosis (or adopt a regulatory phenotype) if they bind peptide-MHC (pMHC) with high affinity. Most DP thymocytes die by neglect, however, as they show no apparent affinity for pMHC molecules. Selected CD4+ or CD8+ thymocytes then migrate through the blood and lymph to populate the lymph nodes and spleen. PI3K controls T-cell biology at multiple junctions of their development (see Fig. 1b).

3.1.1 A Surprising Redundancy Between p110δ and p110γ During T-Cell Development

Both p110δ and p85 double knockout (p85α deleted in T cells, p85β germline deleted) (dko) mice produce near-normal number of thymocytes, whereas p110γ-deficient mice have reduced numbers of thymocytes, which was attributed to increased apoptosis of DP thymocytes (Okkenhaug et al. 2002; Deane et al. 2007; Sasaki et al. 2000; Rodriguez-Borlado et al. 2003). Strikingly, p110δ–p110γ double-deficient mice show a dramatic reduction in thymocyte numbers caused by a block at the DN3–DN4 stage of development in addition to reduced DP cell survival (Webb et al. 2005; Swat et al. 2006). Pre-TcR signaling as this stage was mostly controlled by p85α and p110δ (Shiroki et al. 2007). The unexpected redundancy between p110δ and p110γ therefore suggested that GPCRs might play a greater role at this developmental stage than previously appreciated. Indeed, further investigations demonstrated that coordinated activation of p110δ by the pre-TcR and p110γ by the chemokine receptor CXCR4 is essential for optimal development of double positive T cells (Janas et al. 2010). A key role for PI3K signaling during thymocyte development has also been revealed from mice with T-cell-specific deletions of Akt or Pdk1 (Hinton et al. 2004; Juntilla et al. 2007; Fayard et al. 2007; Mao et al. 2007). In each case, DP cell numbers were dramatically reduced. By contrast, deletion of Pten in pre-T cells led to the development of double positive T cells in the absence of pre-TcR expression, suggesting that unrestrained PI3K signaling can substitute for the signals that normally drive β selection (Hagenbeek et al. 2004). One key observation was that the transition from the DN3 to DN4 stage was accompanied by increased metabolic activity and cell size as well as by the expression of nutrient receptors such as CD71 (the transferrin receptor) and CD98 (an amino acid transporter subunit), presumably required to support the increased growth and proliferative burst (Ciofani and Zuniga-Pflucker 2005; Kelly et al. 2007). Pdk1$^{-/-}$ thymocytes failed to upregulate these receptors, and Pdk1$^{-/-}$ DN4 thymocytes were smaller than normal (Kelly et al. 2007). There is also a requirement for Notch signaling at the DN3 to DP thymocyte transition and PI3K and Akt were required for Notch-dependent increase in metabolism and cell growth at this stage (Juntilla et al. 2007; Ciofani and Zuniga-Pflucker 2005; Kelly et al. 2007). Notch is unlikely to directly regulate the recruitment of PI3K to the plasma membrane. Instead, it has been proposed that Notch can increase PI3K signaling by suppressing the expression of Pten (Palomero et al. 2007). This would be consistent with the observation that deletion of Pten was sufficient to promote DP thymocyte development in the absence of TcR signaling (Hagenbeek et al. 2004). Interestingly, Pten deficiency was also permissive for DP T-cell development in the absence of Pdk1, thus implicating additional PI3K-dependent signaling pathways in early T-cell development (Finlay et al. 2009).

3.2 TcR Signaling and Costimulation

The mature T-cell receptor activates PI3K signaling by recruiting p85/p110 to the immune synapse (Fabre et al. 2005). Precisely how PI3K becomes recruited to the TcR is less clear than it is in the case of BcR signaling. The major TcR-associated tyrosine kinase substrates CD3 and Lat lack obvious p85 docking sites. T cells do express the transmembrane adapter protein TRIM, which contains YxxM motifs that become phosphorylated by ZAP-70 and to which PI3K could dock (Bruyns et al. 1998). However, Akt phosphorylation was unimpaired in TRIM-deficient T cells (Kolsch et al. 2006). Vav also contributes to Akt activation in T cells by a mechanism that requires the catalytic activity of Vav, thus implying a potential role for Rac upstream of PI3K activity also in T cells (Saveliev et al. 2009). CD28 has often been considered to play an analogous role in T cells as CD19 does in B cells; that is, by acting as a docking site for PI3K after phosphorylation by Lck or other TcR-associated kinases (Rudd and Schneider 2003). Consistent with this model is the presence of a p85 SH2 binding motif (Y^{170}MNM) in the CD28 cytoplasmic domain. However, while $CD28^{-/-}$ T cells showed reduced PI3K activity in response to stimulation by pMHC on antigen presenting cells (APCs), retrogenic expression of a $CD28^{Y170F}$ mutant rescued CD28-dependent PI3K activity despite the lack of association between p85-SH2 and $CD28^{Y170F}$ (Garcon et al. 2008). In addition, recent imaging experiments that followed the segregation of the TcR and CD28 into different microclusters revealed that in the mature synapse, PI3K colocalized with the TcR, whereas, surprisingly, protein kinase C-theta (PKCθ) colocalized with CD28 (Yokosuka et al. 2008). Indeed, PKCθ recruitment to the immune synapse is more diffuse in T cells lacking CD28 (Garcon et al. 2008; Sanchez-Lockhart et al. 2004, 2008). Thus, while initial models proposed that CD28-dependent PI3K activity would complement TcR-dependent PKC activity, recent genetic and imaging experiments implicate the TcR as the main activator of PI3K (helped by CD28 through undefined mechanisms) and that CD28 instead contributes to sustained or enhanced PKC activity (again as initiated by the TcR). It remains possible that subsequent to initial activation and in certain T cell lines, CD28 may act more independently of the TcR which may have contributed to the original notion that CD28 is a major activator of PI3K. Indeed, CD28-dependent migration of memory T cells to peripheral tissues was dependent both on the PI3K-binding motif in CD28 and on p110δ (Jarmin et al. 2008; Mirenda et al. 2007). Where CD28 does bind p85, there appears to be a preference for the association with the p85β isoform (Alcazar et al. 2009a). The CD28-related protein ICOS, expressed mainly by primed T cells, is also an important PI3K activator. ICOS contains a p85-SH2 binding motif that shows a higher affinity than CD28 for p85 (Parry et al. 2003). Recent evidence also indicates that ICOS may recruit the smaller p50α subunit, which appears to correlate with greater PI3K enzyme activation (Fos et al. 2008). A tyrosine to phenylalanine mutation in ICOS mimics the ICOS-deficient phenotype, showing that PI3K engagement is essential for ICOS function. Thus, $ICOS^{Y181F}$ mice lacked FO helper T cells, which support the germinal cell reaction that leads to class switching and high-affinity antibody production (Gigoux et al. 2009). Mice lacking p110δ activity have a similar phenotype to ICOS-deficient mice with regard to reduced IL-10 production by T cells, reduced Th2 responses, and impaired germinal center formation in the spleen (Okkenhaug et al. 2002, 2006; Nashed et al. 2007; Dong et al. 1999; Hutloff et al. 1999; Patton et al. 2006) (see Sect. 3.3). Thus, PI3K may be an important component of T-cell costimulatory signaling, but the rules of engagement are different than initially assumed.

In addition to the tyrosine-based recruitment mechanisms described here, p110δ has recently also been shown to bind the atypical Ras family member Rras2 (also known as Tc21) which localizes to the T-cell and B-cell antigen receptor complexes (Delgado et al. 2009). Thus, p85α and p110δ recruitment to the BcR and TcR was impaired in Rras2$^{-/-}$ B cells and T cells, and aspects of the Rras2$^{-/-}$ phenotype, such as the lack of MZ B cells, were reminiscent of that observed in the context of p85α or p110δ deficiency (Delgado et al. 2009). To what extent Rras2 complements or supplements tyrosine-based recruitment mechanisms remains to be further explored.

3.2.1 PI3K Regulates Some, But Not All, TcR-Dependent Signaling Pathways

The activation of Akt is entirely dependent on p85s and p110δ in peripheral T cells stimulated through the antigen receptor with or without CD28 (Okkenhaug et al. 2002, 2006; Deane et al. 2007; Garcon et al. 2008). PI3K deficiency also has a particularly strong impact on Erk phosphorylation in T cells, but a more modest effect on Ca^{2+} release into the cytosol (Okkenhaug et al. 2002; Deane et al. 2007). The modest impact on Ca^{2+} flux suggests that PI3K may act on pathways downstream of, or in parallel to, PLCγ. In this context, it is of interest to note that the PI3K effector Bam32 was required for full Erk activation in T cells (Sommers et al. 2008). PI3K may also contribute to NF-κB nuclear translocation, though this is a subject of considerable debate. Several mechanisms exist for stimulating the nuclear translocation and activation of NF-κB (Schulze-Luehrmann and Ghosh 2006). In response to TcR signaling, the main pathway leading to NF-κB activation involves the activation of PKCθ and sequential engagement of Carma, Bcl10, and Malt1, leading to the activation of IκB kinase (Thome 2004). Where PI3K has been proposed to contribute, it has been suggested to be either via the capacity of Pdk1 to interact with and phosphorylate PKCθ or by Akt-dependent phosphorylation and activation of Cot (also known as Tpl2 or Map3k8) (Lee et al. 2005; Park et al. 2009; Kane et al. 2002; Narayan et al. 2006). The capacity of Pdk1 to phosphorylate AGC kinases (other than Akt, but including the PKCs, is generally thought to be PI3K independent, so a requirement for Pdk1 in activating PKCθ does not necessarily imply a requirement for PI3K. Indeed, a knock-in mutation in the PH domain of Pdk1 that renders Pdk1 insensitive to PI3K signaling and that attenuates Akt phosphorylation did not appear to have any effect on PKCθ activation (Waugh et al. 2009). In addition, the extent to which the putative Pdk1 phosphorylation site in PKCθ regulates downstream signaling events has been questioned (Gruber et al. 2006). Although Akt-dependent phosphorylation of Cot may be critical in some cells, NF-κB activation and IL-2 production are unimpaired in Map3k8-deficient T cells, suggesting that Akt-dependent Cot phosphorylation is also nonessential in this context (Sriskantharajah et al. 2009; Tsatsanis et al. 2008). Pertinently, PKCθ recruitment to the synapse and NF-κB translocation to the nucleus were intact in antigen-stimulated p110δ-deficient T cells (Okkenhaug et al. 2006; Garcon et al.

2008). These data suggest that PI3K contribution to NF-κB signaling in T cells is nonessential. However, PI3K signaling does play a key role in regulating the Foxo family of transcription factors, as these are direct targets for phosphorylation by Akt. Although this may in part relieve T cells from antiproliferative and apoptosis-inducing signals (Coffer and Burgering 2004), more recent data suggest a key role for Foxo family proteins in the regulation of T-cell trafficking, which is discussed further in Sect. 3.6.

3.3 T-Helper Cell Differentiation

Both p110δ-deficient and p85 dko T cells show reduced proliferation and cytokine production (Okkenhaug et al. 2002, 2006; Deane et al. 2007; Nashed et al. 2007; Patton et al. 2006; Ji et al. 2007; Liu et al. 2009; Oak et al. 2006). In general, the defect in cytokine production is more pronounced for the effector cytokines interferon-gamma (IFN-γ) and IL-4, which are mainly produced by antigen-experienced differentiated T helper type-1 (Th1) or T helper type-2 (Th2) cells than it is for interleukin-2 (IL-2), which can be produced by naïve T cells. Thus, PI3K may contribute to Th lineage-specific differentiation programs. In addition, p110δ-inhibitors blocked cytokine production by both mouse and human memory T cells, suggesting that, even after a T-helper cell has differentiated, it remains dependent on PI3K activity for optimal cytokine secretion (Soond et al. 2010). However, defective cytokine secretion has not been observed in all experimental models. For instance, under Th2 skewing conditions in vitro or in vivo, IFN-γ production by p110δ-deficient T cells was enhanced, perhaps reflecting reduced interleukin-10 (IL-10) production, which strongly antagonizes IFN-γ production (Zhang et al. 2008; Nashed et al. 2007). Similarly, mice with a T-cell-specific deletion of SHIP showed enhanced Th1 but reduced Th2 responses under Th2 skewing immunization conditions (Tarasenko et al. 2007b). In addition, both p110δ deficiency and SHIP deficiency are associated with reduced IL-17 production by T cells (Soond et al. 2010; Locke et al. 2009). These are curious observations, as one would anticipate that p110δ deficiency and SHIP deficiency would have opposing effects on T-cell function, as the former is associated with reduced PIP3 levels, whereas the latter is associated with elevated PIP3 levels. However, it should be noted that PtdIns(3,4)P_2 produced by SHIP-mediated hydrolysis of PIP3 may contribute to T-cell activation by mechanisms that have yet to be fully appreciated (Parry et al. 2010).

As a consequence of impaired T-cell development in the thymus, p110δ–p110γ double deficient mice were lymphopenic, particularly with respect to T-cell numbers (Webb et al. 2005; Swat et al. 2006). The T cells found in the periphery secreted elevated amounts of Th2 cytokines. The serum from these mice was also enriched in IgE, had normal levels of IgG1, and reduced levels of IgG2a (Ji et al. 2007). The enhanced Th2-like phenotype in the p110γ–p110δ double deficient mice seems paradoxical considering the reduced Th2 responses observed in p110δ-deficient mice (Nashed et al. 2007), but is most likely explained by stress caused by lymphopenia rather than any role for p110δ or p110γ in suppressing Th2 responses. The requirement for PI3K in promoting Th1 differentiation may be mitigated by a role for PI3K in the negative regulation of TLR signaling in dendritic cells (DCs). Thus, PI3K-deficient DCs secreted larger amounts of IL-12, which promoted Th1 differentiation in p85α knockout mice (Fukao et al. 2002). Thus, while PI3K seems to play a positive and T-cell-intrinsic role in promoting cytokine production, in the context of additional stimuli this defect may not always be apparent.

3.4 Survival and Glucose Homeostasis in Mature T cells

PI3K can also regulate glucose uptake in T cells by stimulating Akt, which controls Glut1 translocation to the membrane (Frauwirth et al. 2002). This was put forward as an alternative mechanism for CD28 to promote costimulation, as it was known that CD28 has minimal impact on the transcription profile of T cells (Riley et al. 2002; Diehn et al. 2002). However, there is no evidence that the CD28 PI3K binding motif was required for this response, and more recent evidence might suggest that CD28 contributes in concert with the TcR to trigger optimal PI3K and Akt activity required for glucose uptake. CD28 and PI3K were also required for the upregulation of Bcl-x_L, which promotes survival of T cells (Okkenhaug et al. 2001; Burr et al. 2001). The serine/threonine Pim kinases, whose expression is regulated by IL-2 in T cells, may also play a partial and PI3K-independent role in stimulating T-cell metabolism and/or survival through mechanisms that remain to be fully defined (Fox et al. 2005). In this context, it is important to note that both Pim kinases and mTOR can be inhibited by the commonly used tool compound LY294002 – thus some studies using this inhibitor alone to probe PI3K function need to be reinterpreted (Jacobs et al. 2005). Therefore, while PI3K may promote cellular fitness by augmenting metabolic and prosurvival responses, other pathways such as Pim and PI3K-independent activation of mTOR may also contribute, which may explain why it is possible to stimulate PI3K-deficient T cells to grow and proliferate in the absence of detectable PI3K activity (Deane et al. 2007).

3.5 PI3K Controls the Development and Suppressive Function of Regulatory T Cells

Regulatory T cells (Tregs) play an essential life-preserving function by reining in the adaptive immune system (Sakaguchi et al. 2008; Zheng and Rudensky 2007). Tregs can develop either in the thymus from immature progenitors or in the periphery from mature CD4 T cells. Peripheral development of Treg requires TGF-β signaling (Li and Flavell 2008). p110δ-deficient mice showed increased proportions of Treg in the thymus, whereas overexpression of myristoylated Akt in thymocytes inhibited

Treg development, suggesting that PI3K negatively regulates thymic development of Tregs via Akt (Patton et al. 2006; Haxhinasto et al. 2008). By contrast, Tregs were found in reduced numbers in the spleen and lymph nodes of p110δ-deficient mice, suggesting a possible role for PI3K in the homeostatic maintenance or peripheral development of Tregs. Consistent with this notion, $SHIP^{-/-}$ mice harbored increased proportions of Treg in the spleen (Locke et al. 2009). Curiously, treatment of naïve T cells with PI3K and mTOR inhibitors 18 h after activation through the antigen receptor led to the induction of Foxp3 expression in the absence of TGF-β (Sauer et al. 2008). In addition, genetic deletion of mTor in T cells led to enhanced Treg development at the expense the Th lineages (Delgoffe et al. 2009). The physiological relevance of how interrupted PI3K and mTOR signaling could lead to induction of Foxp3 expression remains to be defined; nonetheless, these results, together with the $p110\delta^{D910A}$ and Akt overexpressing thymocytes, show that PI3K signaling is intimately linked to the production and maintenance of $Foxp3^{+}$ Tregs.

Treg from p110δ mice are defective in their capacity to suppress T-helper cell proliferation in vitro and to secrete IL-10 (Patton et al. 2006). One apparent consequence of reduced Treg function in p110δ mice is that they were shown to clear Leishmania infections more efficiently than WT mice despite impaired Th1 responses (Liu et al. 2009). However, acute or chronic depletion of Treg cells can also lead to massive lymphoproliferation, lymphocyte tissue infiltration, and multiple organ failure (Kim et al. 2007a). More subtle defects in Treg function or Treg depletion in the absence of intact adaptive immunity, frequently leads to colon inflammation stimulated by commensal flora in the gut (Izcue et al. 2009). Indeed, p110δ-deficient Treg failed to prevent disease in an experimental model of colitis (Patton et al. 2006). Accordingly, $p110\delta^{D910A}$ mice showed signs of inflammatory bowel disease, but no other autoimmune symptoms were evident (Okkenhaug et al. 2002). By contrast, p85 dko mice showed symptoms reminiscent of Sjogren's syndrome, including dry eyes, lacrimal gland destruction, and autoantibodies (Oak et al. 2006). Whether the different symptoms described in the different mice represent real differences in how p85 vs p110δ deficiency affects Treg function or whether other environmental or tissue-specific factors explain the different symptoms is not yet clear. PI3K signaling in response to IL-2 stimulation was reduced in Treg compared to conventional CD4 T cells (Bensinger et al. 2004). Deletion of Pten resulted in elevated PI3K signaling and enhanced proliferation of Treg in response to IL-2 – even in the absence of TcR signaling which is required for the proliferation of conventional Treg (Walsh et al. 2006). However, deletion of SHIP or Pten had no effect on Treg-mediated suppression (Locke et al. 2009; Walsh et al. 2006). Moreover, one study showed that the expression of transgenic Akt in mouse Treg enhanced suppression (Pierau et al. 2009) whereas another study showed that in human Tregs basal PI3K signaling was also lower than in conventional T cells and that ectopic expression of myristoylated Akt interfered with Treg-mediated suppression (Crellin et al. 2007). Thus, PI3Ks regulate many facets of Treg development, homeostasis, and suppression through mechanisms that are still incompletely understood. What seems clear is that the timing and magnitude of the PI3K and mTOR signals have strong impacts on the homeostasis and function of Treg.

3.6 T-Cell Chemotaxis and Migration in Lymph Nodes: Not All About PI3K

The role of p110γ in peripheral T cells is debated. While some investigators observe reduced proliferation and cytokine production by p110$\gamma^{-/-}$ T cells (Sasaki et al. 2000; Alcazar et al. 2007), others observe that p110$\gamma^{-/-}$ T cells proliferate normally (Martin et al. 2008; Thomas et al. 2008). Recent evidence suggests that the TcR can form complexes with chemokine receptors, especially in memory T cells, and may hence influence TcR-dependent events under some circumstances (Molon et al. 2005; Kumar et al. 2006). Other factors, such as the health status of the mice, cell purification protocols, and media components, may also influence to what extent p110γ contributes to T-cell activation by antigen receptor stimulation in vitro by altering the levels of available GPCR ligands. Perhaps surprisingly, chemotactic responses to CCL19, CCL21, and CXCL12 were only moderately affected by p110γ deficiency despite the capacity of these agonists to stimulate PI3K activity in a p110γ-dependent manner (Reif et al. 2004; Nombela-Arrieta et al. 2004; Jarmin et al. 2008). Instead, the Rac guanine nucleotide exchange factor (GEF) Dock2 was required for T-cell chemotaxis in a mostly PI3K-independent manner (Nombela-Arrieta et al. 2004, 2007). More recently, the role of PI3Ks in regulating T-cell migration within the lymph nodes has been investigated using intravital, two-photon microscopy or by fluorescent microscopy of lymph node slices (Matheu et al. 2007; Nombela-Arrieta et al. 2007; Asperti-Boursin et al. 2007). Constitutive migration of T cells within lymph nodes is promoted by chemokines secreted by and adhered to fibroblastic reticular cells (Bajenoff et al. 2006). However, the effects of inhibiting PI3K were more subtle. Interestingly, by some criteria, the loss of the p85 adapter subunits had a stronger effect than inhibition by wortmannin. These observations suggest that p85 may regulate some aspects of cell motility independently of PI3K in T cells as had previously been shown in fibroblasts (Matheu et al. 2007; Jimenez et al. 2000).

In addition to promoting chemotaxis, CCL19 and CCL21 are important for T-cell homeostasis and survival (Link et al. 2007). Thus an intriguing possibility is that chemokines use p110γ in T cells to promote survival or metabolic fitness rather than to detect chemotactic gradients. Consistent with this interpretation is the observation that p110γ deficiency could alleviate autoimmunity caused by the overexpression of an activating form of p85 (referred to as p65) and that this alleviation was correlated with reduced accumulation of memory T cells rather than by reduced infiltration of T cells into affected organs (Barber et al. 2006). Chemokines can also provide costimulatory signals to T cells to enhance proliferation, but this is not p110γ or p110δ dependent (Gollmer et al. 2009). More recent data have provided evidence for a key role for p110γ in mediating chemotactic signaling in previously activated T cells in response to inflammatory chemokine CCL5 or the lipid chemoattractant leukotriene B4 (Martin et al. 2008). This was correlated with a defective recruitment of p110$\gamma^{-/-}$ T cells to the site of vaccinia virus infection and diminished ability to resist infection. Similarly, p110γ$^{-/-}$ CD4 T cells showed reduced response to the chemokine CCL22 (Thomas et al. 2008).

3.7 T-Cell Trafficking

Several lines of evidence have indicated key roles for class IA PI3Ks in regulating lymphocyte motility and trafficking. p85-deficient T cells show reduced motility in intact lymph nodes, and p110δ-deficient T cells show reduced recruitment to sites of inflammation or skin allografts (Matheu et al. 2007; Jarmin et al. 2008). These defects do not appear to reflect a role for p85 or p110δ in chemokine signaling, as PI3K plays a minor role in T cell chemotaxis. Where there is a role for PI3K, p110γ appears to be the relevant isoform. Instead, PI3K, via its activation of Akt, controls the exclusion of Foxo from the nucleus and hence prevents Foxo from driving the transcription of Krüppel-like factor-2 (KLF2) and a number of genes that are critical for lymphocyte trafficking and homing, such as L-selectin (CD62L), CCR7, and receptors for sphingosine-1-phosphate (S1P) and interleukin-7 (IL-7) (Kerdiles et al. 2009; Ouyang et al. 2009; Sinclair et al. 2008; Fabre et al. 2008). L-selectin, CCR7, and S1P receptor expression can also be inhibited by rapamycin, suggesting a novel capacity of mTORC1 to inhibit transcription (Sinclair et al. 2008). CD62L is required for the movement of lymphocytes across the high endothelial venules during entry into lymph nodes. CCR7 contributes to the recruitment of T cells into the T-cell areas of lymph nodes, whereas S1P promotes the exit of T cells from lymph nodes. IL-7 is a key regulator of homeostatic survival of T cells, and it is found in limiting amounts in lymph nodes. Because the effects of the defined Foxo targets are pleiotropic and can promote recruitment or exit from lymph nodes, the effect of PI3K inhibition (and hence sustained expression of these trafficking molecules) may not always be easily predicted.

In memory T cells, the TcR and CD28 can independently stimulate T-cell trafficking to peripheral tissues by mechanisms that remain to be defined but which are absolutely dependent on p110δ activity (Jarmin et al. 2008; Mirenda et al. 2007). The cells used in these studies had been generated by repeated immunizations followed by in vitro cultures, and hence both CCR7 and CD62L expression was low also on the p110δ^{D910A} T cells, resulting in few T cells of either genotype being recruited to the lymph nodes in these studies (Jarmin et al. 2008). Circulating T cells interact with the capillary and postcapillary endothelium and translocate through the endothelial barrier if they detect the expression of foreign antigen or presence of costimulatory ligands. How PI3K is required for this process is presently unclear but may involve the clustering or otherwise alteration of selected integrin affinity or avidity (Mirenda et al. 2007) or additional Foxo or KLF2 targets that have yet to be fully characterized. In this context, it is of interest to note that KLF2-deficient lymphocytes have been found to accumulate preferentially in non-lymphoid tissues (Sebzda et al. 2008). Given that PI3K suppresses KLF2 (Sinclair et al. 2008), this suggests that p110δ-deficient cells may lack the expression of certain genes that would facilitate entry into non-lymphoid organs. These observations have potentially important clinical implications, as they indicate that p110δ inhibitors could reduce the recruitment of T cells to sites of inflammation where the T cells otherwise might increase the inflammatory response.

3.8 Nonessential Role for PI3K in Cytotoxic Responses

p85 or p110δ do not appear to be essential for CD8 T-cell-mediated killing in vivo of either virally infected cells or of hematopoietic allografts (Deane et al. 2007; Garcon et al. 2008). Also, the capacity of NK cells to kill targets was either unaffected or modestly affected in p110δ- or p110γ-deficient mice (Kim et al. 2007b; Tassi et al. 2007; Zebedin et al. 2008; Guo et al. 2008). These results were of particular interest because the NKG2D receptor, which recognizes proteins on cancer and virally infected cells, is linked to PI3K via the adapter DAP10 (Zompi et al. 2003). PI3K-deficient NK cells showed defective IFN-γ secretion, and both p110γ and p110δ played an important role in the recruitment of NK cells to sites of infection and to tumors (Kim et al. 2007b; Tassi et al. 2007; Saudemont et al. 2009a). The precise role of PI3K in CD8 T cells and NK cells remain to be fully understood, but the retention of significant cytotoxicity in the absence of PI3K activity does suggest that, while PI3K inhibitors may limit inflammation and autoimmune responses, they will not necessarily compromise resistance to common viral infections.

Figure 1b outlines key actions of PI3K in T-cell development and function, as highlighted in this section.

4 Prospects for PI3K Inhibitors in Inflammation and Autoimmunity

As will be discussed in other chapters in this book, PI3K inhibitors have been used with notable success in preclinical models of lupus, arthritis, asthma, and allergy. While much of this effort has focused on the important capacity of PI3K inhibitors to reduce neutrophil recruitment and reactive oxygen species generation, p110δ inhibitors may block T-cell- and B-cell-dependent inflammation, autoimmunity, or graft rejection with minimal impact on PI3K signaling in other tissues. By a similar argument, the therapeutic potential of p110γ inhibitors may in part be mediated by suppressing the survival of memory T cells, or they may prevent T cells from being recruited to sites of inflammation. To what extent p110δ or p110γ inhibition will compromise resistance to infection is currently not known, but this will be an important area of investigation as inhibitors of these isoforms enter clinical trials. In addition, attention needs to be paid to the possibility that PI3K inhibition can enhance certain immune responses by blocking Tregs or by enhancing DC function or B-cell class switch recombination.

Acknowledgments Work cited from our laboratories has been funded by grants from the BBSRC (to KO) and from the National Institutes of health and the American Cancer Society (to DAF). We are grateful to Dalya Soond and Dan Patton for constructive comments on the manuscript.

References

Aiba Y, Kameyama M, Yamazaki T, Tedder TF, Kurosaki T (2008) Regulation of B-cell development by BCAP and CD19 through their binding to phosphoinositide 3-kinase. Blood 111:1497–1503

Al-Alwan MM, Okkenhaug K, Vanhaesebroeck B, Hayflick JS, Marshall AJ (2007) Requirement for phosphoinositide 3-kinase p110delta signaling in B cell antigen receptor-mediated antigen presentation. J Immunol 178:2328–2335

Alcazar I, Marques M, Kumar A, Hirsch E, Wymann M, Carrera AC, Barber DF (2007) Phosphoinositide 3-kinase gamma participates in T cell receptor-induced T cell activation. J Exp Med 204:2977–2987

Alcazar I, Cortes I, Zaballos A, Hernandez C, Fruman DA, Barber DF, Carrera AC (2009) p85beta phosphoinositide 3-kinase regulates CD28 coreceptor function. Blood 113:3198–3208

Ali K, Bilancio A, Thomas M, Pearce W, Gilfillan AM, Tkaczyk C, Kuehn N, Gray A, Giddings J, Peskett E, Fox R, Bruce I, Walker C, Sawyer C, Okkenhaug K, Finan P, Vanhaesebroeck B (2004) Essential role for the p110delta phosphoinositide 3-kinase in the allergic response. Nature 431:1007–1011

Ali K, Camps M, Pearce WP, Ji H, Ruckle T, Kuehn N, Pasquali C, Chabert C, Rommel C, Vanhaesebroeck B (2008) Isoform-specific functions of phosphoinositide 3-kinases: p110 delta but not p110 gamma promotes optimal allergic responses in vivo. J Immunol 180:2538–2544

Amin RH, Schlissel MS (2008) Foxo1 directly regulates the transcription of recombination-activating genes during B cell development. Nat Immunol 9:613–622

Asperti-Boursin F, Real E, Bismuth G, Trautmann A, Donnadieu E (2007) CCR7 ligands control basal T cell motility within lymph node slices in a phosphoinositide 3-kinase-independent manner. J Exp Med 204:1167–1179

Bajenoff M, Egen JG, Koo LY, Laugier JP, Brau F, Glaichenhaus N, Germain RN (2006) Stromal cell networks regulate lymphocyte entry, migration, and territoriality in lymph nodes. Immunity 25:989–1001

Barber DF, Bartolome A, Hernandez C, Flores JM, Redondo C, Fernandez-Arias C, Camps M, Ruckle T, Schwarz MK, Rodriguez S, Martinez AC, Balomenos D, Rommel C, Carrera AC (2005) PI3Kgamma inhibition blocks glomerulonephritis and extends lifespan in a mouse model of systemic lupus. Nat Med 11:933–935

Barber DF, Bartolome A, Hernandez C, Flores JM, Fernandez-Arias C, Rodriguez-Borlado L, Hirsch E, Wymann M, Balomenos D, Carrera AC (2006) Class IB-phosphatidylinositol 3-kinase (PI3K) deficiency ameliorates IA-PI3K-induced systemic lupus but not T cell invasion. J Immunol 176:589–593

Bensinger SJ, Walsh PT, Zhang J, Carroll M, Parsons R, Rathmell JC, Thompson CB, Burchill MA, Farrar MA, Turka LA (2004) Distinct IL-2 receptor signaling pattern in $CD4^{+}CD25^{+}$ regulatory T cells. J Immunol 172:5287–5296

Bilancio A, Okkenhaug K, Camps M, Emery JL, Ruckle T, Rommel C, Vanhaesebroeck B (2006) Key role of the p110delta isoform of PI3K in B-cell antigen and IL-4 receptor signaling: comparative analysis of genetic and pharmacologic interference with p110delta function in B cells. Blood 107:642–650

Bluestone JA, St Clair EW, Turka LA (2006) CTLA4Ig: bridging the basic immunology with clinical application. Immunity 24:233–238

Brauweiler A, Tamir I, Dal Porto J, Benschop RJ, Helgason CD, Humphries RK, Freed JH, Cambier JC (2000) Differential regulation of B cell development, activation, and death by the src homology 2 domain-containing 5′ inositol phosphatase (SHIP). J Exp Med 191:1545–1554

Browne CD, Del Nagro CJ, Cato MH, Dengler HS, Rickert RC (2009) Suppression of phosphatidylinositol 3,4,5-trisphosphate production is a key determinant of B cell anergy. Immunity 31:749–760

Bruyns E, Marie-Cardine A, Kirchgessner H, Sagolla K, Shevchenko A, Mann M, Autschbach F, Bensussan A, Meuer S, Schraven B (1998) T cell receptor (TCR) interacting molecule (TRIM), a novel disulfide-linked dimer associated with the TCR-CD3-zeta complex, recruits intracellular signaling proteins to the plasma membrane. J Exp Med 188:561–575

Burr JS, Savage ND, Messah GE, Kimzey SL, Shaw AS, Arch RH, Green JM (2001) Cutting edge: distinct motifs within CD28 regulate T cell proliferation and induction of Bcl-XL. J Immunol 166:5331–5335

Camps M, Ruckle T, Ji H, Ardissone V, Rintelen F, Shaw J, Ferrandi C, Chabert C, Gillieron C, Francon B, Martin T, Gretener D, Perrin D, Leroy D, Vitte PA, Hirsch E, Wymann MP, Cirillo R, Schwarz MK, Rommel C (2005) Blockade of PI3Kgamma suppresses joint inflammation and damage in mouse models of rheumatoid arthritis. Nat Med 11:936–943

Ciofani M, Zuniga-Pflucker JC (2005) Notch promotes survival of pre-T cells at the beta-selection checkpoint by regulating cellular metabolism. Nat Immunol 6:881–888

Clayton E, Bardi G, Bell SE, Chantry D, Downes CP, Gray A, Humphries LA, Rawlings D, Reynolds H, Vigorito E, Turner M (2002) A crucial role for the p110delta subunit of phosphatidylinositol 3-kinase in B cell development and activation. J Exp Med 196:753–763

Coffer PJ, Burgering BM (2004) Forkhead-box transcription factors and their role in the immune system. Nat Rev Immunol 4:889–899

Crabbe T, Welham MJ, Ward SG (2007) The PI3K inhibitor arsenal: choose your weapon! Trends Biochem Sci 32:450–456

Crellin NK, Garcia RV, Levings MK (2007) Altered activation of AKT is required for the suppressive function of human $CD4^{+}CD25^{+}$ T regulatory cells. Blood 109:2014–2022

Deane JA, Trifilo MJ, Yballe CM, Choi S, Lane TE, Fruman DA (2004) Enhanced T cell proliferation in mice lacking the p85beta subunit of phosphoinositide 3-kinase. J Immunol 172:6615–6625

Deane JA, Kharas MG, Oak JS, Stiles LN, Luo J, Moore TI, Ji H, Rommel C, Cantley LC, Lane TE, Fruman DA (2007) T cell function is partially maintained in the absence of class IA phosphoinositide 3-kinase signaling. Blood 109:2894–2902

Delgado P, Cubelos B, Calleja E, Martinez-Martin N, Cipres A, Merida I, Bellas C, Bustelo XR, Alarcon B (2009) Essential function for the GTPase TC21 in homeostatic antigen receptor signaling. Nat Immunol 10:880–888

Delgoffe GM, Kole TP, Zheng Y, Zarek PE, Matthews KL, Xiao B, Worley PF, Kozma SC, Powell JD (2009) The mTOR kinase differentially regulates effector and regulatory T cell lineage commitment. Immunity 30:832–844

Dengler HS, Baracho GV, Omori SA, Bruckner S, Arden KC, Castrillon DH, DePinho RA, Rickert RC (2008) Distinct functions for the transcription factor Foxol at various stages of B cell differentiation. Nat Immunol 9:1388–1398

Depoil D, Fleire S, Treanor BL, Weber M, Harwood NE, Marchbank KL, Tybulewicz VL, Batista FD (2008) CD19 is essential for B cell activation by promoting B cell receptor-antigen microcluster formation in response to membrane-bound ligand. Nat Immunol 9:63–72

Diehn M, Alizadeh AA, Rando OJ, Liu CL, Stankunas K, Botstein D, Crabtree GR, Brown PO (2002) Genomic expression programs and the integration of the CD28 costimulatory signal in T cell activation. Proc Natl Acad Sci USA 99:11796–11801

Doi T, Obayashi K, Kadowaki T, Fujii H, Koyasu S (2008) PI3K is a negative regulator of IgE production. Int Immunol 20:499–508

Donahue AC, Fruman DA (2007) Distinct signaling mechanisms activate the target of rapamycin in response to different B-cell stimuli. Eur J Immunol 37:2923–2936

Donahue AC, Hess KL, Ng KL, Fruman DA (2004) Altered splenic B cell subset development in mice lacking phosphoinositide 3-kinase p85alpha. Int Immunol 16:1789–1798

Dong H, Zhu G, Tamada K, Chen L (1999) B7-H1, a third member of the B7 family, co-stimulates T-cell proliferation and interleukin-10 secretion. Nat Med 5:1365–1369

Doughty CA, Bleiman BF, Wagner DJ, Dufort FJ, Mataraza JM, Roberts MF, Chiles TC (2006) Antigen receptor-mediated changes in glucose metabolism in B lymphocytes: role of phosphatidylinositol 3-kinase signaling in the glycolytic control of growth. Blood 107:4458–4465

Edwards JC, Leandro MJ, Cambridge G (2004) B lymphocyte depletion therapy with rituximab in rheumatoid arthritis. Rheum Dis Clin North Am 30:393–403, viii

Fabre S, Lang V, Harriague J, Jobart A, Unterman TG, Trautmann A, Bismuth G (2005) Stable activation of phosphatidylinositol 3-kinase in the T cell immunological synapse stimulates Akt signaling to FoxO1 nuclear exclusion and cell growth control. J Immunol 174:4161–4171

Fabre S, Carrette F, Chen J, Lang V, Semichon M, Denoyelle C, Lazar V, Cagnard N, Dubart-Kupperschmitt A, Mangeney M, Fruman DA, Bismuth G (2008) FOXO1 regulates L-selectin and a network of human T cell homing molecules downstream of phosphatidylinositol 3-kinase. J Immunol 181:2980–2989

Fayard E, Gill J, Paolino M, Hynx D, Hollander GA, Hemmings BA (2007) Deletion of PKBalpha/Akt1 affects thymic development. PLoS One 2:e992

Feldmann M, Maini SR (2008) Role of cytokines in rheumatoid arthritis: an education in pathophysiology and therapeutics. Immunol Rev 223:7–19

Finlay DK, Sinclair LV, Feijoo C, Waugh CM, Hagenbeek TJ, Spits H, Cantrell DA (2009) Phosphoinositide-dependent kinase 1 controls migration and malignant transformation but not cell growth and proliferation in PTEN-null lymphocytes. J Exp Med 206:2441–2454

Floto RA, Clatworthy MR, Heilbronn KR, Rosner DR, MacAry PA, Rankin A, Lehner PJ, Ouwehand WH, Allen JM, Watkins NA, Smith KG (2005) Loss of function of a lupus-associated FcgammaRIIb polymorphism through exclusion from lipid rafts. Nat Med 11: 1056–1058

Fos C, Salles A, Lang V, Carrette F, Audebert S, Pastor S, Ghiotto M, Olive D, Bismuth G, Nunes JA (2008) ICOS ligation recruits the p50alpha PI3K regulatory subunit to the immunological synapse. J Immunol 181:1969–1977

Fox CJ, Hammerman PS, Thompson CB (2005) The Pim kinases control rapamycin-resistant T cell survival and activation. J Exp Med 201:259–266

Frauwirth KA, Riley JL, Harris MH, Parry RV, Rathmell JC, Plas DR, Elstrom RL, June CH, Thompson CB (2002) The CD28 signaling pathway regulates glucose metabolism. Immunity 16:769–777

Fruman DA, Bismuth G (2009) Fine tuning the immune response with PI3K. Immunol Rev 228:253–272

Fruman DA, Snapper SB, Yballe CM, Davidson L, Yu JY, Alt FW, Cantley LC (1999) Impaired B cell development and proliferation in absence of phosphoinositide 3-kinase p85alpha. Science 283:393–397

Fruman DA, Satterthwaite AB, Witte ON (2000) Xid-like phenotypes: a B cell signalosome takes shape. Immunity 13:1–3

Fukao T, Tanabe M, Terauchi Y, Ota T, Matsuda S, Asano T, Kadowaki T, Takeuchi T, Koyasu S (2002) PI3K-mediated negative feedback regulation of IL-12 production in DCs. Nat Immunol 3:875–881

Garcon F, Patton DT, Emery JL, Hirsch E, Rottapel R, Sasaki T, Okkenhaug K (2008) CD28 provides T-cell costimulation and enhances PI3K activity at the immune synapse independently of its capacity to interact with the p85/p110 heterodimer. Blood 111:1464–1471

Gigoux M, Shang J, Pak Y, Xu M, Choe J, Mak TW, Suh WK (2009) Inducible costimulator promotes helper T-cell differentiation through phosphoinositide 3-kinase. Proc Natl Acad Sci USA 106:20371–20376

Gollmer K, Asperti-Boursin F, Tanaka Y, Okkenhaug K, Vanhaesebroeck B, Peterson JR, Fukui Y, Donnadieu E, Stein JV (2009) CCL21 mediates CD4$^+$ T cell costimulation via a DOCK2/Rac-dependent pathway. Blood 114:580–588

Gruber T, Freeley M, Thuille N, Heit I, Shaw S, Long A, Baier G (2006) Comment on "PDK1 nucleates T cell receptor-induced signaling complex for NF-kappaB activation". Science 312:55, author reply 55

Guo H, Samarakoon A, Vanhaesebroeck B, Malarkannan S (2008) The p110 delta of PI3K plays a critical role in NK cell terminal maturation and cytokine/chemokine generation. J Exp Med 205:2419–2435

Hagenbeek TJ, Naspetti M, Malergue F, Garcon F, Nunes JA, Cleutjens KB, Trapman J, Krimpenfort P, Spits H (2004) The loss of PTEN allows TCR alphabeta lineage thymocytes to bypass IL-7 and pre-TCR-mediated signaling. J Exp Med 200:883–894

Hardy RR, Kincade PW, Dorshkind K (2007) The protean nature of cells in the B lymphocyte lineage. Immunity 26:703–714

Haxhinasto S, Mathis D, Benoist C (2008) The AKT–mTOR axis regulates de novo differentiation of CD4$^+$Foxp3$^+$ cells. J Exp Med 205:565–574

Henley T, Kovesdi D, Turner M (2008) B-cell responses to B-cell activation factor of the TNF family (BAFF) are impaired in the absence of PI3K delta. Eur J Immunol 38:3543–3548

Herzog S, Hug E, Meixlsperger S, Paik JH, DePinho RA, Reth M, Jumaa H (2008) SLP-65 regulates immunoglobulin light chain gene recombination through the PI(3)K-PKB-Foxo pathway. Nat Immunol 9:623–631

Herzog S, Reth M, Jumaa H (2009) Regulation of B-cell proliferation and differentiation by pre-B-cell receptor signalling. Nat Rev Immunol 9:195–205

Hess KL, Donahue AC, Ng KL, Moore TI, Oak J, Fruman DA (2004) Frontline: the p85alpha isoform of phosphoinositide 3-kinase is essential for a subset of B cell receptor-initiated signaling responses. Eur J Immunol 34:2968–2976

Hinton HJ, Alessi DR, Cantrell DA (2004) The serine kinase phosphoinositide-dependent kinase 1 (PDK1) regulates T cell development. Nat Immunol 5:539–545

Hirsch E, Katanaev VL, Garlanda C, Azzolino O, Pirola L, Silengo L, Sozzani S, Mantovani A, Altruda F, Wymann MP (2000) Central role for G protein-coupled phosphoinositide 3-kinase gamma in inflammation. Science 287:1049–1053

Hutloff A, Dittrich AM, Beier KC, Eljaschewitsch B, Kraft R, Anagnostopoulos I, Kroczek RA (1999) ICOS is an inducible T-cell co-stimulator structurally and functionally related to CD28. Nature 397:263–266

Inabe K, Ishiai M, Scharenberg AM, Freshney N, Downward J, Kurosaki T (2002) Vav3 modulates B cell receptor responses by regulating phosphoinositide 3-kinase activation. J Exp Med 195:189–200

Izcue A, Coombes JL, Powrie F (2009) Regulatory lymphocytes and intestinal inflammation. Annu Rev Immunol 27:313–338

Jacobs MD, Black J, Futer O, Swenson L, Hare B, Fleming M, Saxena K (2005) Pim-1 ligand-bound structures reveal the mechanism of serine/threonine kinase inhibition by LY294002. J Biol Chem 280:13728–13734

Janas ML, Hodson D, Stamataki Z, Hill S, Welch K, Gambardella L, Trotman LC, Pandolfi PP, Vigorito E, Turner M (2008) The effect of deleting p110delta on the phenotype and function of PTEN-deficient B cells. J Immunol 180:739–746

Janas ML, Varano G, Gudmundsson K, Noda M, Nagasawa T, Turner M (2010) Thymic development beyond beta-selection requires phosphatidylinositol 3-kinase activation by CXCR4. J Exp Med 207(247–261):S241–S242

Jarmin SJ, David R, Ma L, Chai JG, Dewchand H, Takesono A, Ridley AJ, Okkenhaug K, Marelli-Berg FM (2008) T cell receptor-induced phosphoinositide-3-kinase p110delta activity is required for T cell localization to antigenic tissue in mice. J Clin Invest 118:1154–1164

Ji H, Rintelen F, Waltzinger C, Bertschy Meier D, Bilancio A, Pearce W, Hirsch E, Wymann MP, Ruckle T, Camps M, Vanhaesebroeck B, Okkenhaug K, Rommel C (2007) Inactivation of

PI3Kgamma and PI3Kdelta distorts T-cell development and causes multiple organ inflammation. Blood 110:2940–2947

Jimenez C, Portela RA, Mellado M, Rodriguez-Frade JM, Collard J, Serrano A, Martinez AC, Avila J, Carrera AC (2000) Role of the PI3K regulatory subunit in the control of actin organization and cell migration. J Cell Biol 151:249–262

Jou ST, Carpino N, Takahashi Y, Piekorz R, Chao JR, Carpino N, Wang D, Ihle JN (2002) Essential, nonredundant role for the phosphoinositide 3-kinase p110delta in signaling by the B-cell receptor complex. Mol Cell Biol 22:8580–8591

Juntilla MM, Wofford JA, Birnbaum MJ, Rathmell JC, Koretzky GA (2007) Akt1 and Akt2 are required for alphabeta thymocyte survival and differentiation. Proc Natl Acad Sci USA 104:12105–12110

Kane LP, Mollenauer MN, Xu Z, Turck CW, Weiss A (2002) Akt-dependent phosphorylation specifically regulates Cot induction of NF-kappa B-dependent transcription. Mol Cell Biol 22:5962–5974

Kelly AP, Finlay DK, Hinton HJ, Clarke RG, Fiorini E, Radtke F, Cantrell DA (2007) Notch-induced T cell development requires phosphoinositide-dependent kinase 1. EMBO J 26: 3441–3450

Kerdiles YM, Beisner DR, Tinoco R, Dejean AS, Castrillon DH, DePinho RA, Hedrick SM (2009) Foxo1 links homing and survival of naive T cells by regulating L-selectin, CCR7 and interleukin 7 receptor. Nat Immunol 10:176–184

Kim JM, Rasmussen JP, Rudensky AY (2007a) Regulatory T cells prevent catastrophic autoimmunity throughout the lifespan of mice. Nat Immunol 8:191–197

Kim N, Saudemont A, Webb L, Camps M, Ruckle T, Hirsch E, Turner M, Colucci F (2007b) The p110delta catalytic isoform of PI3K is a key player in NK-cell development and cytokine secretion. Blood 110:3202–3208

Kishimoto H, Ohteki T, Yajima N, Kawahara K, Natsui M, Kawarasaki S, Hamada K, Horie Y, Kubo Y, Arase S, Taniguchi M, Vanhaesebroeck B, Mak TW, Nakano T, Koyasu S, Sasaki T, Suzuki A (2007) The Pten/PI3K pathway governs the homeostasis of Valpha14iNKT cells. Blood 109:3316–3324

Kolsch U, Arndt B, Reinhold D, Lindquist JA, Juling N, Kliche S, Pfeffer K, Bruyns E, Schraven B, Simeoni L (2006) Normal T-cell development and immune functions in TRIM-deficient mice. Mol Cell Biol 26:3639–3648

Kumar A, Humphreys TD, Kremer KN, Bramati PS, Bradfield L, Edgar CE, Hedin KE (2006) CXCR4 physically associates with the T cell receptor to signal in T cells. Immunity 25: 213–224

Lee KY, D'Acquisto F, Hayden MS, Shim JH, Ghosh S (2005) PDK1 nucleates T cell receptor-induced signaling complex for NF-kappaB activation. Science 308:114–118

Lee KS, Lee HK, Hayflick JS, Lee YC, Puri KD (2006) Inhibition of phosphoinositide 3-kinase delta attenuates allergic airway inflammation and hyperresponsiveness in murine asthma model. FASEB J 20:455–465

Li MO, Flavell RA (2008) TGF-beta: a master of all T cell trades. Cell 134:392–404

Li Z, Jiang H, Xie W, Zhang Z, Smrcka AV, Wu D (2000) Roles of PLC-beta2 and -beta3 and PI3Kgamma in chemoattractant-mediated signal transduction. Science 287:1046–1049

Lin L, Hron JD, Peng SL (2004) Regulation of NF-kappaB, Th activation, and autoinflammation by the forkhead transcription factor Foxo3a. Immunity 21:203–213

Link A, Vogt TK, Favre S, Britschgi MR, Acha-Orbea H, Hinz B, Cyster JG, Luther SA (2007) Fibroblastic reticular cells in lymph nodes regulate the homeostasis of naive T cells. Nat Immunol 8:1255–1265

Liu D, Zhang T, Marshall AJ, Okkenhaug K, Vanhaesebroeck B, Uzonna JE (2009) The p110delta isoform of phosphatidylinositol 3-kinase controls susceptibility to Leishmania major by regulating expansion and tissue homing of regulatory T cells. J Immunol 183:1921–1933

Llorian M, Stamataki Z, Hill S, Turner M, Martensson IL (2007) The PI3K p110delta is required for down-regulation of RAG expression in immature B cells. J Immunol 178:1981–1985

Locke NR, Patterson SJ, Hamilton MJ, Sly LM, Krystal G, Levings MK (2009) SHIP regulates the reciprocal development of T regulatory and Th17 cells. J Immunol 183:975–983

Mao C, Tili EG, Dose M, Haks MC, Bear SE, Maroulakou I, Horie K, Gaitanaris GA, Fidanza V, Ludwig T, Wiest DL, Gounari F, Tsichlis PN (2007) Unequal contribution of Akt isoforms in the double-negative to double-positive thymocyte transition. J Immunol 178:5443–5453

Martin AL, Schwartz MD, Jameson SC, Shimizu Y (2008) Selective regulation of CD8 effector T cell migration by the p110 gamma isoform of phosphatidylinositol 3-kinase. J Immunol 180:2081–2088

Matheu MP, Deane JA, Parker I, Fruman DA, Cahalan MD (2007) Class IA phosphoinositide 3-kinase modulates basal lymphocyte motility in the lymph node. J Immunol 179:2261–2269

Mirenda V, Jarmin SJ, David R, Dyson J, Scott D, Gu Y, Lechler RI, Okkenhaug K, Marelli-Berg FM (2007) Physiologic and aberrant regulation of memory T-cell trafficking by the costimulatory molecule CD28. Blood 109:2968–2977

Molon B, Gri G, Bettella M, Gomez-Mouton C, Lanzavecchia A, Martinez AC, Manes S, Viola A (2005) T cell costimulation by chemokine receptors. Nat Immunol 6:465–471

Narayan P, Holt B, Tosti R, Kane LP (2006) CARMA1 is required for Akt-mediated NF-kappaB activation in T cells. Mol Cell Biol 26:2327–2336

Nashed BF, Zhang T, Al-Alwan M, Srinivasan G, Halayko AJ, Okkenhaug K, Vanhaesebroeck B, Hayglass KT, Marshall AJ (2007) Role of the phosphoinositide 3-kinase p110delta in generation of type 2 cytokine responses and allergic airway inflammation. Eur J Immunol 37:416–424

Nombela-Arrieta C, Lacalle RA, Montoya MC, Kunisaki Y, Megias D, Marques M, Carrera AC, Manes S, Fukui Y, Martinez AC, Stein JV (2004) Differential requirements for DOCK2 and phosphoinositide-3-kinase gamma during T and B lymphocyte homing. Immunity 21:429–441

Nombela-Arrieta C, Mempel TR, Soriano SF, Mazo I, Wymann MP, Hirsch E, Martinez AC, Fukui Y, von Andrian UH, Stein JV (2007) A central role for DOCK2 during interstitial lymphocyte motility and sphingosine-1-phosphate-mediated egress. J Exp Med 204:497–510

Oak JS, Deane JA, Kharas MG, Luo J, Lane TE, Cantley LC, Fruman DA (2006) Sjogren's syndrome-like disease in mice with T cells lacking class 1A phosphoinositide-3-kinase. Proc Natl Acad Sci USA 103:16882–16887

Oak JS, Chen J, Peralta RQ, Deane JA, Fruman DA (2009) The p85beta regulatory subunit of phosphoinositide 3-kinase has unique and redundant functions in B cells. Autoimmunity 442:447–458

Okkenhaug K, Wu L, Garza KM, La Rose J, Khoo W, Odermatt B, Mak TW, Ohashi PS, Rottapel R (2001) A point mutation in CD28 distinguishes proliferative signals from survival signals. Nat Immunol 2:325–332

Okkenhaug K, Bilancio A, Farjot G, Priddle H, Sancho S, Peskett E, Pearce W, Meek SE, Salpekar A, Waterfield MD, Smith AJ, Vanhaesebroeck B (2002) Impaired B and T cell antigen receptor signaling in p110delta PI 3-kinase mutant mice. Science 297:1031–1034

Okkenhaug K, Patton DT, Bilancio A, Garcon F, Rowan WC, Vanhaesebroeck B (2006) The p110delta isoform of phosphoinositide 3-kinase controls clonal expansion and differentiation of Th cells. J Immunol 177:5122–5128

Okkenhaug K, Ali K, Vanhaesebroeck B (2007) Antigen receptor signalling: a distinctive role for the p110delta isoform of PI3K. Trends Immunol 28:80–87

Omori SA, Cato MH, Anzelon-Mills A, Puri KD, Shapiro-Shelef M, Calame K, Rickert RC (2006) Regulation of class-switch recombination and plasma cell differentiation by phosphatidylinositol 3-kinase signaling. Immunity 25:545–557

Ouyang W, Beckett O, Flavell RA, Li MO (2009) An essential role of the Forkhead-box transcription factor Foxo1 in control of T cell homeostasis and tolerance. Immunity 30:358–371

Palomero T, Sulis ML, Cortina M, Real PJ, Barnes K, Ciofani M, Caparros E, Buteau J, Brown K, Perkins SL, Bhagat G, Agarwal AM, Basso G, Castillo M, Nagase S, Cordon-Cardo C, Parsons R, Zuniga-Pflucker JC, Dominguez M, Ferrando AA (2007) Mutational loss of PTEN induces resistance to NOTCH1 inhibition in T-cell leukemia. Nat Med 13:1203–1210

Park SG, Schulze-Luehrman J, Hayden MS, Hashimoto N, Ogawa W, Kasuga M, Ghosh S (2009) The kinase PDK1 integrates T cell antigen receptor and CD28 coreceptor signaling to induce NF-kappaB and activate T cells. Nat Immunol 10:158–166

Parry RV, Rumbley CA, Vandenberghe LH, June CH, Riley JL (2003) CD28 and inducible costimulatory protein Src homology 2 binding domains show distinct regulation of phosphatidylinositol 3-kinase, Bcl-xL, and IL-2 expression in primary human CD4 T lymphocytes. J Immunol 171:166–174

Parry RV, Harris SJ, Ward SG (2010) Fine tuning T lymphocytes: a role for the lipid phosphatase SHIP-1. Biochim Biophys Acta 1804:592–597

Patton DT, Garden OA, Pearce WP, Clough LE, Monk CR, Leung E, Rowan WC, Sancho S, Walker LS, Vanhaesebroeck B, Okkenhaug K (2006) Cutting edge: the phosphoinositide 3-kinase p110 delta is critical for the function of $CD4^{+}CD25^{+}Foxp3^{+}$ regulatory T cells. J Immunol 177:6598–6602

Phee H, Rodgers W, Coggeshall KM (2001) Visualization of negative signaling in B cells by quantitative confocal microscopy. Mol Cell Biol 21:8615–8625

Pierau M, Engelmann S, Reinhold D, Lapp T, Schraven B, Bommhardt UH (2009) Protein kinase B/Akt signals impair Th17 differentiation and support natural regulatory T cell function and induced regulatory T cell formation. J Immunol 183:6124–6134

Randis TM, Puri KD, Zhou H, Diacovo TG (2008) Role of PI3Kdelta and PI3Kgamma in inflammatory arthritis and tissue localization of neutrophils. Eur J Immunol 38: 1215–1224

Reif K, Okkenhaug K, Sasaki T, Penninger JM, Vanhaesebroeck B, Cyster JG (2004) Cutting edge: differential roles for phosphoinositide 3-kinases, p110gamma and p110delta, in lymphocyte chemotaxis and homing. J Immunol 173:2236–2240

Riley JL, Mao M, Kobayashi S, Biery M, Burchard J, Cavet G, Gregson BP, June CH, Linsley PS (2002) Modulation of TCR-induced transcriptional profiles by ligation of CD28, ICOS, and CTLA-4 receptors. Proc Natl Acad Sci USA 99:11790–11795

Roberts TM, Ilić N (2010) Comparing the roles of the p110a and p110b isoforms of PI3K in signaling and cancer. Curr Top Microbiol Immunol doi 10.1007/82_2010_63

Rodriguez-Borlado L, Barber DF, Hernandez C, Rodriguez-Marcos MA, Sanchez A, Hirsch E, Wymann M, Martinez AC, Carrera AC (2003) Phosphatidylinositol 3-kinase regulates the CD4/CD8 T cell differentiation ratio. J Immunol 170:4475–4482

Rommel C, Camps M, Ji H (2007) PI3K delta and PI3K gamma: partners in crime in inflammation in rheumatoid arthritis and beyond? Nat Rev Immunol 7:191–201

Ruckle T, Schwarz MK, Rommel C (2006) PI3Kgamma inhibition: towards an 'aspirin of the 21st century'? Nat Rev Drug Discov 5:903–918

Rudd CE, Schneider H (2003) Unifying concepts in CD28, ICOS and CTLA4 co-receptor signalling. Nat Rev Immunol 3:544–556

Sakaguchi S, Yamaguchi T, Nomura T, Ono M (2008) Regulatory T cells and immune tolerance. Cell 133:775–787

Sanchez-Lockhart M, Marin E, Graf B, Abe R, Harada Y, Sedwick CE, Miller J (2004) Cutting edge: CD28-mediated transcriptional and posttranscriptional regulation of IL-2 expression are controlled through different signaling pathways. J Immunol 173:7120–7124

Sanchez-Lockhart M, Graf B, Miller J (2008) Signals and sequences that control CD28 localization to the central region of the immunological synapse. J Immunol 181:7639–7648

Sasaki T, Irie-Sasaki J, Jones RG, Oliveira-dos-Santos AJ, Stanford WL, Bolon B, Wakeham A, Itie A, Bouchard D, Kozieradzki I, Joza N, Mak TW, Ohashi PS, Suzuki A, Penninger JM (2000) Function of PI3Kgamma in thymocyte development, T cell activation, and neutrophil migration. Science 287:1040–1046

Saudemont A, Garcon F, Yadi H, Roche-Molina M, Kim N, Segonds-Pichon A, Martin-Fontecha A, Okkenhaug K, Colucci F (2009) p110gamma and p110delta isoforms of phosphoinositide 3-kinase differentially regulate natural killer cell migration in health and disease. Proc Natl Acad Sci USA 106:5795–5800

Sauer S, Bruno L, Hertweck A, Finlay D, Leleu M, Spivakov M, Knight ZA, Cobb BS, Cantrell D, O'Connor E, Shokat KM, Fisher AG, Merkenschlager M (2008) T cell receptor signaling controls Foxp3 expression via PI3K, Akt, and mTOR. Proc Natl Acad Sci USA 105: 7797–7802

Saveliev A, Vanes L, Ksionda O, Rapley J, Smerdon SJ, Rittinger K, Tybulewicz VLJ (2009) Function of the nucleotide exchange activity of vav1 in T cell development and activation. Sci Signal 2:ra83

Schulze-Luehrmann J, Ghosh S (2006) Antigen-receptor signaling to nuclear factor kappa B. Immunity 25:701–715

Sebzda E, Zou Z, Lee JS, Wang T, Kahn ML (2008) Transcription factor KLF2 regulates the migration of naive T cells by restricting chemokine receptor expression patterns. Nat Immunol 9:292–300

Shiroki F, Matsuda S, Doi T, Fujiwara M, Mochizuki Y, Kadowaki T, Suzuki H, Koyasu S (2007) The p85alpha regulatory subunit of class IA phosphoinositide 3-kinase regulates beta-selection in thymocyte development. J Immunol 178:1349–1356

Sinclair LV, Finlay D, Feijoo C, Cornish GH, Gray A, Ager A, Okkenhaug K, Hagenbeek TJ, Spits H, Cantrell DA (2008) Phosphatidylinositol-3-OH kinase and nutrient-sensing mTOR pathways control T lymphocyte trafficking. Nat Immunol 9:513–521

Sommers CL, Gurson JM, Surana R, Barda-Saad M, Lee J, Kishor A, Li W, Gasser AJ, Barr VA, Miyaji M, Love PE, Samelson LE (2008) Bam32: a novel mediator of Erk activation in T cells. Int Immunol 20:811–818

Soond DR, Bjorgo E, Moltu K, Dale VQ, Patton DT, Torgersen KM, Galleway F, Twomey B, Clark J, Gaston JH, Tasken K, Bunyard P, Okkenhaug K (2010) PI3K p110{delta} regulates T cell cytokine production during primary and secondary immune responses in mice and humans. Blood 115:2203–2213

Srinivasan L, Sasaki Y, Calado DP, Zhang B, Paik JH, DePinho RA, Kutok JL, Kearney JF, Otipoby KL, Rajewsky K (2009) PI3 kinase signals BCR-dependent mature B cell survival. Cell 139:573–586

Sriskantharajah S, Belich MP, Papoutsopoulou S, Janzen J, Tybulewicz V, Seddon B, Ley SC (2009) Proteolysis of NF-kappaB1 p105 is essential for T cell antigen receptor-induced proliferation. Nat Immunol 10:38–47

Suzuki H, Terauchi Y, Fujiwara M, Aizawa S, Yazaki Y, Kadowaki T, Koyasu S (1999) Xid-like immunodeficiency in mice with disruption of the p85alpha subunit of phosphoinositide 3-kinase. Science 283:390–392

Suzuki H, Matsuda S, Terauchi Y, Fujiwara M, Ohteki T, Asano T, Behrens TW, Kouro T, Takatsu K, Kadowaki T, Koyasu S (2003) PI3K and Btk differentially regulate B cell antigen receptor-mediated signal transduction. Nat Immunol 4:280–286

Swat W, Montgrain V, Doggett TA, Douangpanya J, Puri K, Vermi W, Diacovo TG (2006) Essential role of PI3Kdelta and PI3Kgamma in thymocyte survival. Blood 107:2415–2422

Tarasenko T, Dean JA, Bolland S (2007a) FcgammaRIIB as a modulator of autoimmune disease susceptibility. Autoimmunity 40:409–417

Tarasenko T, Kole HK, Chi AW, Mentink-Kane MM, Wynn TA, Bolland S (2007b) T cell-specific deletion of the inositol phosphatase SHIP reveals its role in regulating Th1/Th2 and cytotoxic responses. Proc Natl Acad Sci USA 104:11382–11387

Tassi I, Cella M, Gilfillan S, Turnbull I, Diacovo TG, Penninger JM, Colonna M (2007) p110gamma and p110delta phosphoinositide 3-kinase signaling pathways synergize to control development and functions of murine NK cells. Immunity 27:214–227

Thomas MS, Mitchell JS, DeNucci CC, Martin AL, Shimizu Y (2008) The p110gamma isoform of phosphatidylinositol 3-kinase regulates migration of effector CD4 T lymphocytes into peripheral inflammatory sites. J Leukoc Biol 84:814–823

Thome M (2004) CARMA1, BCL-10 and MALT1 in lymphocyte development and activation. Nat Rev Immunol 4:348–359

Tsatsanis C, Vaporidi K, Zacharioudaki V, Androulidaki A, Sykulev Y, Margioris AN, Tsichlis PN (2008) Tpl2 and ERK transduce antiproliferative T cell receptor signals and inhibit transformation of chronically stimulated T cells. Proc Natl Acad Sci USA 105:2987–2992

Tze LE, Schram BR, Lam KP, Hogquist KA, Hippen KL, Liu J, Shinton SA, Otipoby KL, Rodine PR, Vegoe AL, Kraus M, Hardy RR, Schlissel MS, Rajewsky K, Behrens TW (2005) Basal immunoglobulin signaling actively maintains developmental stage in immature B cells. PLoS Biol 3:e82

Verkoczy L, Duong B, Skog P, Ait-Azzouzene D, Puri K, Vela JL, Nemazee D (2007) Basal B cell receptor-directed phosphatidylinositol 3-kinase signaling turns off RAGs and promotes B cell-positive selection. J Immunol 178:6332–6341

Vigorito E, Bardi G, Glassford J, Lam EW, Clayton E, Turner M (2004) Vav-dependent and vav-independent phosphatidylinositol 3-kinase activation in murine B cells determined by the nature of the stimulus. J Immunol 173:3209–3214

Wagle NM, Faassen AE, Kim JH, Pierce SK (1999) Regulation of B cell receptor-mediated MHC class II antigen processing by FcgammaRIIB1. J Immunol 162:2732–2740

Walmsley MJ, Ooi SK, Reynolds LF, Smith SH, Ruf S, Mathiot A, Vanes L, Williams DA, Cancro MP, Tybulewicz VL (2003) Critical roles for Rac1 and Rac2 GTPases in B cell development and signaling. Science 302:459–462

Walsh PT, Buckler JL, Zhang J, Gelman AE, Dalton NM, Taylor DK, Bensinger SJ, Hancock WW, Turka LA (2006) PTEN inhibits IL-2 receptor-mediated expansion of $CD4^+$ CD25+ Tregs. J Clin Invest 116:2521–2531

Waugh C, Sinclair L, Finlay D, Bayascas JR, Cantrell D (2009) Phosphoinositide (3,4,5)-triphosphate binding to phosphoinositide-dependent kinase 1 regulates a protein kinase B/Akt signaling threshold that dictates T-cell migration, not proliferation. Mol Cell Biol 29:5952–5962

Webb LM, Vigorito E, Wymann MP, Hirsch E, Turner M (2005) Cutting edge: T cell development requires the combined activities of the p110gamma and p110delta catalytic isoforms of phosphatidylinositol 3-kinase. J Immunol 175:2783–2787

Weber M, Treanor B, Depoil D, Shinohara H, Harwood NE, Hikida M, Kurosaki T, Batista FD (2008) Phospholipase C-gamma2 and vav cooperate within signaling microclusters to propagate B cell spreading in response to membrane-bound antigen. J Exp Med 205:853–868

Yokosuka T, Kobayashi W, Sakata-Sogawa K, Takamatsu M, Hashimoto-Tane A, Dustin ML, Tokunaga M, Saito T (2008) Spatiotemporal regulation of T cell costimulation by TCR-CD28 microclusters and protein kinase C theta translocation. Immunity 29:589–601

Zebedin E, Simma O, Schuster C, Putz EM, Fajmann S, Warsch W, Eckelhart E, Stoiber D, Weisz E, Schmid JA, Pickl WF, Baumgartner C, Valent P, Piekorz RP, Freissmuth M, Sexl V (2008) Leukemic challenge unmasks a requirement for PI3Kdelta in NK cell-mediated tumor surveillance. Blood 112:4655–4664

Zhang TT, Okkenhaug K, Nashed BF, Puri KD, Knight ZA, Shokat KM, Vanhaesebroeck B, Marshall AJ (2008) Genetic or pharmaceutical blockade of p110delta phosphoinositide 3-kinase enhances IgE production. J Allergy Clin Immunol 122(811–819):e812

Zheng Y, Rudensky AY (2007) Foxp3 in control of the regulatory T cell lineage. Nat Immunol 8:457–462

Zompi S, Hamerman JA, Ogasawara K, Schweighoffer E, Tybulewicz VL, Di Santo JP, Lanier LL, Colucci F (2003) NKG2D triggers cytotoxicity in mouse NK cells lacking DAP12 or Syk family kinases. Nat Immunol 4:565–572

The Regulation of Class IA PI 3-Kinases by Inter-Subunit Interactions

Jonathan M. Backer

Contents

Abstract Phosphoinositide 3-kinases (PI 3-kinases) are activated by growth factor and hormone receptors, and regulate cell growth, survival, motility, and responses to changes in nutritional conditions (Engelman et al. 2006). PI 3-kinases have been classified according to their subunit composition and their substrate specificity for phosphoinositides (Vanhaesebroeck et al. 2001). The class IA PI 3-kinase is a heterodimer consisting of one regulatory subunit (p85α, p85β, p55α, p50α, or p55γ) and one 110-kDa catalytic subunit (p110α, β or δ). The Class IB PI 3-kinase is also a dimer, composed of one regulatory subunit (p101 or p87) and one catalytic subunit (p110γ) (Wymann et al. 2003). Class I enzymes will utilize PI, PI[4]P, or PI[4,5]P2 as substrates *in vitro*, but are thought to primarily produce PI[3,4,5]P3 in cells.

J.M. Backer
Department of Molecular Pharmacology, Albert Einstein College of Medicine, 1300 Morris Park Avenue, Bronx, NY 10461, USA
e-mail: jonathan.backer@einstein.yu.edu

C. Rommel et al. (eds.). *Phosphoinositide 3-kinase in Health and Disease*, Volume 1
Current Topics in Microbiology and Immunology 346, DOI 10.1007/82_2010_52

The crystal structure of the Class IB PI 3-kinase catalytic subunit p110γ was solved in 1999 (Walker et al. 1999), and crystal or NMR structures of the Class IA p110α catalytic subunit and all of the individual domains of the Class IA p85α regulatory subunit have been solved (Booker et al. 1992; Günther et al. 1996; Hoedemaeker et al. 1999; Huang et al. 2007; Koyama et al. 1993; Miled et al. 2007; Musacchio et al. 1996; Nolte et al. 1996; Siegal et al. 1998). However, a structure of an intact PI 3-kinase enzyme has remained elusive. In spite of this, studies over the past 10 years have lead to important insights into how the enzyme is regulated under physiological conditions. This chapter will specifically discuss the regulation of Class IA PI 3-kinase enzymatic activity, focusing on regulatory interactions between the p85 and p110 subunits and the modulation of these interactions by physiological activators and oncogenic mutations. The complex web of signaling downstream from Class IA PI 3-kinases will be discussed in other chapters in this volume.

1 Structural Organization of p85α/p110α

The p85 regulatory and p110 catalytic domains of Class IA PI 3-kinase are both multi-domain proteins (Fig. 1). The crystal structure of p110α (Huang et al. 2007) shows an N-terminal Adapter Binding Domain (ABD), which binds to the coiled-coil domain of p85 (the iSH2 domain). The ABD has an ubiquitin-like fold that makes a large hydrophobic contact with the iSH2 domain (Huang et al. 2007; Miled et al. 2007). The remainder of p110α consists of a Ras-binding domain, a C2 domain, a helical domain (found in all PI 3-kinase and formerly referred to as the PIK domain), and a kinase domain with a two-lobed structure typical of protein kinases; this organization is similar to that found in p110γ (Huang et al. 2007; Walker et al. 1999).

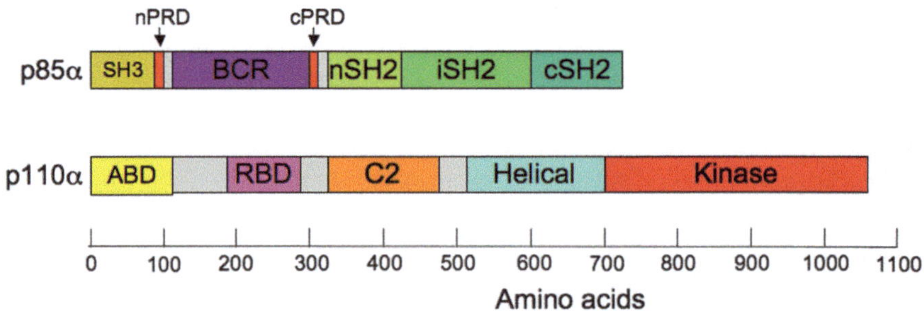

Fig. 1 *Domain structure of p85α and p110α.* The p85α regulatory subunit contains Src-homology 3 (SH3), BCR-homology (BCR), Proline rich (PRD), and Src homology 2 (SH2) and domains. The inter-SH2 domain (iSH2) is an antiparallel coiled-coil that links the two SH2 domains. The p110α catalytic subunit contains the adapter binding domain (ABD), which binds to the iSH2 domain of p85, a Ras-binding domain (RBD), a C2 domain, a helical domain (formally called the PIK homology domain), and a kinase domain

The full-length p85 isoforms, p85α and p85β, consist of an SH3 domain, a Rac-binding domain (the BCR homology domain) flanked by two proline-rich domains (nPRD and cPRD), and two SH2 domains (nSH2 and cSH2) linked by the inter-SH2 (iSH2) domain (Vanhaesebroeck et al. 2001). Three shorter regulatory subunits, p55γ and the p85α splice variants p55α and p50α, lack the SH3, nPRD, and BCR domains. The two p85 nSH2 and cSH2 domains share a similar specificity for tyrosine-phosphorylated motifs of the form YxxM (Songyang et al. 1993). While individual domains of p85α have been structurally characterized, as yet there is no structural data concerning the spatial organization of these domains within p85.

The iSH2 domain was predicted to be a long antiparallel coiled coil, based on sequence analysis (Dhand et al. 1994a) and subsequent site-specific spin labeling/electron paramagnetic resonance spectroscopy and molecular modeling of the nSH2-iSH2 fragment of p85α (Fu et al. 2003). A crystal structure of the isolated iSH2 domain (residues 431–600) bound to the ABD of p110α (Miled et al. 2007) shows a 115-Å long rod comprising two major helices (440–515 and 518–587). The C-terminal end of the iSH2 domain forms a short third helix (588–599); the extent of this third helix in intact p85α is not yet known. The presence of the p85α N- and C-terminal SH2 domains at the ends of an antiparallel coiled-coil suggests that the domains are relatively close to each other. The linker between the nSH2 domain and the iSH2 domain is only 10 amino acids long, and recent EPR data suggest that this short tether limits the range of mobility of the nSH2 domain relative to the iSH2 domain (Sen et al. 2010; see below). In contrast, the linker region between structurally defined helical regions of the iSH2 domain and the cSH2 domain (23 amino acids) is long enough to allow for considerable uncertainty as to their relative orientation. Isothermal calorimetry studies showed that the two SH2 domains of p85α can bind simultaneously to a peptide containing phosphotyrosine residues separated by 11 amino acids (O'Brien et al. 2000), and the nSH2 domain can bind to a phosphorylated Tyr688 in the C-terminal SH2 domain in the context of a p85/p110α dimer (Cuevas et al. 2001). While these studies suggest that the phosphotyrosine binding sites of the nSH2 and cSH2 domains can in some cases be close together, the range of motion and orientation of the two domains in the p85α/p110 dimer is not known.

The crystal structure of p110α bound to an nSH2-iSH2 fragment of p85α (Huang et al. 2007) is shown in Fig. 1 in both space-filling and schematic views, from two orientations. Structures of the p110α ABD or intact p110α bound to the iSH2 domain of p85α show extensive hydrophobic interactions between the ABD and the iSH2 domain, involving seven helical turns in iSH2 helix 1 and three turns of helix 2 (Miled et al. 2007). The C2 domain and kinase domains of p110α drape over the iSH2 domain, with predicted hydrogen bonding between residue N345 in the C2 domain with residues D560 and N564 in helix 2 of the iSH2 domain. The helical domain contacts the kinase and C2 domains, and is located near the p85 SH2 domains and away from the ABD-iSH2 domain contacts. In the orientation shown in Fig. 2a, b, the Ras-binding domain is on top, separated from the iSH2 domain and making contacts with the kinase domain. The position of the p85α nSH2 domain is not seen in the crystal structure of wild-type p110α, but is visible in

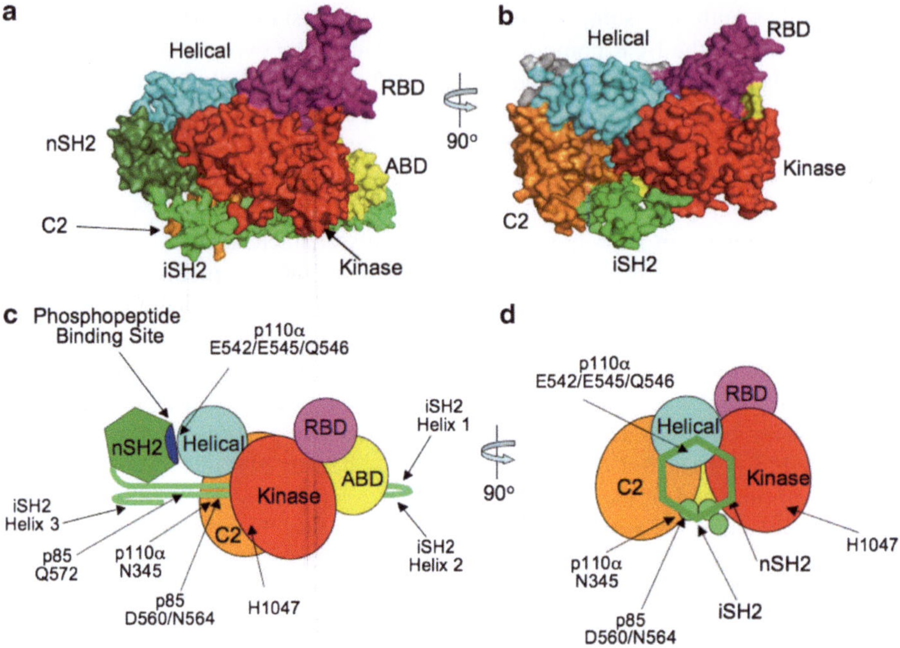

Fig. 2 Space-filing views of the nSH2-iSH2-p110α[H1047R] crystal structure (Huang et al. 2007; Mandelker et al. 2009), facing the iSH2 domain from the side (**a**) or down its barrel with the nSH2 domain removed (**b**). (**c**, **d**) Schematic models of the same orientations. In (**d**), the nSH2 domain is shown only in outline, to allow the rest of the iSH2 domain and p110α to be seen. The sites of the oncogenic mutations in p85α (D560, N564 and Q572) and p110α (E542/E545/Q546 and H1047) are shown, as is the phosphopeptide binding site of the nSH2 domain

the structure of the p85 nSH2-iSH2 fragment dimerized with the H1047R mutant of p110α (Huang et al. 2007; Mandelker et al. 2009). The nSH2 domain makes contacts with the helical domain of p110α, as previously predicted (Miled et al. 2007), as well as with the kinase and C2 domains.

2 p85α Inhibits and Stabilizes p110

The p85α and p85β regulatory and p110α catalytic subunits of Class IA PI 3K were cloned in 1991 and 1992 (Escobedo et al. 1991; Hiles et al. 1992; Otsu et al. 1991; Skolnik et al. 1991). There was initially some confusion as to whether p85α activated or inhibited p110α. The catalytic subunit is active as a monomer when expressed in insect cells (Hiles et al. 1992). In contrast, p85α is required for p110α activity in mammalian cells; based on these latter data, Williams and coworkers proposed that p85α is a p110α activator (Klippel et al. 1994). Studies in yeast suggested that p85α is a p110α inhibitor, as expression of p85α rescues the lethal effects of p110α expression in *S. pombe* (Kodaki et al. 1994). Similarly, using recombinant monomeric p110α produced in insect cells, it was shown that p85α binding inhibits the activity of monomeric p110α by as much as 80% (Yu et al. 1998a). These data were reconciled with the discovery that monomeric p110α is heat labile, and is stabilized by dimerization with p85α (Yu et al. 1998a). Monomeric p110α rapidly loses activity when incubated at 37°C, and it undergoes rapid degradation when expressed as a monomer in mammalian cells. However, the p110α monomer is active when expressed in insect cells, which grow at 25°C. Similarly, the specific activity of overexpressed monomeric p110α in mammalian cells is increased 15-fold by culturing the cells at reduced temperature. These data explain the discrepancies in the earlier literature: the apparent activation of p110α by its co-expression with p85α in mammalian cells in fact reflects the stabilization of p110α in an inhibited, low activity state. The heat labile nature of monomeric p110α also explains the later observation that the homozygous deletion of p85α and p85β in MEFs leads to parallel losses of p110 expression (Fruman et al. 2000).

The stabilization of the p110α subunit by binding to p85α is not as yet understood. Several groups showed that the N-terminus of p110, the adapter-binding domain or ABD, binds to the coiled-coil domain of p85α, the iSH2 domain (Holt et al. 1994; Hu et al. 1993; Hu and Schlessinger 1994; Klippel et al. 1993). This interaction is necessary and sufficient for p110-p85α dimerization and for stabilization of p110α in mammalian cells, although it does not replicate physiological regulation of p110α (see below). Surprisingly, the role of regulatory subunit binding in p110α stabilization is supplanted by the linkage of epitope tags to the N-terminus of p110α; the degree of stabilization correlates with the size of the tag (Yu et al. 1998a). This explains the finding that a fusion of the p85α iSH2 domain with the N-terminus of p110α (the commonly used p110*) is constitutively active in mammalian cells (Hu et al. 1995). Based on more recent biochemical and structural data, this construct is unlikely to replicate ABD-iSH2 interactions. The ABD of p110α binds to residues near the hinge region of the rod-like iSH2 antiparallel coiled-coil, at the end furthest away from the two SH2 domains (Huang et al. 2007; Miled et al. 2007) (Fig. 2). In contrast, the p110* chimera links helix 3 of the iSH2 domain to the N-terminus of p110α, placing the p110α ABD at some distance from its normal binding site in the iSH2 domain. Thus, the iSH2 domain in p110* presumably stabilizes p110α by acting as a bulky N-terminal tag, and not by replicating ABD-iSH2 domain interactions. Consistent with this idea, Vogt and colleagues have found that an oncogenic mutant of p110α, p110α-H1047R, is not stabilized by a p110*-like linkage to the iSH2 domain (Zhao and Vogt 2008), whereas we find that p110α H1047R is stabilized by co-expression with the iSH2 domain in trans (J.M. Backer, unpublished observations).

The stabilization of p110α catalytic subunits (and presumably also p110β and p110δ) poses a problem for overexpression studies, since N-terminally tagged p110 will show a higher stability, and hence a higher constitutive activity, than wild-type p110 (Yu et al. 1998a). Whereas expression levels of wild type p110α are limited by the amount of available p85, this is not true for N-terminally tagged p110α. Unfortunately, recent data suggest that some C-terminal tags may inhibit the activity of p110α toward PI[4,5]P2 *in vivo* (Bart Vanhaesebroeck, personal communication). Thus, the definition of an activity-neutral tag for the study of p110 isoforms continues to pose a significant experimental problem.

The inhibition of p110 by p85 is an example of a widely used regulatory scheme in eukaryotic ells, in which regulatory subunits of kinases maintain enzyme activity at a low level, with subsequent activation of the enzyme by a release of inhibition. The best-studied example of this scheme is PKA, where R1 or R2 subunits inhibit the activity of the C subunits (Taylor et al. 2005). In PKA, the mechanism of disinhibition is dissociative: in the presence of cAMP, the binding between the regulatory and catalytic subunits is disrupted. However, for Class IA PI 3-kinase, this mechanism of disinhibition is impossible for two reasons. First, as discussed above, monomeric p110 is heat labile and would rapidly lose activity should p85 dissociate. Second, p85–p110α binding is extremely tight, and is essentially irreversible under cellular conditions (Woscholski et al. 1994). Thus, activation of p85/p110 dimers by upstream regulators must occur by an intramolecular conformational change, in which the subunits remain bound together and activator binding to p85 causes a conformational change in p110.

3 Activation of p85α/p110α Dimers by Phosphopeptides

Early studies on Class IA PI 3-kinases demonstrated an increase in PI 3-kinase activity in anti-phosphotyrosine or anti-receptor tyrosine kinase immunoprecipitates upon growth factor stimulation (reviewed in Cantley et al. 1991). With the demonstration that SH2 domains are binding sites for tyrosine phosphorylated proteins (Matsuda et al. 1991; Mayer et al. 1991; Moran et al. 1990), as well as the identification of relatively PI 3-kinase-specific phosphotyrosine motifs in receptor tyrosine kinases (Fantl et al. 1992; Songyang et al. 1993), it became clear that the production of PIP3 *in vivo* is mediated at least in part by the recruitment of PI 3-kinases to activated receptor tyrosine kinases or their substrates. This led to the demonstration that constitutive targeting of p110α to cell membranes by N-terminal myristylation or C-terminal isoprenylation is sufficient to produce constitutive increases in PIP3 production (Klippel et al. 1996). The preferred tyrosine phosphorylated peptide binding site for the p85α SH2 domains was defined by degenerate peptide library binding studies to be (P)YxxM (Songyang et al. 1993). A crystal structure of a phosphopeptide-bound p85α nSH2 domain shows a two-pronged socket binding mechanism, similar to that seen in the Src SH2 domain, in which the negatively-charged phosphotyrosine fits into a positively charged pocket, and the bulky methionine fits into a hydrophobic pocket (Nolte et al. 1996; Waksman et al. 1993).

With the development of antibodies against endogenous p85α, as well as methods to produce recombinant p85α and p110α in baculovirus (Gout et al. 1992; Woscholski et al. 1994), it was possible to directly examine the effect of p85α SH2 domain occupancy on the specific activity of PI 3-kinase, independently of effects on PI 3-kinase membrane recruitment. Initial studies showed that binding of p85α/p110α dimers to tyrosine phosphorylated IRS-1 or tyrosine phosphopeptides

peptides from IRS-1, Polyoma middle T or the PDGF receptor causes a two- to threefold activation *in vitro* (Backer et al. 1992; Carpenter et al. 1993). Bis-phosphopeptides activate PI 3-kinase with higher potency than corresponding monophosphopeptides (Carpenter et al. 1993; Herbst et al. 1994). p85α contains two SH2 domains, and a number of tyrosine-phosphorylated proteins that bind p85α are known to have two or more tyrosine phosphorylated YXXM motifs (e.g., PDRF-R, IRS-1). This suggested that dual occupancy of both domains might lead to increased PI 3-kinase activation. Consistent with this model, individual mutation of the conserved FLRVES motif in the N-terminal or C-terminal SH2 domains of p85α leads to partial loss of phosphopeptide activation; mutation of both SH2 domains leads to a complete loss (Rordorf-Nikolic et al. 1995).

While bis-phosphopeptides lead to more potent activation of p85α/p110α dimers, the idea that the N- and C-terminal SH2 domains of p85α bind simultaneously to a single tyrosine-phosphorylated protein has been challenged. Biochemical studies showed that a bis-phosphorylated peptide from Polyoma MT binds to the p85α nSH2 domain, but not to the cSH2 domain, with enhanced affinity relative to a monophosphopeptide; a second noncanonical binding site for phosphotyrosine in the nSH2 domain was defined using NMR (Günther et al. 1996). Bis-phosphopeptides were also found to induce formation of higher order dimers between p85α/p110α holoeznymes; dimerization is independent of peptide concentration, suggesting that it does not involve direct bridging of SH2 domains in distinct p85α/p110α dimers (Layton et al. 1998). Finally, isothermal calorimetry experiments suggested that the mode of p85α binding to receptors containing two potential binding sites (e.g., the PDGF-R receptor) is concentration dependent: two distinct p85α molecules bind to a single bis-phosphopeptide when p85α is in excess, whereas both SH2 domains from a single p85α molecule bind to a single bis-phosphopeptide when peptide is in excess (O'Brien et al. 2000). It remains unclear whether PI 3-kinase binding to ligands like the tyrosine bis-phosphorylated PDGF receptor *in vivo* involves divalent interactions.

4 A Mechanism for Phosphopeptide Activation

A model for the regulation of p85α/p110α dimers by phosphopeptides emerged from biochemical studies, in which active monomeric p110α (produced in insect cells) was reconstituted with fragments of p85α, and its activity was measured in the absence and presence of tyrosine phosphopeptides (Yu et al. 1998b). These studies showed that although the iSH2-ABD contact is the major interaction responsible for formation of the p85α/p110α dimer, as previously shown by several groups (Holt et al. 1994; Hu et al. 1993; Hu and Schlessinger 1994; Klippel et al. 1993), the iSH2 domain is not involved in the regulation of p110α activity. First, although the iSH2 domain is sufficient to bind the p110α ABD, this binding has no affect on p110α activity. Regulation of p110α by tyrosine phosphopeptides was instead shown to specifically require the presence of the N-terminal SH2 domain:

nSH2-iSH2 fragments of p85α (p85ni) inhibit p110 and the p85ni/p110α dimer is activated by phosphopeptide. In contrast, iSH2, iSH2-cSH2, and cSH2-iSH2 fragments bind but do not inhibit p110. Thus, p85ni is the minimal fragment of p85α that confers phosphopeptide regulation of p110 (Yu et al. 1998b).

Second, NMR and EPR studies on the structure and mobility of the iSH2 domain showed that the iSH2 domain is rigid and conformationally unresponsive to SH2 domain occupancy (Fu et al. 2003, 2004; O'Brien et al. 2000). Based on these data, it was proposed that the iSH2 domain is a rigid tether for p110, holding the subunits together and enabling a second regulatory contact between the nSH2 and an unidentified region of p110α (Fu et al. 2004). Subsequent activation of the p85α/p110α dimer would involve a disruption of this inhibitory interface by binding of a phosphoprotein to the SH2 domain.

While these data suggested an inhibitory nSH2-iSH2 contact, the site of this contact was not known. A potential clue to the site of interaction emerged from studies by Samuels et al., which identified oncogenic mutants of p110α from human tumor samples (discussed in more detail below) (Samuels et al. 2004). Sporadic mutations are scattered over the length of p110α, but two hotspots account for nearly 80% of the mutations: an H1047R mutant in the C-terminus of p110α, and a cluster of 3 change of-charge mutations (E542K, E545K, Q546K) in the helical domain of p110α. Could these hotspots indicate the site of the inhibitory nSH2-p110α contact? If so, then it seemed likely that the mutations would be at residues conserved between Class IA catalytic subunit isoforms (p110α/β/δ), since p85α binds to and inhibits all three isoforms. However, whereas the acidic cluster in the helical domain is conserved in all three p110 isoforms, H1047 is not found in the kinase domains p110β or p110δ, where residues corresponding to $H^{1046}H^{1047}$ are replaced with LR. Thus, the helical domain acidic cluster seemed the more likely site of nSH2-p110α contact.

This hypothesis was tested using an *in vitro* reconstitution approach (Miled et al. 2007). Helical domain mutants of p110α (E542K, E545K, E546Q) bind p85ni but are not inhibited, nor are p85α/p110α-E545K dimers additionally activated by phosphopeptide. In contrast, although the H1047R mutant has increased specific activity as compared to wild-type p110α, p85α/p110α-H1047R dimers are additionally activated by phosphopeptide (Carson et al. 2008), consistent with the observation that p110α-H1047R is inhibited by p85ni (J.M. Backer and R.L. Williams, unpublished observations). Thus, the helical domain appeared to be the site of the inhibitory nSH2 domain contact. Miled et al. next reasoned that the acidic patch in the helical domain was likely to interact with basic residues on the surface of the nSH2 domain. If so, then a basic-to-acidic mutation in the nSH2 domain might reduce the inhibition of wild-type p110, but restore the inhibition of the helical domain mutants. In fact, it was shown that K379E and R340E mutants of p85ni exhibited a loss of inhibition of wild-type p110. Additional experiments showed that the K379E mutant of p85ni restores the inhibition of the E545K mutant of p110, indicative of contacts between basic residues in the nSH2 domain and the acidic patch in the helical domain.

These data shed light on both the physiological activation of p85α/p110α by tyrosyl phosphoproteins, as well as the mechanism of oncogenic activation in the helical domain mutants (Fig. 3). The nSH2 basic residues whose mutation to Glu reduce the inhibition of wild-type p110α (K379 and R340) lie adjacent to the phosphopeptide-binding site of the nSH2 domain. One can readily see how the binding of the tyrosine-phosphorylated proteins to the nSH2 domain would disrupt contacts between these residues and the helical domain of p110α. In the cancer-related p110α helical domain mutations, this contact is already disrupted. Thus, the helical domain mutants of p110α found in human cancer mimic the physiological disruption of p110α inhibition caused by phosphotyrosine protein binding to the nSH2 domain of p85α. More recent data suggest that helical domain mutations may have additional effects in the context of transformation (Zhao and Vogt 2008) (discussed below).

Given that the nSH2 phosphopeptide binding site contacts the E542/E545/Q546 acidic patch in the helical domain of p110α, the nSH2 domain presumably must move away from the helical domain to accommodate the binding of tyrosine phosphorylated activators. Recent studies suggest that the nSH2 domain shows considerable mobility relative to the iSH2 domain, at least in the context of an isolated p85α (Sen et al. 2010; Wu et al. 2009; see below). Biochemical data also suggests that the ability of nSH2 domain to move about the proximal end of the iSH2 domain is necessary for PI 3-kinase activation, as introduction of an engineered disulfide

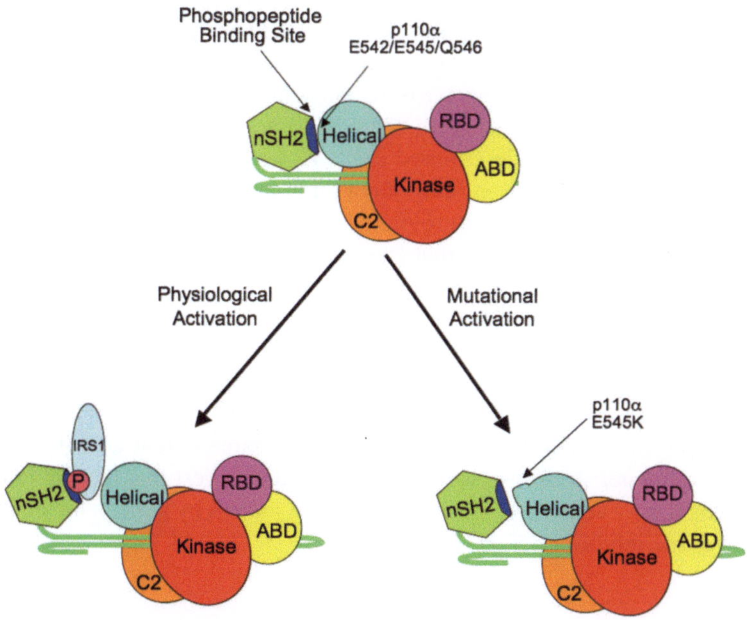

Fig. 3 *Models of physiological and oncogenic activation of p85/p110 dimers*. The models show a schematic of p110α bound to the nSH2-iSH2 fragment of p85. In normal cells, phosphoprotein binding to the nSH2 domain disrupts the inhibitory interface with the helical domain of p110α. In transformed cells, mutations of residues in the acidic patch in the helical domain (for example E545K) constitutively disrupt nSH2-helical domain interactions, leading to the deregulation of the enzyme

bond between the nSH2 and iSH2 domains produces a mutant than inhibits p110, but cannot be activated by phosphopeptides (H. Wu and J.M. Backer, unpublished observations). A similar model has been proposed by Hale et al. (2008) for the activation of p85β/p110α dimers by the NS1 protein from Influenza A. NS1 binds to the C-terminal end of the p85β iSH2 domain, including helix 3, and the authors propose that binding of NS1 to this region may sterically force the nSH2 domain into a conformational that disrupts its inhibitory contacts with the helical domain of p110.

As noted above, maximal activation of intact p85α/p110α dimers by phosphopeptides requires functional nSH2 and cSH2 domains (Rordorf-Nikolic et al. 1995). Given that an iSH2-cSH2 fragment of p85α binds but does not inhibit p110α (Yu et al. 1998b), the mechanisms of activation by the C-terminal SH2 domain must be different. Interestingly, phosphopeptide binding to the cSH2 domain contributes to p110α activation only within the context of full-length p85α/p110α dimers, but not when p110α is dimerized with the nSH2-iSH2-cSH2 construct; a disabling mutation of the nSH2 domain only partially abolishes phosphopeptide activation of p85α/p110, whereas it completely abolishes activation of p110α bound to the nSH2-iSH2-cSH2 construct (Yu et al. 1998b). This means that activation of p110α dimers by occupancy of the cSH2 domain requires the presence of the N-terminal domains of p85α (the SH3, proline rich and BCR-homology domains). Presumably, these domains position the cSH2 domain so that it makes direct contacts with p110, or affect inhibitory contacts between p110α and other domains of p85α. An example of this latter mechanism was described by Cuevas et al. (2001), who suggested that the tyrosine phosphorylation of a site in the cSH2 domain (Y688) could interact with the nSH2 domain, leading to PI 3-kinase signaling to Akt activation in cultured cells. However, in this case the activation occurs through phosphorylation rather than through cSH2 domain occupancy.

A mechanistically independent role for the cSH2 domain was also suggested by Chan et al. (2002), who showed that Ras-dependent Akt activation is blocked by expression of a p85α fragment consisting of helix 3 of the iSH2 domain, the iSH2-cSH2 linker, and the cSH2 domain. A truncation mutant identified in a Hodgkin's disease patient (a frame shift that truncates p85α at residue 635 in the cSH2 domain, and adds 25 unrelated amino acids) also leads to increased Akt activation and transformation in cultured cells (Jucker et al. 2002), despite the fact that this construct appears to inhibit p110α *in vitro* to the same extent as wild type p85α (JM Backer, unpublished observations). These data suggest that the deletion of the cSH2 domain may activate PI 3-kinase signaling *in vivo* by a mechanism that does not involve the inhibition of p110 by p85α.

5 Activation of p85/p110 by GTPases

Class I PI 3-kinases are activated by the binding of activated (GTP-bound) small GTPases. Activated Ras binds directly to the RBD of the catalytic subunit of both Class IA and IB PI-3 kinase. While all four p110 isoforms can bind Ras, only p110α and p110γ have been clearly shown to be activated by Ras *in vitro* (Jimenez et al. 2002; Pacold et al. 2000; Rodriguez-Viciana et al. 1996; Rubio et al. 1997). This does not mean that Ras binding is not involved in signaling by p110β and p110δ, since an increase in membrane association could lead to increased PIP3 production *in vivo* that might not be detected *in vitro*. For example, p110β and p110δ but not p110α are recruited to Ras in redox-stimulated cells (Deora et al. 1998). In the case of p110γ, the crystal structure of the GTP-Ras-p110γ complex shows that activated Ras binding to the RBD involves both the Switch I and II domains of Ras. Ras binding causes a substantial shift in the position of a number of helices near the activation loop, consistent with an allosteric activation mechanism of p110γ activation (Pacold et al. 2000). A study of knock-in mice expressing RBD mutants of p110α has clearly shown that Ras-p110α binding is required for normal lymphatic development, and for tumorigenesis by activated mutants of K-ras (Gupta et al. 2007; Ramjaun and Downward 2007).

Binding of activated (GTP-bound) forms of the small GTPases Cdc42 or Rac activates p85α/p110α dimers *in vitro* (Bokoch et al. 1996; Tolias et al. 1995; Zheng et al. 1994); activation of p85α/p110β or p85α/p110δ dimers has not been assessed. Rac and Cdc42 bind to the BCR homology domain of p85α (Beeton et al. 1999), but the mechanism by which this binding affects p85α/p110α activity is not known. The BCR homology domain is so named because of homology to the Rho-GAP domain of the BCR gene product (Diekmann et al. 1991). A mutagenesis analysis of the BCR homolog β-chimaerin identified a number of residues necessary for Rho-GAP activity that are not conserved in the p85α BCR domain (Ahmed et al. 1994). Alternatively, another group has found that p85α does have GTPase activity toward Rac and Cdc42, as well as toward Rab5 and Rab4, but not Rab 11 (Chamberlain et al. 2004).

The original classification scheme for Class I PI 3-kinases suggested that p85α dimers with p110α, p110β, or p110δ are regulated primarily by receptor tyrosine kinases, whereas dimers of p101 or p87 with p110γ are regulated by G-protein coupled receptors (GPCRs). This division of labors has been recently questioned, with the demonstration that p85α/p110β dimers are also regulated primarily by GPCRs (Guillermet-Guibert et al. 2008).

The activation of p85/p110β dimers by Gβγ subunits was first noted in the 1997 by Kurosu et al. (1997), who identified a Gβγ-stimulated PI 3-kinase from rat liver as p85/p110β. The *in vitro* activation of p110β by Gβγ occurs in the absence or presence of p85; in the latter case, Gβγ activation is synergistic with phosphopeptide activation (Maier et al. 1999). In a strain of NIH 3T3 cells expressing undetectable levels of p110β, expression of p110β is required for LPA-stimulated activation of Akt (Murga et al. 2000). While these early studies suggested that p85/p110β dimers could respond to Gβγ, recent evidence suggests that p85/p110β dimers may in fact be unresponsive to receptor tyrosine kinases and primarily regulated by GPCRs. In MEFS expressing a kinase-dead p110β knock-in, or in the livers of conditional p110β knockout animals, signaling is normal in response to ligands for receptor tyrosine kinases, such as insulin or IGF-I, but defective in response to GPCR ligands such as LPA or S1P (Ciraolo et al. 2008; Jia et al. 2008). In macrophages, which also express the GPCR-regulated p110γ, both p85/p110β and p101/p110γ dimers respond to a similar range of GPCR ligands, and knockout or inhibition of both isoforms is required to completely inhibit GPCR-mediated activation of Akt (Guillermet-Guibert et al. 2008). The finding that p85/p110β dimers are responsive to GPCRs explains a number of data in the recent literature using knockouts, kinase-dead knock-ins, or isoform-specific PI 3-kinase inhibitors to evaluate specific signaling by distinct p110 isoforms. For example, in myotubes or adipocyte cell lines, similar amounts of p110α and p110β are bound to IRS-1, but pharmacological inhibition of p110β has minimal affect on IRS-1-associated PI 3-kinase activity or insulin signaling (Knight et al. 2006). Similarly, p110α is recruited preferentially to IRS-1 as compared to p110β in insulin-stimulated fat, muscle and liver (Foukas et al. 2006). These data are consistent with the idea that p110β is primarily regulated by GPCRs.

From a mechanistic point of view, however, we have very little insight into why p110β is regulated differently from p110α, or presumably p110δ. p110β is inhibited by p85 *in vitro* (H. Dbouk and JM Backer, unpublished observations), and p85/p110β dimers are activated by tyrosine phosphopeptides *in vitro* (Maier et al. 1999). The helical domain acidic patch required for regulation of p110α (Miled et al. 2007) is conserved in p110β. Thus, all the elements for SH2-mediated regulation of p110β would seem to be in place. This suggests that p110β is inhibited by nSH2-helical domain contacts, making it surprising that p85/p110β dimers are not activated by binding to tyrosine-phosphorylated IRS-1 (Knight et al. 2006). The failure of p110β-bound p85 to bind to tyrosine phosphorylated IRS-1, even in tissues where p110β is abundant, is also unexplained (Foukas et al. 2006).

Finally, little is known about how Gβγ interacts with p85/p110β. As mentioned above, Gβγ activation of Akt has been observed in cells expressing p110β without p85, and p110β is activated in the absence or p85 *in vitro*, suggesting that Gβγ can interact directly with the p110β catalytic subunit (Maier et al. 1999; Murga et al. 2000). p110β-mediated activation of Akt is blocked by overexpression of a construct comprising residues 34–349 of p110β, which contains the RBD flanked by truncated ABD and C2 domains (Kubo et al. 2005), but a more thorough definition of the Gβγ-p110γ binding site is still wanting.

p110β was identified in a proteomics analysis of proteins that bind to the activated form of Rab5, the early endosomal small GTPases (Christoforidis et al. 1999). The binding site for this interaction in p110β has not been established, nor has it been shown whether Rab5 binding to p85/p110β dimers increases PI 3-kinase activity. A possible endosome-specific function for p110β was suggested by the finding that early endosomal PIP3 can be converted by the sequential action of phosphoinositide phosphatase to the functionally important early endosomal lipid PI[3]P (Shin et al. 2005). In addition, a role in clathrin-mediated endocytosis has been suggested by studies showing a marked internalization defect in MEFs derived from mice in which p110β expression is knocked out or significantly decreased (Ciraolo et al. 2008; Jia et al. 2008). This may be consistent with the fact that Rab5-associated p110β is found in clathrin coated vesicles but not early endosomes

(Christoforidis et al. 1999). It is intriguing to note that the internalization defect in p110β-deficient MEFs is rescued by kinase-dead mutant of p110β (Ciraolo et al. 2008; Jia et al. 2008), suggesting that the Rab5-p110β interaction may not involve regulation of p110β lipid kinase activity, but rather the regulation of an as yet undefined scaffolding function of p110β.

6 Activation by Binding to p85 SH3 and Proline-Rich Domains

The SH3 domain of p85α binds to proline-rich ligands of the general form RXLPPRPXX or XXXPPLPXR; the two types of ligand may bind in opposite orientations (Koyama et al. 1993). At present, nothing is known about the position of the p85α SH3 domain relative to the rest of p85α or p110, although *in vitro* binding studies show that the N-terminal proline rich sequences from p85α can interact with the p85α SH3 domain (Kapeller et al. 1994). This suggests that the SH3 domain may fold forward toward the BCR domain, which is flanked by the two proline-rich domains. If this hypothesis is correct, then the disruption of intramolecular SH3-proline-rich domain interactions might provide an activation mechanism for PI 3-kinase. Several studies have suggested that p85α/p110 dimers are activated by either binding of proline-rich peptides to the p85α SH3 domain, including the proline-rich motifs from the FAK tyrosine kinase (Guinebault et al. 1995) and the influenza virus protein NS1 (Shin et al. 2007a, b). However, as discussed above, activation of PI 3-kinase by influenza virus may also involve binding between NS1 and the C-terminal end of the p85β iSH2 domain (Hale et al. 2008). Similarly, addition of recombinant SH3 domains from Src-family kinases activates p85/p110 dimers *in vitro* (Pleiman et al. 1994), and PI 3-kinase activation by the binding the SH3 domains of Grb2 has been demonstrated in IGF-1-stimulated vascular smooth muscle cells (Imai and Clemmons 1999). These data suggest that binding of either the SH3 or proline rich domains of p85α lead to PI 3-kinase activation, perhaps by disrupting an inhibitory intramolecular SH2-PRD interaction.

7 Regulation by Autophosphorylation

Class I PI 3-kinases possess protein as well as lipid kinase activity. Ser608 in the iSH2-cSH2 linker region of p85α is a substrate for p110-mediated phosphorylation, which inhibits PI 3-kinase activity *in vitro* (Dhand et al. 1994b). This inter-subunit phosphorylation presumably represents a form of negative feedback regulation, as Ser608 phosphorylation is increased in response to stimulation by agonists like insulin or PDGF that increase Class IA activity (Foukas et al. 2004). However, evaluation of the role of this residue *in vivo* has been difficult, as mutation to either Ala or Glu decreases PI 3-kinase activity (Foukas et al. 2004). Interestingly, p85α-Ser608 phosphorylation is more robust in dimers with p110α as opposed to p110β (Beeton et al. 2000; Foukas et al. 2004); phosphorylation by p110δ was not tested.

Phosphorylation of the corresponding residue (Ser602) in p85β has not been studied. In contrast, both p110β and p110δ undergo inhibitory C-terminal autophosphorylation, at residues S1070 and S1039, respectively (Czupalla et al. 2003; Vanhaesebroeck et al. 1999), whereas p110α autophosphorylation is not robust (Boldyreff et al. 2008; Dhand et al. 1994b). Thus, distinct Class IA PI 3-kinase isoforms all undergo inhibitory autophosphorylation, although at different sites and on different subunits.

8 Oncogenic Mutation of p85α and p110

Studies over the past five years have identified mutations in p85α and p110α that clearly define PI 3-kinase as an oncogene in human cancer. The landmark study by Samuels et al. examined lipid kinases in human colon, brain, gastric, breast and lung tumors (Samuels et al. 2004). Mutations were commonly found in p110α, but not in p110β, p110δ or p110γ. The vast majority of sites involve mutation of H1047 to Arg in the catalytic domain (33%), and E542, E5445 or Q546 to Lys in the helical domain (47%). Less frequent sites of mutation are in the C2 domain and the ABD. Multiple subsequent studies have identified these mutants in p110α from an ever-widening number of malignancies (reviewed in Liu et al. 2009). In a summary of series examining more than 100 patient samples, p110α mutations are found in 25% of breast tumors, 22% of endometrial tumors%, 17% of urinary tract tumors, 14% of large intestine tumors, 10% of cervical and upper aerodigestive tract tumors, and at lower frequencies in other tumors types (http://www.sanger.ac.uk/genetics/CGP/cosmic).

For the common E545K and H1047R mutants, as well as for the less common G106V, C420R, E453Q, E542K, and M1043I mutants, an increase in catalytic activity has been demonstrated *in vitro* (Carson et al. 2008; Ikenoue et al. 2005). Introduction of p110α mutants into chick fibroblasts or human mammary endothelial cells causes increased levels of Akt phosphorylation, transformation (Ikenoue et al. 2005; Isakoff et al. 2005; Kang et al. 2005; Zhao et al. 2005), tumor formation in the chick chorioallantoic membrane assay and in mice (Bader et al. 2006; Samuels et al. 2005; Zhao et al. 2005), and growth-factor independent growth in hematopoietic cells (Horn et al. 2008). In chick fibroblasts, a gradient of transforming efficiency was noted, and it was possible to subdivide p110α mutants into strongly (H1047R, N345K, C420R, P539R, E542K, E545K, E545G, Q546K, Q546P, H1047L, H1047R), moderately (E545A, T1025S, M1043I, M1043V, H1047Y), and weakly (R38H, K111N) transforming (Gymnopoulos et al. 2007).

It is surprising that mutants in other p110 catalytic subunits have not been identified. Two studies on the transforming potential of p110β,δ, and γ reached divergent conclusions: Kang et al. found that, in contrast to p110α, overexpression of wild type p110β,δ or γ is sufficient to cause transformation of avian fibroblasts (Kang et al. 2006). p110γ already contains an Arg residue at the site homologous to H1047R in p110α, but its mutation to His does not appear to alter p110γ activity

(Kang et al. 2006). The authors proposed that given the transforming capacity of these isoforms in wild type state, their contribution to oncogenesis might be primarily via changes in expression as opposed to mutation. In contrast, Zhao et al. found that p110β is poorly transforming in mammary epithelial cells, even when the equivalent of the p110α E545K hotspot mutant (E552K) is introduced (Zhao et al. 2005). Given the differences in the systems used, the potential for transformation by modulation of p110β, δ and γ expression deserves further study.

Mutations in the p85α regulatory subunit have been described in murine and human cancers. The first mutants to be discovered were deletions or truncations in the C-terminal end of the iSH2 domain. Jimenez et al. described the fusion of p85α with a non-catalytic fragment of the Eph tyrosine kinase, which results in the termination of p85α at residue 571 (Jimenez et al. 1998). Expression of either the p85α-Eph fusion or the truncated p85α(1-571) leads to increased PI 3-kinase signaling and a modest increase in transformation in cultured cells. Philp et al. described the replacement of residues Met582–Asp605 of p85α with Ile (Philp et al. 2001), and Jucker et al. described a missense mutation resulting in the substitution of 25 unrelated amino acids after residue 635, in the cSH2 domain of p85α (Jucker et al. 2002); these mutations also lead to increased Akt activation and PIP3 production in cultured cells. Recent sequencing efforts in glioblastoma and colon cancer have identified additional point mutants and small deletions, most clustered in the iSH2 domain and the cSH2 domain (CGART 2008; Jaiswal et al. 2009; Parsons et al. 2008). A number of these mutants lead to increased PI 3-kinase activity, and Akt signaling, survival and transformation in BaF3 cells (Jaiswal et al. 2009).

8.1 Helical Domain Mutants: Disruption of the nSH2-Helical Domain Interface

The mechanism by which helical domain mutants lead to activation of p110α has been discussed above in the section on phosphopeptide activation of p85α/p110α dimers. Charge change mutations in a patch of acidic residues on the helical domain surface (E542K, E545K and Q546K) are proposed to disrupt an inhibitory contact with the nSH2 domain of p85α, mimicking the disruption of this interface that would occur with physiological SH2 domain occupancy (Miled et al. 2007). Transformation by helical domain is blocked by an RBD mutant (K277E) that disrupts Ras binding (Zhao and Vogt 2008), which is consistent with a previous study showing that phosphopeptide activation of p85α/p110α facilitates subsequent activation by Ras (Jimenez et al. 2002). Surprisingly, the E545K mutant causes increased transformation even in the context of an ABD truncation (residues 1–72) that abolishes p85α binding, suggesting that this mutation might have some effects independent of the disruption of inhibition by p85α (Zhao and Vogt 2008). However, it is important to note that the ABD truncation also causes a significant increase in transformation by wild type p110α, to a level that is only slightly less than that seen with the truncated helical domain mutant. This would suggest that in the context of the ABD truncation, the additional activating effect of the helical domain mutant is small relative to its effect on wild type p110α.

8.2 The Kinase Domain Mutant H1047R

The H1047R mutant of p110α is inhibited by p85α (Y. Yan and J.M. Backer, unpublished observations), and p85α/p110-H1047R dimers are activated by phosphopeptides (Carson et al. 2008), suggesting that this mutation is mechanistically distinct from helical domain mutants. Consistent with these data, the H1047R mutant is additive with the E545K mutant for Akt activation and transformation in avian fibroblasts (Zhao and Vogt 2008). Unlike the E545K mutant, the H1047R mutant is unaffected by mutation of the Ras-binding domain, and is not activated by Ras-GTP binding (Chaussade et al. 2009; Zhao and Vogt 2008). These data led to the proposal that the H1047R mutant mimics the effect of Ras-GTP binding to the RBD of p110α. Surprisingly, the transforming effects of the H1047R mutant are abolished by deletion of the ABD, to levels much lower than were seen in a similarly deleted wild type p110α (Zhao and Vogt 2008). The reason for this requirement for the p85-binding domain, which is not seen with wild type or E545K p110α, is not clear.

A crystal structure of the H1047R hotspot mutant of p110α has been recently solved (Mandelker et al. 2009). The R1047 residue is oriented at a 90° angle from the position of H1047 in the wild type enzyme, and leads to changes in the conformation of two loops (864–874 and 1050–1062). The net effect of the mutation is predicted to increase the ability of the kinase to interact with cellular membranes; this is also suggested by biochemical data showing a greater sensitivity to variations in membrane lipid composition than wild type p110α during *in vitro* kinase assays. If a major effect of Ras binding to p110α is to enhance its association with cellular membranes (see below), then this model for the enhanced activity of p110αH1047R would be consistent with its insensitivity to mutations in the Ras binding domain.

8.3 p110α Mutations Disrupting the C2-iSH2 Domain Interface

The crystal structure of p110α identified a previously unappreciated contact between the C2 domain of p110α and the iSH2 domain of p85α (Huang et al. 2007). In particular, hydrogen bonding was predicted between p110α residues N345 and p85α iSH2 domain residues D560 and N564. Mutation of p110α N345 to Lys had previously be shown to induce transformation in avian fibroblasts, suggesting that disruption of the contact would lead to activation of p110α (Gymnopoulos et al. 2007). More recently, mutations at residues D560 and N564, as well other p85α residues near the C2-iSH2 domain interface, were discovered in human glioblastoma and colon cancer samples (CGART 2008; Jaiswal et al. 2009;

Parsons et al. 2008). Analysis of the naturally occurring D560Y, N564D, D569Y/N564D, and ΔYQL^{579} mutants show that they bind but fail to inhibit p110α, β or δ, and lead to enhanced cell survival, Akt activation, anchorage independent cell growth, and oncogenesis (Jaiswal et al. 2009).

These data all point to the C2-iSH2 contact as an important inhibitory interface, whose mutation leads to increased constitutive PI 3-kinase activity and cellular transformation. However, unlike the nSH2-helical domain interface that is modulated during phosphoprotein regulation of PI 3-kinase, it is not yet known whether the C2-iSH2 domain interface is regulated under normal conditions. Furthermore, it is not yet known whether the loss of inhibition due to mutations at this site are due to direct conformational effects on p110α, as opposed to indirect effects due to a loosening of the iSH2-p110α coupling that might secondarily affect nSH2-helical domain interactions.

8.4 *p85α Mutations Disrupting the C2-iSH2 Domain Interface*

Interestingly, recent data suggest that the iSH2 truncation and deletion mutants described by Jimenez et al. (1998) and Philp et al. (2001), discussed above, may also exert their effects by disrupting the iSH2-C2 domain interface. A number of hypotheses as to the mechanism of PI 3-kinase activation by these mutations have been proposed. Foukas et al. (2004) noted that these mutants cause the loss of Ser608, whose transphosphorylation by p110α had been shown to cause an inhibition of PI 3-kinase lipid kinase activity (Dhand et al. 1994b). The authors proposed that the loss of this inhibitory phosphorylation site would lead to constitutive elevated activity *in vivo*. However, Shekar et al. compared the inhibition of p110α *in vitro* by an nSH2-iSH2 construct (residues 321–600) as opposed to a truncated version of this construct (residues 321–572) (Shekar et al. 2005). Whereas the nSH2-iSH2 construct inhibits p110α, even though it did not contain Ser608, truncation of the iSH2 domain at residue 572 causes a significant loss of p110α inhibition. Thus, while the loss of the inhibitory Ser608 site might modulate the level of PI 3-kinase activation *in vivo*, truncation of the iSH2 domain at residue 572 causes a loss of inhibition independent of the removal of Ser608.

A second hypothesis was based on the finding that introduction of spin labels into the C-terminal end of the iSH2 domain, the region that is deleted in the p85α (1-572) truncation, has pronounced relaxation effects on a single face of the nSH2 domain (Shekar et al. 2005). The authors interpreted the data as showing a contact between the nSH2 domain and the C-terminal iSH2 domain, and proposed that these contacts were required to orient the nSH2 domain so as to make inhibitory contacts with p110α. However, subsequent studies on the nSH2-iSH2 fragment of p85α show that the nSH2 domain is in fact highly mobile with respect to the iSH2 domain. ^{15}N NMR relaxation methods were used to measure the rotational dynamics of the nSH2 domain within the nSH2-iSH2 construct, and defined an apparent rotational correlation time of 12.7 ± 0.7 ns for the nSH2 domain (Wu et al. 2009).

These data are similar to experimental measurements for isolated SH2 domains (6.5–9.2 ns) and quite different from hydrodynamic calculations for several proposed nSH2-iSH2 geometries (45–52 ns) (Bernado et al. 2002; Farrow et al. 1994; Fushman et al. 1999; Zhang et al. 1998). Pulsed EPR measurements and molecular modeling of spin-labeled nSH2-iSH2 showed that the motion of the nSH2 domain is restrained to a torus-like distribution around the proximal end of the iSH2 domain (Sen et al. 2010). Thus, the nSH2 domain moves in a hinge-like manner, and not like a ball on a string; this restricted motion presumably explains why relaxation effects from iSH2 spin labels were only seen on one face of the nSH2 domain in the earlier study (Shekar et al. 2005). Similar results were obtained for the motion of the nSH2 domain in the context of full-length p85α (Sen IK, Wu H, Backer JM and Gerfen GJ, unpublished observations). Taken together, these data do not support a model in which truncation of the iSH2 domain relieves a conformational constraint on the nSH2 domain.

A third hypothesis was proposed by Huang et al., based on the discovery of a close contact between the C2 domain of p110α and the iSH2 domain of p85α (discussed above) (Huang et al. 2007). The authors proposed that the loss of residues 572–600 in the truncation mutant described by Jimenez et al. (1998) might destabilize the iSH2 domain coiled-coil at the C2 domain contact sites (near residues D560 and N564). This hypothesis was directly tested by Wu et al. (2009), who showed that wild type p110α is inhibited efficiently by the wild type nSH2-iSH2 construct but not by a construct truncated at residue Q572. In contrast, the p110α-N345K mutant shows the same degree of partial inhibition by either full-length nSH2-iSH2 or a construct truncated at residue Q572. These data suggest that the Q572 truncation has no phenotype when paired with a mutant p110α that is incapable of forming the C2-iSH2 domain contact. *In vivo* data further supported this hypothesis. Either truncation of the nSH2-iSH2 fragment at Q572, or mutation of D560 and N565 to Lys, leads to constitutive activation of Akt and S6K1 and increased transformation, but no additive effects are seen in a truncated D560K/N565K construct. Similarly, the p1110α-N345K mutation leads to increased rates of transformation as compared to wild type p110α (Gymnopoulos et al. 2007), but cells co-transfected with p110α-N345K plus wild-type, truncated, D560K/D565K, or truncated D560K/D565K mutants all show similar rates of transformation. These data suggest that truncation of p85α at residue Q572, or mutation of p85α residues D560K/N565K, both lead to constitutive PI 3-kinase activation by disrupting the inhibitory C2-iSH2 domain interface identified by Huang et al. (2007).

9 Regulation of p85/p110 *In vivo*: Activation Versus Translocation

PI 3-kinase must interact with cellular membranes in order to reach their phospholipid substrates. Constitutive membrane association of p110α catalytic subunits linked to N-terminal myristylation motifs or C-terminal CAAX motifs, leads to large increases in intracellular PIP3, Akt phosphorylation, and cellular transformation (Klippel et al. 1996). Since Class I PI 3-kinase has substantial activity under basal conditions (Beeton et al. 2000), membrane recruitment of PI 3-kinase leads to increased PIP3 production even in the absence of changes in PI 3-kinase specific activity. However, as described above, PI 3-kinases are also directly activated by allosteric interactions with tyrosine phosphorylated proteins and GTP-bound small Rac (Backer et al. 1992; Bokoch et al. 1996; Carpenter et al. 1993; Tolias et al. 1995; Zheng et al. 1994). These activations can be measured *in vitro*, independently of membrane recruitment, and reflect an increase in enzyme specific activity.

The allosteric activation of PI 3-kinases *in vitro* is not large (in the range of two- to fivefold). Given the magnitude of these changes in activity, there has been some question as to whether membrane translocation is sufficient to drive most PI 3-kinase dependent processes, and whether increases in catalytic activity play an important role in PI 3-kinase signaling. In this regard, the oncogenic potential of the helical domain p110α is informative. The helical domain mutants disrupt the inhibitory interface that is regulated by SH2 domain occupancy, and lead to an approximately twofold increase in activity *in vitro* (Carson et al. 2008). These mutants have normal SH2 domains and still interact with Ras (Zhao and Vogt 2008), and would therefore be expected to undergo normal ligand-stimulated translocation to cell membranes. Nonetheless, these mutations have potent transforming activity. These data strongly suggest that the relatively small increases in Class IA PI 3-kinase specific activity caused by SH2 domain occupancy are critical for function *in vivo*.

The effect of Ras binding on Class IA PI 3-kinase activity is somewhat more complex. As discussed above, Ras binding to p110γ causes changes in the catalytic cleft, consistent with direct allosteric regulation of activity (Pacold et al. 2000). However, the original studies on Ras activation only saw increases in PI 3-kinase activity with prenylated Ras, despite the fact that non-prenylated GTP-loaded Ras can still bind to p110α (Rodriguez-Viciana et al. 1996). Thus, Ras is likely to activate by Class I PI 3-kinase by a combination of translocation and activation. This point was directly supported by studies showing that the transforming capacity of overexpressed p110β and δ is abolished by mutation of the Ras-binding site, but rescued by N-terminal myristylation (Denley et al. 2008). Similarly, the H1047R mutant of p110α, whose transforming potential is insensitive to mutation of the Ras binding sites, appears to act by enhancing p110α interactions with cellular membranes (Mandelker et al. 2009; Zhao and Vogt 2008).

10 Unanswered Mechanistic Questions on p85–p110 Interactions

Despite the concerted efforts of numerous laboratories, many very basic questions about p85–p110 interactions remain unanswered. With regard to structural questions, the structure of the nSH2-iSH2 fragment of p85 dimerized with p110α has been solved (Mandelker et al. 2009), but we know virtually nothing about how the rest of p85 (the cSH2 domain, as well as the SH3, proline-rich and BCR homology domains) is organized in space. Hopefully, continued advances using biochemical and biophysical methods will lead to new information on these issues in the near future, and will undoubtedly increase our understanding of how interactions between these domains and upstream activators regulate the catalytic activity of p110.

An even more complex question is that of the specificity of p85–p110 pairing and regulation. Mass spectrometry studies have suggested that p85 and p110 expression levels are matched in multiple cell types and tissues (Geering et al. 2007), although this point has been controversial (Brachmann et al. 2005). However, we know very little about whether there is selectivity for the interaction of distinct p110 isoforms with distinct p85/p55 isoforms, since any one of the p110 ABD domains should be able to bind to all the p85/p50/p55 iSH2 domains. Given that knockout, knock-in and transgenic studies in mice have clearly demonstrated distinct physiological roles for the isoforms of p85 and p110 (reviewed in Fruman 2007; Liu et al. 2009; Vanhaesebroeck et al. 2005), one would think that the pairing of different Class IA regulatory and catalytic subunits would be a regulated process with significant physiological consequences. If so, then we have not even begun to figure out the rules for this sorting process, nor are the mechanisms for selectivity apparent. This question is likely to converge on the broader question of the isoform specificity of PI 3-kinase signaling *in vivo*, which has been a major focus of recent work in the field (Marone et al. 2008; Rommel et al. 2007; Stuart and Sellers 2009).

A significant caveat with regard to our current understanding of p85–p110 interactions is that nearly all the *in vitro* experiments have been done with p85α and p110α. For example, it will be very important to analyze mutants such as the acidic helical domain mutants of p110α (e.g., E545K) in the context of p110β and p10δ, to see if they have similar effects on the regulation of these isoforms. This is also true for recently described point mutants of p85α. New biochemical studies on p110β should be particularly interesting, since this isoform appears to have all the p85-related tools in place for regulation by receptor tyrosine kinases, yet somehow is instead activated downstream of GPCRs. Finally, very little biochemical analysis has been done on the shorter splice variants and isoforms of p85, p55α, p50α and p55γ. These variants contain the same nSH2-iSH2-cSH2 organization found in the so-called p85nic construct that has been extensively studied *in vitro* (Wu et al. 2007), but p55γ and p55α contain homologous 30 amino acid N-terminal extensions that contains a YXXM motif (Dey et al. 1998; Inukai et al. 1996) and has been suggested to associate with tubulin and with the retinoblastoma gene product Rb (Inukai et al. 2000; Xia et al. 2003). The potential role of these additional sequences in the regulation of Class IA dimers is not known. A better understanding of the enzymatic regulation of p110 catalytic subunits by these isoforms will be important in deciphering their differential regulation and physiological functions *in vivo*.

The biochemical analysis of these large and complex enzymes remains a challenging enterprise. Given the important role of the PI 3-kinases in human physiology, it is to be hoped that an enhanced understanding of how the enzymes work will lead to new ideas about the pharmacological modulation of their activity.

Acknowledgments This work was supported by NIH grants GM55692, DK07069, and 1 PO1 CA100324, and grants from American Diabetes Association and the Janey Fund. I thank Dr. Mirvat El-Sibai, Beirut University, for critical reading of the manuscript. I also thank Drs. Mark Girvin, Gary Gerfen, Steve Almo, and George Orr (deceased) at Albert Einstein College of Medicine, and Dr. Roger Williams at the MRC, Cambridge, for collaboration and hours of discussion over the past decade. Finally, I thank all my graduate students and postdocs for their scientific contributions to many of the studies cited here.

References

Ahmed S, Lee J, Wen L-P, Zhao Z, Ho J, Best A, Kozma R, Lim L (1994) Breakpoint Cluster Region gene product-related domain of n-Chimaerin. J Biol Chem 269:17642–17648

Backer JM, Myers MG Jr, Shoelson SE, Chin DJ, Sun XJ, Miralpeix M, Hu P, Margolis B, Skolnik EY, Schlessinger J et al (1992) Phosphatidylinositol 3′-kinase is activated by association with IRS-1 during insulin stimulation. EMBO J 11(9):3469–3479

Bader AG, Kang S, Vogt PK (2006) Cancer-specific mutations in PIK3CA are oncogenic *in vivo*. Proc Natl Acad Sci USA 103(5):1475–1479

Beeton CA, Das P, Waterfield MD, Shepherd PR (1999) The SH3 and BH domains of the p85alpha adapter subunit play a critical role in regulating Class Ia phosphoinositide 3-kinase function. Mol Cell Biol Res Comm 1:153–157

Beeton CA, Chance EM, Foukas LC, Shepherd PR (2000) Comparison of the kinetic properties of the lipid- and protein-kinase activities of the p110alpha and p110beta catalytic subunits of class-Ia phosphoinositide 3-kinases. Biochem J 350(Pt 2):353–359

Bernado P, Garcia de la Torre J, Pons M (2002) Interpretation of 15N NMR relaxation data of globular proteins using hydrodynamic calculations with HYDRONMR. J Biomol NMR 23(2):139–150

Bokoch GM, Vlahos CJ, Wang Y, Knaus UG, Traynor-Kaplan AE (1996) Rac GTPase interacts specifically with phosphatidylinositol 3- kinase. Biochem J 315:775–779

Boldyreff B, Rasmussen TL, Jensen HH, Cloutier A, Beaudet L, Roby P, Issinger OG (2008) Expression and purification of PI3 kinase alpha and development of an ATP depletion and an alphascreen PI3 kinase activity assay. J Biomol Screen 13(10):1035–1040

Booker GW, Breeze AL, Downing AK, Panayotou G, Gout I, Waterfield MD, Campbell LD (1992) Structure of an SH2 domain of the p85α subunit of phosphatidylinositol-3 OH-kinase. Nature 358:684–687

Brachmann SM, Ueki K, Engelman JA, Kahn RC, Cantley LC (2005) Phosphoinositide 3-kinase catalytic subunit deletion and regulatory subunit deletion have opposite effects on insulin sensitivity in mice. Mol Cell Biol 25(5):1596–1607

Cantley LC, Auger KR, Carpenter C, Duckworth B, Graziani A, Kapeller R, Soltoff S (1991) Oncogenes and signal transduction. Cell 64:231–302

Carpenter CL, Auger KR, Chanudhuri M, Yoakim M, Schaffhausen B, Shoelson S, Cantley LC (1993) Phosphoinositide 3-kinase is activated by phosphopeptides that bind to the SH2 domains of the 85-kDa subunit. J Biol Chem 268(13):9478–9483

Carson JD, Van Aller G, Lehr R, Sinnamon RH, Kirkpatrick RB, Auger KR, Dhanak D, Copeland RA, Gontarek RR, Tummino PJ, Luo L (2008) Effects of oncogenic p110alpha subunit mutations on the lipid kinase activity of phosphoinositide 3-kinase. Biochem J 409(2):519–524

CGART (2008) Comprehensive genomic characterization defines human glioblastoma genes and core pathways. Nature 455:1061–1068

Chamberlain MD, Berry TR, Pastor MC, Anderson DH (2004) The p85alpha subunit of phosphatidylinositol 3′-kinase binds to and stimulates the GTPase activity of Rab proteins. J Biol Chem 279(47):48607–48614
Chan TO, Rodeck U, Chan AM, Kimmelman AC, Rittenhouse SE, Panayotou G, Tsichlis PN (2002) Small GTPases and tyrosine kinases coregulate a molecular switch in the phosphoinositide 3-kinase regulatory subunit. Cancer Cell 1(2):181–191
Chaussade C, Cho K, Mawson C, Rewcastle GW, Shepherd PR (2009) Functional differences between two classes of oncogenic mutation in the PIK3CA gene. Biochem Biophys Res Commun 381(4):577–581
Christoforidis S, Miaczynska M, Ashman K, Wilm M, Zhao L, Yip S-C, Waterfield MD, Backer JM, Zerial M (1999) Phosphatidylinositol-3-OH kinases are Rab5 effectors. Nature Cell Biology 1:249–252
Ciraolo E, Iezzi M, Marone R, Marengo S, Curcio C, Costa C, Azzolino O, Gonella C, Rubinetto C, Wu H, Dastru W, Martin EL, Silengo L, Altruda F, Turco E, Lanzetti L, Musiani P, Ruckle T, Rommel C, Backer JM, Forni G, Wymann MP, Hirsch E (2008) Phosphoinositide 3-kinase p110beta activity: key role in metabolism and mammary gland cancer but not development. Sci Signal 1(36):ra3
Cuevas BD, Lu Y, Mao M, Zhang J, LaPushin R, Siminovitch K, Mills GB (2001) Tyrosine phosphorylation of p85 relieves its inhibitory activity on phosphatidylinositol 3-kinase. J Biol Chem 276(29):27455–27461
Czupalla C, Culo M, Muller EC, Brock C, Reusch HP, Spicher K, Krause E, Nurnberg B (2003) Identification and characterization of the autophosphorylation sites of phosphoinositide 3-kinase isoforms beta and gamma. J Biol Chem 278(13):11536–11545
Denley A, Kang S, Karst U, Vogt PK (2008) Oncogenic signaling of class I PI3K isoforms. Oncogene 27(18):2561–2574
Deora AA, Win T, Vanhaesebroeck B, Lander HM (1998) A redox-triggered ras-effector interaction. Recruitment of phosphatidylinositol 3′-kinase to Ras by redox stress. J Biol Chem 273 (45):29923–29928
Dey BR, Furlanetto RW, Nissley SP (1998) Cloning of human p55 gamma, a regulatory subunit of phosphatidylinositol 3-kinase, by a yeast two-hybrid library screen with the insulin-like growth factor-I receptor. Gene 209(1–2):175–183
Dhand R, Hara K, Hiles I, Bax B, Gout I, Panayotou G, Fry MJ, Yonezawa K, Kasuga M, Waterfield MD (1994a) PI 3-kinase: Structural and functional analysis of intersubunit interactions. EMBO J 13:511–521
Dhand R, Hiles I, Panayotou G, Roche S, Fry MJ, Gout I, Totty NF, Truong O, Vicendo P, Yonezawa K, Kasuga M, Courtneidge SA, Waterfield MD (1994b) PI 3-kinase is a dual specificity enzyme: Autoregulation by an intrinsic protein-serine kinase activity. EMBO J 13:522–533
Diekmann D, Brill S, Garrett MD, Totty N, Hsuan J, Monfries C, Hall C, Lim L, Hall A (1991) Bcr encodes a GTPase-activating protein for p21-rac. Nature 351:400–402
Engelman JA, Luo J, Cantley LC (2006) The evolution of phosphatidylinositol 3-kinases as regulators of growth and metabolism. Nat Rev Genet 7(8):606–619
Escobedo JA, Navankasattusas S, Kavanaugh WM, Milfay D, Fried VA, Williams LT (1991) cDNA cloning of a novel 85 kd protein that has SH2 domains and regulates binding of PI3-kinase to the PDGF beta-receptor. Cell 65(1):75–82
Fantl WJ, Escobedo JA, Martin GA, Turck CW, del Rosario M, McCormick F, Williams LT (1992) Distinct Phosphotyrosines on a Growth Factor Receptor Bind to Specific Molecules that Mediate Different Signalling Pathways. Cell 69:413–423
Farrow NA, Muhandiram R, Singer AU, Pascal SM, Kay CM, Gish G, Shoelson SE, Pawson T, Forman-Kay JD, Kay LE (1994) Backbone dynamics of a free and phosphopeptide-complexed Src homology 2 domain studied by 15N NMR relaxation. Biochemistry (Mosc) 33(19):5984–6003
Foukas LC, Beeton CA, Jensen J, Phillips WA, Shepherd PR (2004) Regulation of phosphoinositide 3-kinase by its intrinsic serine kinase activity *in vivo*. Mol Cell Biol 24(3):966–975

Foukas LC, Claret M, Pearce W, Okkenhaug K, Meek S, Peskett E, Sancho S, Smith AJ, Withers DJ, Vanhaesebroeck B (2006) Critical role for the p110alpha phosphoinositide-3-OH kinase in growth and metabolic regulation. Nature 441(7091):366–370

Fruman DA (2007) The role of class I phosphoinositide 3-kinase in T-cell function and autoimmunity. Biochem Soc Trans 35(Pt 2):177–180

Fruman DA, Mauvais-Jarvis F, Pollard DA, Yballe CM, Brazil D, Bronson RT, Kahn CR, Cantley LC (2000) Hypoglycaemia, liver necrosis and perinatal death in mice lacking all isoforms of phosphoinositide 3-kinase p85 alpha. Nat Genet 26(3):379–382

Fu Z, Aronoff-Spencer E, Backer JM, Gerfen GJ (2003) The structure of the inter-SH2 domain of class IA phosphoinositide 3-kinase determined by site-directed spin labeling EPR and homology modeling. Proc Natl Acad Sci USA 100(6):3275–3280

Fu Z, Aronoff-Spencer E, Wu H, Gerfen GJ, Backer JM (2004) The iSH2 domain of PI 3-kinase is a rigid tether for p110 and not a conformational switch. Arch Biochem Biophys 432(2):244–251

Fushman D, Xu R, Cowburn D (1999) Direct determination of changes of interdomain orientation on ligation: use of the orientational dependence of 15N NMR relaxation in Abl SH(32). Biochemistry (Mosc) 38(32):10225–10230

Geering B, Cutillas PR, Nock G, Gharbi SI, Vanhaesebroeck B (2007) Class IA phosphoinositide 3-kinases are obligate p85-p110 heterodimers. Proc Natl Acad Sci USA 104(19):7809–7814

Gout I, Dhand R, Panayotou G, Fry MJ, Hiles I, Otsu M, Waterfield MD (1992) Expression and characterization of the p85 subunit of the phosphatidylinositol 3-kinase complex and a related p85 beta protein by using the baculovirus expression system. Biochem J 288(Pt 2):395–405

Guillermet-Guibert J, Bjorklof K, Salpekar A, Gonella C, Ramadani F, Bilancio A, Meek S, Smith AJ, Okkenhaug K, Vanhaesebroeck B (2008) The p110beta isoform of phosphoinositide 3-kinase signals downstream of G protein-coupled receptors and is functionally redundant with p110gamma. Proc Natl Acad Sci USA 105(24):8292–8297

Guinebault C, Payrastre B, Racaud-Sultan C, Mazarguil H, Breton M, Mauco G, Plantavid M, Chap H (1995) Integrin-dependent translocation of phosphoinositide 3-kinase to the cytoskeleton of thrombin-activated platelets involves specific interactions of p85 alpha with actin filaments and focal adhesion kinase. J Cell Biol 129(3):831–842

Günther UL, Liu YX, Sanford D, Bachovchin WW, Schaffhausen B (1996) NMR analysis of interactions of a phosphatidylinositol 3′-kinase SH2 domain with phosphotyrosine peptides reveals interdependence of major binding sites. Biochemistry 35:15570–15581

Gupta S, Ramjaun AR, Haiko P, Wang Y, Warne PH, Nicke B, Nye E, Stamp G, Alitalo K, Downward J (2007) Binding of ras to phosphoinositide 3-kinase p110alpha is required for ras-driven tumorigenesis in mice. Cell 129(5):957–968

Gymnopoulos M, Elsliger MA, Vogt PK (2007) Rare cancer-specific mutations in PIK3CA show gain of function. Proc Natl Acad Sci USA 104(13):5569–5574

Hale BG, Batty IH, Downes CP, Randall RE (2008) Binding of influenza A virus NS1 protein to the inter-SH2 domain of p85 suggests a novel mechanism for phosphoinositide 3-kinase activation. J Biol Chem 283(3):1372–1380

Herbst JJ, Andrews G, Contillo L, Lamphere L, Gardner J, Lienhard GE, Gibbs EM (1994) Potent activation of phosphatidylinositol 3′-kinase by simple phosphotyrosine peptides derived from insulin receptor substrate 1 containing two YMXM motifs for binding SH2 domains. Biochemistry 33:9376–9381

Hiles ID, Otsu M, Volinia S, Fry MJ, Gout I, Dhand R, Panayotou G, Ruiz-Larrea F, Thompson A, Totty NF, Hsuan JJ, Courtneidge SA, Parker PJ, Waterfield MD (1992) Phosphatidylinositol 3-Kinase: Structure and Expression of the 110 kd Catalytic Subunit. Cell 70:419–429

Hoedemaeker FJ, Siegal G, Roe SM, Driscoll PC, Abrahams JP (1999) Crystal structure of the C-terminal SH2 domain of the p85alpha regulatory subunit of phosphoinositide 3-kinase: an SH2 domain mimicking its own substrate. J Mol Biol 292(4):763–770

Holt KH, Olson AL, Moye-Rowley WS, Pessin JE (1994) Phosphatidylinositol 3-kinase activation is mediated by high-affinity interactions between distinct domains within the p110 and p85 subunits. Mol Cell Biol 14:42–49

Horn S, Bergholz U, Jucker M, McCubrey JA, Trumper L, Stocking C, Basecke J (2008) Mutations in the catalytic subunit of class IA PI3K confer leukemogenic potential to hematopoietic cells. Oncogene 27(29):4096–4106
Hu P, Schlessinger J (1994) Direct association of p110 beta phosphatidylinositol 3-kinase with p85 is mediated by an N-terminal fragment of p110 beta. Mol Cell Biol 14(4):2577–2583
Hu P, Mondino A, Skolnik EY, Schlessinger J (1993) Cloning of a novel, ubiquitously expressed human phosphatidylinositol 3-kinase and identification of its binding site on p85. Mol Cell Biol 13:7677–7688
Hu Q, Klippel A, Muslin AJ, Fantl WJ, Williams LT (1995) Ras-dependent induction of cellular responses by constitutively active phosphatidylinositol-3 kinase. Science 268(5207): 100–102
Huang CH, Mandelker D, Schmidt-Kittler O, Samuels Y, Velculescu VE, Kinzler KW, Vogelstein B, Gabelli SB, Amzel LM (2007) The structure of a human p110alpha/p85alpha complex elucidates the effects of oncogenic PI3Kalpha mutations. Science 318(5857): 1744–1748
Ikenoue T, Kanai F, Hikiba Y, Obata T, Tanaka Y, Imamura J, Ohta M, Jazag A, Guleng B, Tateishi K, Asaoka Y, Matsumura M, Kawabe T, Omata M (2005) Functional analysis of PIK3CA gene mutations in human colorectal cancer. Cancer Res 65(11):4562–4567
Imai Y, Clemmons DR (1999) Roles of phosphatidylinositol 3-kinase and mitogen-activated protein kinase pathways in stimulation of vascular smooth muscle cell migration and deoxyriboncleic acid synthesis by insulin-like growth factor-I. Endocrinology 140(9):4228–4235
Inukai K, Anai M, Van Breda E, Hosaka T, Katagiri H, Funaki M, Fukushima Y, Ogihara T, Yazaki Y, Kikuchi YO, Asano T (1996) A novel 55-kDa regulatory subunit for phosphatidylinositol 3-kinase structurally similar to p55PIK Is generated by alternative splicing of the p85alpha gene. J Biol Chem 271(10):5317–5320
Inukai K, Funaki M, Nawano M, Katagiri H, Ogihara T, Anai M, Onishi Y, Sakoda H, Ono H, Fukushima Y, Kikuchi M, Oka Y, Asano T (2000) The N-terminal 34 residues of the 55 kDa regulatory subunits of phosphoinositide 3-kinase interact with tubulin. Biochem J 346 (Pt 2):483–489
Isakoff SJ, Engelman JA, Irie HY, Luo J, Brachmann SM, Pearline RV, Cantley LC, Brugge JS (2005) Breast cancer-associated PIK3CA mutations are oncogenic in mammary epithelial cells. Cancer Res 65(23):10992–11000
Jaiswal BS, Janakiraman V, Kljavin NM, Chaudhari S, Stern HM, Wang W, Kan Z, Dbouk HA, Peters BA, Waring P, Dela Vega T, Kenski DM, Bowman K, Lorenzo M, Wu J, Modrusan Z, Stinson J, Eby M, Yue P, Kaminker J, de Sauvage FJ, Backer JM, Seshagiri S (2009) Somatic mutations in p85α promote tumorigenesis through class IA PI3K activation. Cancer Cell 16(6):463–474
Jia S, Liu Z, Zhang S, Liu P, Zhang L, Lee SH, Zhang J, Signoretti S, Loda M, Roberts TM, Zhao JJ (2008) Essential roles of PI(3)K-p110beta in cell growth, metabolism and tumorigenesis. Nature 454(7205):776–779
Jimenez C, Jones DR, Rodríguez-Viciana P, Gonzalez-García A, Leonardo E, Wennström S, Von Kobbe C, Toran JL, Borlado LR, Calvo V, Copin SG, Albar JP, Gaspar ML, Diez E, Marcos MAR, Downward J, Martinez C, Mérida I, Carrera AC (1998) Identification and characterization of a new oncogene derived from the regulatory subunit of phosphoinositide 3-kinase. EMBO J 17:743–753
Jimenez C, Hernandez C, Pimentel B, Carrera AC (2002) The p85 regulatory subunit controls sequential activation of phosphoinositide 3-kinase by Tyr kinases and Ras. J Biol Chem 277 (44):41556–41562
Jucker M, Sudel K, Horn S, Sickel M, Wegner W, Fiedler W, Feldman RA (2002) Expression of a mutated form of the p85alpha regulatory subunit of phosphatidylinositol 3-kinase in a Hodgkin's lymphoma-derived cell line (CO). Leukemia 16(5):894–901
Kang S, Bader AG, Vogt PK (2005) Phosphatidylinositol 3-kinase mutations identified in human cancer are oncogenic. Proc Natl Acad Sci USA 102(3):802–807

Kang S, Denley A, Vanhaesebroeck B, Vogt PK (2006) Oncogenic transformation induced by the p110beta, -gamma, and -delta isoforms of class I phosphoinositide 3-kinase. Proc Natl Acad Sci USA 103(5):1289–1294

Kapeller R, Prasad KVS, Janssen O, Hou W, Schaffhausen BS, Rudd CE, Cantley LC (1994) Identification of two SH3-binding motifs in the regulatory subunit of phosphatidylinositol 3-kinase. J Biol Chem 269:1927–1933

Klippel A, Escobedo JA, Hu Q, Williams LT (1993) A region of the 85-kilodalton (kDa) subunit of phosphatidylinositol 3-kinase binds the 110-kDa catalytic subunit *in vivo*. Mol Cell Biol 13:5560–5566

Klippel A, Escobedo JA, Hirano M, Williams LT (1994) The interaction of small domains between the subunits of phosphatidylinositol 3-kinase determines enzyme activity. Mol Cell Biol 14(4):2675–2685

Klippel A, Reinhard C, Kavanaugh WM, Apell G, Escobedo MA, Williams LT (1996) Membrane localization of phosphatidylinositol 3-kinase is sufficient to activate multiple signal-transducing kinase pathways. Mol Cell Biol 16:4117–4127

Knight ZA, Gonzalez B, Feldman ME, Zunder ER, Goldenberg DD, Williams O, Loewith R, Stokoe D, Balla A, Toth B, Balla T, Weiss WA, Williams RL, Shokat KM (2006) A pharmacological map of the PI3-K family defines a role for p110alpha in insulin signaling. Cell 125(4):733–747

Kodaki T, Woscholski R, Hallberg B, Rodriguez-Viciana P, Downward J, Parker PJ (1994) The activation of phosphatidylinositol 3-kinase by Ras. Curr Biol 4:798–806

Koyama S, Yu H, Dalgarno DC, Shin TB, Zydowsky LD, Schreiber SL (1993) Structure of the PI3K SH3 domain and analysis of the SH3 family. Cell 72:945–952

Kubo H, Hazeki K, Takasuga S, Hazeki O (2005) Specific role for p85/p110beta in GTP-binding-protein-mediated activation of Akt. Biochem J 392(Pt 3):607–614

Kurosu H, Maehama T, Okada T, Yamamoto T, Hoshino S, Fukui Y, Ui M, Hazeki O, Katada T (1997) Heterodimeric phosphoinositide 3-kinase consisting of p85 and p110β is synergistically activated by the βgamma subunits of G proteins and phosphotyrosyl peptide. J Biol Chem 272:24252–24256

Layton MJ, Harpur AG, Panayotou G, Bastiaens PIH, Waterfield MD (1998) Binding of a diphosphotyrosine-containing peptide that mimics activated platelet-derived growth factor receptor β induces oligomerization of phosphatidylinositol 3-kinase. J Biol Chem 273(50):33379–33385

Liu P, Cheng H, Roberts TM, Zhao JJ (2009) Targeting the phosphoinositide 3-kinase pathway in cancer. Nat Rev Drug Discov 8(8):627–644

Maier U, Babich A, Nürnberg B (1999) Roles of non-catalytic subunits in Gβgamma-induced activation of class I phosphoinositide 3-kinase isoforms β and gamma. J Biol Chem 274 (41):29311–29317

Mandelker D, Gabelli SB, Schmidt-Kittler O, Zhu J, Cheong I, Huang CH, Kinzler KW, Vogelstein B, Amzel LM (2009) A frequent kinase domain mutation that changes the interaction between PI3K{alpha} and the membrane. Proc Natl Acad Sci USA 106: 16996–17001

Marone R, Cmiljanovic V, Giese B, Wymann MP (2008) Targeting phosphoinositide 3-kinase: moving towards therapy. Biochim Biophys Acta 1784(1):159–185

Matsuda M, Mayer BJ, Hanafusa H (1991) Identification of domains of the v-crk oncogene product sufficient for association with phosphotyrosine-containing proteins. Mol Cell Biol 11(3):1607–1613

Mayer BJ, Jackson PK, Baltimore D (1991) The noncatalytic src homology region 2 segment of abl tyrosine kinase binds to tyrosine-phosphorylated cellular proteins with high affinity. Proc Natl Acad Sci USA 88(2):627–631

Miled N, Yan Y, Hon WC, Perisic O, Zvelebil M, Inbar Y, Schneidman-Duhovny D, Wolfson HJ, Backer JM, Williams RL (2007) Mechanism of two classes of cancer mutations in the phosphoinositide 3-kinase catalytic subunit. Science 317(5835):239–242

Moran MF, Koch CA, Anderson D, Ellis C, England L, Martin GS, Pawson T (1990) Src homology region 2 domains direct protein-protein interactions in signal transduction. Proc Natl Acad Sci USA 87(21):8622–8626

Murga C, Fukuhara S, Gutkind JS (2000) A novel role for phosphatidylinositol 3-kinase β in signaling from G protein-coupled receptors to Akt. J Biol Chem 275(16):12069–12073

Musacchio A, Cantley LC, Harrison SC (1996) Crystal structure of the breakpoint cluster region homology domain from phosphoinositide 3-kinase p85α subunit. Proc Natl Acad Sci USA 93:14373–14378

Nolte RT, Eck MJ, Schlessinger J, Shoelson SE, Harrison SC (1996) Crystal structure of the PI 3-kinase p85 amino-terminal SH2 domain and its phosphopeptide complexes. Nat Struct Biol 3(4):364–374

O'Brien R, Rugman P, Renzoni D, Layton M, Handa R, Hilyard K, Waterfield MD, Driscoll PC, Ladbury JE (2000) Alternative modes of binding of proteins with tandem SH2 domains. Protein Sci 9(3):570–579

Otsu M, Hiles I, Gout I, Fry MJ, Ruiz-Larrea F, Panayotou G, Thompson A, Dhand R, Hsuan J, Totty N, Smith AD, Morgan SJ, Courtneidge SA, Parker PJ, Waterfield MD (1991) Characterization of two 85 kD proteins that associate with receptor tyrosine kinases, middle-T/pp 60c-src complexes and PI3-kinase. Cell 65:91–104

Pacold ME, Suire S, Perisic O, Lara-Gonzalez S, Davis CT, Walker EH, Hawkins PT, Stephens L, Eccleston JF, Williams RL (2000) Crystal structure and functional analysis of Ras binding to its effector phosphoinositide 3-kinase gamma. Cell 103(6):931–943

Parsons DW, Jones S, Zhang X, Lin JC, Leary RJ, Angenendt P, Mankoo P, Carter H, Siu IM, Gallia GL, Olivi A, McLendon R, Rasheed BA, Keir S, Nikolskaya T, Nikolsky Y, Busam DA, Tekleab H, Diaz LA Jr, Hartigan J, Smith DR, Strausberg RL, Marie SK, Shinjo SM, Yan H, Riggins GJ, Bigner DD, Karchin R, Papadopoulos N, Parmigiani G, Vogelstein B, Velculescu VE, Kinzler KW (2008) An integrated genomic analysis of human glioblastoma multiforme. Science 321(5897):1807–1812

Philp AJ, Campbell IG, Leet C, Vincan E, Rockman SP, Whitehead RH, Thomas RJ, Phillips WA (2001) The phosphatidylinositol 3′-kinase p85alpha gene is an oncogene in human ovarian and colon tumors. Cancer Res 61(20):7426–7429

Pleiman CM, Hertz WM, Cambier JC (1994) Activation of phosphatidylinositol-3′ kinase by Src-family kinase SH3 binding to the p85 subunit. Science 263:1609–1612

Ramjaun AR, Downward J (2007) Ras and phosphoinositide 3-kinase: partners in development and tumorigenesis. Cell Cycle 6(23):2902–2905

Rodriguez-Viciana P, Warne PH, Vanhaesebroeck B, Waterfield MD, Downward J (1996) Activation of phosphoinositide 3-kinase by interaction with Ras and by point mutation. EMBO J 15:2442–2451

Rommel C, Camps M, Ji H (2007) PI3K delta and PI3K gamma: partners in crime in inflammation in rheumatoid arthritis and beyond? Nat Rev Immunol 7(3):191–201

Rordorf-Nikolic T, Van Horn DJ, Chen D, White MF, Backer JM (1995) Regulation of phosphatidylinositol 3′-kinase by tyrosyl phosphoproteins. Full activation requires occupancy of both SH2 domains in the 85-kDa regulatory subunit. J Biol Chem 270:3662–3666

Rubio I, Rodriguez-Viciana P, Downward J, Wetzker R (1997) Interaction of Ras with phosphoinositide 3-kinase gamma. Biochem J 326:891–895

Samuels Y, Wang Z, Bardelli A, Silliman N, Ptak J, Szabo S, Yan H, Gazdar A, Powell SM, Riggins GJ, Willson JK, Markowitz S, Kinzler KW, Vogelstein B, Velculescu VE (2004) High frequency of mutations of the PIK3CA gene in human cancers. Science 304(5670): 554

Samuels Y, Diaz LA Jr, Schmidt-Kittler O, Cummins JM, Delong L, Cheong I, Rago C, Huso DL, Lengauer C, Kinzler KW, Vogelstein B, Velculescu VE (2005) Mutant PIK3CA promotes cell growth and invasion of human cancer cells. Cancer Cell 7(6):561–573

Sen KI, Wu H, Backer JM, Gerfen GJ (2010) The Structure of p85ni in Class IA Phosphoinositide 3-Kinase Exhibits Interdomain Disorder. Biochemistry 49:2159–2166

Shekar SC, Wu H, Fu Z, Yip SC, Cahill SM, Girvin ME, Backer JM (2005) Mechanism of constitutive PI 3-kinase activation by oncogenic mutants of the p85 regulatory subunit. J Biol Chem 280:27850–27855

Shin HW, Hayashi M, Christoforidis S, Lacas-Gervais S, Hoepfner S, Wenk MR, Modregger J, Uttenweiler-Joseph S, Wilm M, Nystuen A, Frankel WN, Solimena M, De Camilli P, Zerial M (2005) An enzymatic cascade of Rab5 effectors regulates phosphoinositide turnover in the endocytic pathway. J Cell Biol 170(4):607–618

Shin YK, Li Y, Liu Q, Anderson DH, Babiuk LA, Zhou Y (2007a) SH3 binding motif 1 in influenza A virus NS1 protein is essential for PI3K/Akt signaling pathway activation. J Virol 81(23):12730–12739

Shin YK, Liu Q, Tikoo SK, Babiuk LA, Zhou Y (2007b) Influenza A virus NS1 protein activates the phosphatidylinositol 3-kinase (PI3K)/Akt pathway by direct interaction with the p85 subunit of PI3K. J Gen Virol 88(Pt 1):13–18

Siegal G, Davis B, Kristensen SM, Sankar A, Linacre J, Stein RC, Panayotou G, Waterfield MD, Driscoll PC (1998) Solution structure of the C-terminal SH2 domain of the p85α regulatory subunit of phosphoinositide 3-kinase. J Mol Biol 276:461–478

Skolnik EY, Margolis B, Mohammadi M, Lowenstein E, Fischer R, Drepps A, Ullrich A, Schlessinger J (1991) Cloning of PI3 kinase-associated p85 utilizing a novel method for expression/cloning of target proteins for receptor tyrosine kinases. Cell 65:83–90

Songyang Z, Shoelson SE, Chaudhuri M, Gish G, Pawson T, Haser WG, King F, Roberts T, Ratnofsky S, Lechleider RJ, Neel BG, Birge RB, Fajardo JE, Chou MM, Hanafusa H, Schaffhausen B, Cantley LC (1993) SH2 domains recognize specific phosphopeptide sequences. Cell 72:767–778

Stuart D, Sellers WR (2009) Linking somatic genetic alterations in cancer to therapeutics. Curr Opin Cell Biol 21(2):304–310

Taylor SS, Kim C, Vigil D, Haste NM, Yang J, Wu J, Anand GS (2005) Dynamics of signaling by PKA. Biochim Biophys Acta 1754(1–2):25–37

Tolias KF, Cantley LC, Carpenter CL (1995) Rho family GTPases bind to phosphoinositide kinases. J Biol Chem 270:17656–17659

Vanhaesebroeck B, Higashi K, Raven C, Welham M, Anderson S, Brennan P, Ward SG, Waterfield MD (1999) Autophosphorylation of p110δ phosphoinositide 3-kinase: a new paradigm for the regulation of lipid kinases *in vitro* and *in vivo*. EMBO J 18(5):1292–1302

Vanhaesebroeck B, Leevers SJ, Ahmadi K, Timms J, Katso R, Driscoll PC, Woscholski R, Parker PJ, Waterfield MD (2001) Synthesis and function of 3-phosphorylated inositol lipids. Annu Rev Biochem 70:535–602

Vanhaesebroeck B, Ali K, Bilancio A, Geering B, Foukas LC (2005) Signalling by PI3K isoforms: insights from gene-targeted mice. Trends Biochem Sci 30(4):194–204

Waksman G, Shoelson SE, Pant N, Cowburn D, Kuriyan J (1993) Binding of a high affinity phosphotyrosyl peptide to the Src SH2 domain: crystal structures of the complexed and peptide-free forms. Cell 72(5):779–790

Walker EH, Perisic O, Ried C, Stephens L, Williams RL (1999) Structural insights into phosphoinositide 3-kinase catalysis and signalling. Nature 402(6759):313–320

Woscholski R, Dhand R, Fry MJ, Waterfield MD, Parker PJ (1994) Biochemical characterization of the free catalytic p110α and the complexed heterodimeric p110α•p85α forms of the mammalian phosphatidylinositol 3-kinase. J Biol Chem 269:25067–25072

Wu H, Yan Y, Backer JM (2007) Regulation of class IA PI3Ks. Biochem Soc Trans 35(Pt 2):242–244

Wu H, Shekar SC, Flinn RJ, El-Sibai M, Jaiswal BS, Sen KI, Janakiraman V, Seshagiri S, Gerfen GJ, Girvin ME, Backer JM (2009) Regulation of Class IA PI 3-kinases: C2 domain-iSH2 domain contacts inhibit p85/p110alpha and are disrupted in oncogenic p85 mutants. Proc Natl Acad Sci USA 106(48):20258–20263

Wymann MP, Bjorklof K, Calvez R, Finan P, Thomast M, Trifilieff A, Barbier M, Altruda F, Hirsch E, Laffargue M (2003) Phosphoinositide 3-kinase gamma: a key modulator in inflammation and allergy. Biochem Soc Trans 31(Pt 1):275–280

Xia X, Cheng A, Akinmade D, Hamburger AW (2003) The N-terminal 24 amino acids of the p55 gamma regulatory subunit of phosphoinositide 3-kinase binds Rb and induces cell cycle arrest. Mol Cell Biol 23(5):1717–1725

Yu J, Zhang Y, McIlroy J, Rordorf-Nikolic T, Orr GA, Backer JM (1998a) Regulation of the p85/p110 phosphatidylinositol 3′-kinase: Stabilization and inhibition of the p110-alpha catalytic subunit by the p85 regulatory subunit. Mol Cell Biol 18:1379–1387

Yu JH, Wjasow C, Backer JM (1998b) Regulation of the p85/p110α Phosphatidylinositol 3′-kinase - Distinct roles for the N-terminal and C-terminal SH2 domains. J Biol Chem 273(46):30199–30203

Zhang W, Smithgall TE, Gmeiner WH (1998) Self-association and backbone dynamics of the hck SH2 domain in the free and phosphopeptide-complexed forms. Biochemistry (Mosc) 37(20):7119–7126

Zhao L, Vogt PK (2008) Helical domain and kinase domain mutations in p110alpha of phosphatidylinositol 3-kinase induce gain of function by different mechanisms. Proc Natl Acad Sci USA 105(7):2652–2657

Zhao JJ, Liu Z, Wang L, Shin E, Loda MF, Roberts TM (2005) The oncogenic properties of mutant p110alpha and p110beta phosphatidylinositol 3-kinases in human mammary epithelial cells. Proc Natl Acad Sci USA 102(51):18443–18448

Zheng Y, Bagrodia S, Cerione RA (1994) Activation of phosphoinositide 3-kinase by Cdc42Hs binding to p85. J Biol Chem 269:18727–18730

Phosphoinositide Signalling Pathways in Metabolic Regulation

Lazaros C. Foukas and Dominic J. Withers

Contents

Abstract The Insulin Receptor/PI 3–kinase (INSR/PI3K) signalling pathway is a key regulator of cell and organismal metabolism. Phosphoinositides generated by PI 3-kinases following insulin and other metabolic hormone receptor activation give rise to signalling cascades involving a multitude of effector molecules. The physiological roles of these molecules have been dissected with the use of both pharmacological and genetic tools. Furthermore, tissue-specific mutagenesis has

L.C. Foukas (✉)
Institute of Healthy Ageing, University College London, Gower Street, London WC1E 6BT, UK
Department of Genetics, Evolution and Environment, University College London, Gower Street, London WC1 6BT, UK
e-mail: l.foukas@ucl.ac.uk

D.J. Withers (✉)
Metabolic Signalling Group, Medical Research Council Clinical Sciences Centre, Imperial College, Du Cane Road, London W12 0NN, UK
e-mail: d.withers@imperial.ac.uk

C. Rommel et al. (eds.). *Phosphoinositide 3-kinase in Health and Disease*, Volume 1
Current Topics in Microbiology and Immunology 346, DOI 10.1007/82_2010_59

revealed the extent to which individual insulin-target organs and signalling molecules contribute to whole-body carbohydrate and lipid homeostasis. These studies have generated important information with respect to the function of these molecules in normal physiology and their implication in the development of metabolic diseases such as type-2 diabetes and obesity.

1 Introduction

A number of organs and their constituent cells take part in the complex process of regulation of whole body energy homeostasis. Along with the classical metabolic tissues such as liver, skeletal muscle and adipose tissue, pancreatic beta cells and a number of specific central nervous system neuronal populations sense and respond to insulin and other key metabolic regulatory hormones. Glucose released in the bloodstream following ingestion of foods stimulates secretion of insulin from pancreatic beta cells. Amongst a plethora of actions, insulin stimulates uptake of glucose in insulin sensitive peripheral tissues, mainly skeletal muscle and adipose tissue, and inhibits hepatic glucose production. Furthermore, insulin as well as leptin, an adipocyte-derived key metabolic hormone, modulate the function of specific neuronal populations in the hypothalamic arcuate nucleus in the central nervous system (CNS) to regulate food intake and energy expenditure and thus whole-body energy balance.

Binding of insulin/IGF1 and leptin to their cognate receptors results in activation of phosphoinositide 3-kinase (PI3K) and production of membrane-associated phosphoinositide phosphatidylinositol-(3,4,5)-trisphosphate (PIP_3), which acts as a second messenger and activates signalling cascades (Niswender et al. 2004; Shepherd et al. 1998). PIP_3 is generated by the enzymatic action of class I phosphoinositide-3 kinases. This class of PI3Ks comprises four catalytic subunits designated p110α, p110β, p110γ and p110δ. p110α and p110β are broadly expressed, whereas p110γ and p110δ are predominantly expressed in leukocytes.

Signalling cascades are initiated following binding of effector molecules bearing PH domains (a protein phosphoinositide-binding module) to PIP_3. A key molecule activated following PIP_3 generation is the Ser/Thr kinase AKT. AKT is a master kinase and a critical node in the insulin signalling network (Taniguchi et al. 2006a). AKT is activated by dual phosphorylation, with phosphoinositide-dependent kinase 1 (PDK1) phosphorylating Thr308 in the activation loop of AKT. PDK1 has also a PH domain and binding to PIP_3 is required for optimal AKT phosphorylation. However, PIP_3 binding is not necessary for phosphorylation of all of its many (at least 23 AGC kinases) substrates (Mora et al. 2004). Most of the actions of insulin in responsive tissues involve AKT activation. However, other phosphoinositide-activated kinases have also been implicated in insulin action, notably isoforms of the atypical PKC (Farese et al. 2005).

In addition to responding to growth factors/hormones, cells sense nutrient availability and respond by adjusting their metabolic activity accordingly. A key molecule involved in nutrient sensing is the mammalian target of rapamycin (mTOR) and its effectors p70 ribosomal S6-kinase (S6K)-1 and -2 (Dann et al. 2007). mTOR exists in two distinct complexes: the rapamycin sensitive mTOR Complex 1, which consists of mTOR, Raptor, PRAS40 and GβL and the insensitive to acute rapamycin treatment mTOR Complex 2, in which mTOR is associated with Rictor, mSIN1 and GβL. Downstream targets of mTOR, S6-kinase (S6K) and 4EBP1 control a multitude of biological functions. 4EBP1 operates in translational control, whereas S6K is a node where nutrient availability and growth factor/hormonal signals are integrated to control a number of metabolic responses.

Here we have attempted to summarise what is known with respect to the function of the various molecules participating in the insulin signalling network with a focus on data validated by phenotyping gene-targeted mouse models (Tables 1 and 2).

2 Role of Insulin Signalling Network Molecules in Metabolic Regulation as Revealed by Global Gene Inactivation in Mice

Insulin Receptor. Insulin is a key hormonal regulator of cellular and organismal metabolism. Consistent with this role, global inactivation of the insulin receptor in mice results in early neonatal lethality (Accili et al. 1996; Joshi et al. 1996). Hence, the role of insulin signalling in each of the major insulin sensitive tissues has been assessed by tissue-specific insulin receptor inactivation (see below).

Insulin Receptor Substrates 1/2. Signals emanating from the insulin receptor are transmitted mainly via insulin receptor substrates (IRS) 1 and 2, which are major adapter proteins recruited and phosphorylated by the insulin receptor tyrosine kinase. Global knock-out of IRS1 results in growth retarded, insulin resistant, but normoglycemic mice (Araki et al. 1994; Tamemoto et al. 1994). Knock-out of IRS2 results in a more severe phenotype, comprising severe insulin resistance, beta cell hypoplasia and reduced insulin secretion, which together lead to progressive diabetes (Withers et al. 1998).

Class I PI3K. Class I PI3K is further divided into two subclasses designated IA and IB. Class IA PI3Ks are heterodimeric enzymes that bind tyrosine phosphorylated IRS proteins and further transmit the insulin signal by producing PIP_3. They consist of a 110 kDa catalytic subunit (p110) and one of various regulatory subunits (collectively referred to as "p85s"). Despite a key role of PI3K in the insulin signalling pathway, mice deficient in the p85 regulatory subunit (either p85α or p85β) display enhanced insulin sensitivity and glucose tolerance (Fruman et al. 2000; Terauchi et al. 1999; Ueki et al. 2002). To explain this paradoxical effect of p85 deletion, it has been suggested that p85 has a number of negative roles in PI3K signalling downstream the insulin receptor, which have previously been discussed elsewhere (Taniguchi et al. 2006a; Vanhaesebroeck et al. 2005). Targeting of the p110 catalytic subunits has provided important insights, consistent with a critical role for class IA PI3K in insulin signalling. Inactivation of p110α by point mutation that renders the protein catalytically inactive results in early embryonic lethality

Table 1 Metabolic phenotypes of mice with global or tissue-specific inactivation of phosphoinositide-regulated signalling molecules in insulin-sensitive peripheral tissues

Target	Function	Tissue	Phenotype	References
Insulin receptor (IR)	Receptor tyrosine kinase	Liver (LIRKO)	Insulin resistance, glucose intolerance, increased hepatic glucose output	Michael et al. (2000)
		Muscle (MIRKO)	Increased adiposity and serum lipids. Normal glucose and insulin levels and tolerance. Reduced glucose uptake in isolated muscle	Bruning et al. (1998)
		Adipose tissue (FIRKO)	Protection from obesity and obesity-related glucose intolerance	Bluher et al. (2002)
		Brown adipose tissue (BATIRKO)	Impaired insulin secretion, glucose intolerance	Guerra et al. (2001))
IRS1	Adaptor	Global	Growth retardation, beta cell hyperplasia, insulin resistance	Araki et al. (1994) and Tamemoto et al. (1994)
IRS2	Adaptor	Global	Beta cell hypoplasia, insulin resistance, diabetes	Withers et al. (1998)
p85α	Class IA PI3K regulatory subunit	Global	All spice variants: Enhanced insulin sensitivity, hypoglycaemia, perinatal lethality	Terauchi et al. (1999)
		Global	Full-length variant: Enhanced insulin sensitivity, hypoglycemia	Fruman et al. (2000)
		Global	Short-length variants (p50α/p55α): Enhanced insulin sensitivity and glucose uptake by muscle, adipocytes, resistant to thioglucose induced obesity	Chen et al. (2004)
p85β	Class IA PI3K regulatory subunit	Global	Enhanced insulin sensitivity	Ueki et al. (2002)
p110α	Class IA PI3K catalytic subunit	Global	Glucose and insulin intolerance	Foukas et al. (2006)
p110β	Class IA PI3K catalytic subunit	Global	Glucose and insulin intolerance, impaired hepatic glucose output	Ciraolo et al. (2008)
		Liver	Glucose and insulin intolerance, impaired hepatic glucose output	Jia et al. (2008)

p110γ	Class IA PI3K catalytic subunit	Global	Impaired insulin secretion from beta cells, mild glucose intolerance, enhanced insulin sensitivity	MacDonald et al. (2004)
PTEN	Lipid and protein phosphatase	Liver	Increased fatty acid synthesis, hepatomegaly, fatty liver. Enhanced liver insulin action, improved systemic glucose tolerance	Stiles et al. (2004)
		Muscle	Protection from high fat diet induced insulin resistance. Enhanced insulin-stimulated glucose uptake and AKT phosphorylation in soleus, but not in extensor digitorum longus of high fat fed mice	Wijesekara et al. (2005)
		Adipose tissue	Improved glucose tolerance and insulin sensitivity. Resistance to streptozotocin-induced diabetes. Decreased serum resistin levels	Kurlawalla-Martinez et al. (2005)
SHIP2	Inositol-5-phosphatase	Global	Protection from high fat diet-induced obesity and obesity-associated diabetes	Sleeman et al. (2005)
PDK1	Ser/Thr kinase	Global, PH domain point mutant (K465E)	Hyperinsulinemia, glucose and insulin intolerance	Bayascas et al. (2008)
		Liver	Impaired glucose tolerance, increased hepatic glucose output, severely reduced liver glycogen content	Mora et al. (2005)
AKT2	Ser/Thr kinase	Global	Impaired glucose and insulin tolerance, impaired glucose uptake, increased hepatic glucose output	Cho et al. (2001)
		Liver	Protection from hepatic lipid accumulation induced by leptin deficiency or high-fat diet feeding	Leavens et al. (2009)
FOXO1	Transcription factor	Liver	Impaired fasting and cAMP induced glycogenolysis and gluconeogenesis	Matsumoto et al. (2007)
S6K1	Ser/Thr kinase	Global	Hypoinsulinaemia, glucose intolerance and diminished beta cell size	Pende et al. (2000)

(*continued*)

Table 1 (continued)

Target	Function	Tissue	Phenotype	References
GSK3α	Ser/Thr kinase	Global KI (S21A)	No significant metabolic phenotypes	McManus et al. (2005)
		Global KO	Enhanced glucose and insulin sensitivity, reduced fat mass, Increased fasted and glucose-stimulated hepatic glycogen content. Increased insulin-stimulated AKT phosphorylation and IRS1 expression in liver	MacAulay et al. (2007)
GSK3β	Ser/Thr kinase	Global KI (S9A)	Impaired activation of glycogen synthase in muscle	McManus et al. (2005)
		Liver	No significant metabolic phenotypes	Patel et al. (2008)
		Muscle	Age-limited improved glucose and insulin tolerance. Enhanced insulin-stimulated glycogen synthase regulation and glycogen deposition	
Raptor	mTORC1 complex constituent	Muscle	Progressive muscle dystrophia, impaired oxidative capacity, and increased glycogen stores	Bentzinger et al. (2008)
		Adipose tissue	Increased energy expenditure, leanness, resistance to high fat diet	Polak et al. (2008)
PKCλ	Ser/Thr kinase	Muscle	Insulin resistance, glucose intolerance, increased adiposity, hyperlipidemia, hepatosteatosis	Farese et al. (2007)
		Liver	Reduced expression of SREBP-1c, decreased triglyceride content, increased insulin sensitivity	Matsumoto et al. (2003)

Table 2 Metabolic phenotypes of mice with global or tissue-specific inactivation of phosphoinositide-regulated signalling molecules in pancreatic beta cells and hypothalamic neurons

Target	Function	Tissue	Phenotype	References
Insulin receptor	Receptor tyrosine kinase	Pancreatic beta cells	Reduced first phase insulin secretion	Kulkarni et al. (1999)
		Brain (NIRKO)	Mild diet-sensitive obesity. Reproductive abnormalities	Bruning et al. (2000)
		Hypothalamic POMC neurons	No discernable metabolic phenotype	Konner et al. (2007)
		Hypothalamic AgRP neurons	Failure in suppression of hepatic glucose production	
IRS2	Adaptor	Pancreatic beta cells	Reduced beta cell mass	Cantley et al. (2007), Choudhury et al. (2005), Kubota et al. (2004) and Lin et al. (2004)
p85α (on a global p85β background)	Class IA PI3K regulatory subunit	Hypothalamic POMC neurons	Defects in acute regulation of food intake by leptin. Blockade of the effects of insulin and leptin on POMC neuron excitability. No long-term alteration in energy homeostasis	Hill et al. (2008)
p110α	Class IA PI3K catalytic subunit	Hypothalamic POMC neurons	Mild obesity	Hill et al. (2009)
			Diet-induced obesity	Al-Qassab et al. (2009)
		Hypothalamic AgRP neurons	No phenotype	Al-Qassab et al. (2009)
p110β	Class IA PI3K catalytic subunit	Hypothalamic POMC neurons	Obesity and central leptin resistance	Al-Qassab et al. (2009)
		Hypothalamic AgRP neurons	Leanness and increased insulin sensitivity	Al-Qassab et al. (2009)
PTEN	Lipid and protein phosphatase	Hypothalamic POMC neurons	Paradoxical diet-induced obesity	Plum et al. (2006)
PDK1	Ser/Thr kinase	Hypothalamic POMC neurons	Mild obesity and adrenocortical failure	Belgardt et al. (2008)
GSK3β (on IRS2 KO background)	Ser/Thr kinase	Pancreatic beta cells	Preservation of beta cell mass	Tanabe et al. (2008)

(Foukas et al. 2006). However, heterozygote mice display small body size, insulin resistance and glucose intolerance, hyperleptinemia and hyperphagia. These studies demonstrated that following insulin stimulation, the p110α PI3K isoform productively engages IRS1/2 complexes. An independent pharmacological study, employing isoform-selective inhibitors, has confirmed a prominent role for p110α in insulin signalling (Knight et al. 2006). In contrast, p110β is not acutely activated by insulin, but a gene targeting strategy aiming to produce a conditionally kinase-dead p110β has reported metabolic phenotypes in these mice (Ciraolo et al. 2008). Mice bearing this inactivating mutation become insulin resistant and glucose intolerant at 6 months of age (Ciraolo et al. 2008). Altered kinetics of AKT phosphorylation was documented both in mutant mouse embryonic fibroblasts (MEFs) and in HepG2 hepatocytes treated with a p110β-selective inhibitor (Ciraolo et al. 2008). These studies suggested that p110β may confer a sustained signal for Akt activation. Interestingly, knockout of p110γ, the single class IB PI3K isoform, which is predominantly expressed in leukocytes and activated downstream G-protein coupled receptors, results in defective insulin secretion from pancreatic beta cells (MacDonald et al. 2004). These animals develop mild only glucose intolerance, presumably due to an as yet unexplained hypersensitivity to insulin.

AKT. The Ser/Thr kinase AKT is a key node in the insulin signalling network (Taniguchi et al. 2006a). Out of three AKT isoforms, AKT2 specifically has been shown to play a role in the regulation of glucose homeostasis (Cho et al. 2001). Disruption of *Akt2* gene in mice results in impaired glucose and insulin tolerance. Impaired glucose uptake mainly in muscle and to a lesser extent in adipocytes as well as a complete failure of insulin to suppress hepatic glucose output underlie the phenotype of these mutants. Importantly, the role of AKT2 in insulin signalling in humans is supported by the discovery of a germline mutation in *AKT2* in a family with severe insulin resistance and diabetes (George et al. 2004).

PDK1. The mode of action of PDK1 in insulin-stimulated Akt phosphorylation has been elegantly demonstrated *in vivo* by mutating the mouse *Pdk1* gene so that the PH domain of the protein can no longer bind PIP_3 [PDK1 (K465E)] (Bayascas et al. 2008). AKT phosphorylation in tissues from these animals is reduced and the mice display hyperinsulinemia and glucose and insulin intolerance. Interestingly, phosphorylation of other substrates of PDK1, such as p90RSK and SGK1 was not affected in these animals. This showed that, in contrast to AKT, phosphorylation of these substrates by PDK1 does not require PIP_3 binding.

GSK3. A key target of AKT in metabolic regulation signalling pathways is GSK3, a kinase that phosphorylates and inactivates glucogen synthase. Phosphorylation of GSK3 by AKT results in inactivation of the former and thus activation of glycogen synthase. Upregulation of GSK3 has been reported in tissues from diabetic rodents and human patients. Hence, GSK3 has been considered a potential therapeutic target in diabetes (Eldar-Finkelman 2002; Nikoulina et al. 2000).

The physiological roles of the GSK3 isoforms, GSK3α and GSK3β, have been precisely dissected both by knock-in (KI) point mutation of the respective AKT phosphorylation sites (McManus et al. 2005) and by knock-out (KO) approaches (MacAulay et al. 2007; Patel et al. 2008). GSK3 KI mice do not display substantial metabolic phenotypes. This approach has revealed that glycogen synthesis in muscle is regulated by GSK-3β rather than GSK3α (McManus et al. 2005). Liver-specific GSK3β KO mice do not display substantial metabolic phenotypes, whereas muscle-specific GSK3β KO mice display improved glucose and insulin tolerance when young, though this improved metabolic phenotype is lost by the age of 6 months (Patel et al. 2008). Knock-out of GSK3α revealed that this isoform is responsible for the regulation of glycogen synthesis in the liver (MacAulay et al. 2007). These mice had also improved glucose and insulin tolerance and increased insulin-stimulated AKT phosphorylation in the liver, presumably as a result of increased IRS1 expression (see also below). These findings are in favour of therapeutically targeting GSK3 in diabetes.

S6K. p70S6K is a molecular node where growth/factor and nutrient availability signals converge. S6K1 knockout mice are hypoinsulinemic and glucose intolerant (Pende et al. 2000). Hypoinsulinemia is the result of a growth defect in beta cells which impairs insulin content and secretion. Muscles of these mice are insulin sensitive (presumably hypersensitive), and this is in line with a negative impact of S6K1 activity on insulin receptor signalling (see below).

3 Role of INSR/PI3K Pathway Components in the Development and Function of Insulin Sensitive Tissues

3.1 *Insulin Sensitive Peripheral Tissues*

Liver. Insulin signalling in the liver has a fundamental role in maintaining glucose homeostasis. The main action of insulin in this tissue is inhibition of hepatic glucose output. Excessive hepatic glucose output is a major cause of hyperglycemia in diabetes. The role of liver insulin signalling in body glucose homeostasis has been demonstrated by liver-specific *Insr* knock-out (LIRKO) in mice. LIRKO mice display severe insulin resistance, glucose intolerance and elevated hepatic glucose production (Michael et al. 2000).

Despite severe metabolic defects and diabetes in mice, with global inactivation of IRS proteins, in particular IRS2 (Withers et al. 1998), there is little, if any, metabolic defect caused by liver-specific deletion of either IRS1 (Dong et al. 2008) or IRS2 (Dong et al. 2008; Simmgen et al. 2006) in mice. This likely reflects a redundant role of hepatic IRS proteins. However, combined inactivation of both IRS1 and IRS2 has a strong negative impact on glucose homeostasis (Dong et al. 2008). Importantly, this phenotype could be rescued by concomitant inactivation of FOXO1, a key transcriptional target of AKT signalling. This shows that insulin signalling via IRS proteins exerts its regulatory role on gluconeogenesis via FOXO1. This mechanism is further supported by studies on liver-specific inactivation of *Foxo1*. Ablation of FOXO1 in mouse liver impairs fasting, cAMP-induced glycogenolysis and gluconeogenesis (Matsumoto et al. 2007).

PI3K activity plays an important role in insulin receptor signalling in liver. A dominant-negative form of p85α expressed via adenoviral delivery in the liver of mice has been shown before to interfere with glucose and lipid homeostasis (Miyake et al. 2002). Likewise, combined liver-specific deletion of p85α on a global p85β KO background resulted in impaired glucose and insulin tolerance increased gluconeogenic gene expression (Taniguchi et al. 2006b). These outcomes were associated with severely reduced PI3K activity and AKT phosphorylation. Moreover, these mice displayed hypolipidemia and decreased expression of SREBP-1c, which was attributable to impaired PKCλ/ζ activation. Indeed, PKCλ liver-specific deletion has been shown before to downregulate SREBP-1c (a major lipogenic transcription factor) expression and reduce triglyceride content in this tissue (Matsumoto et al. 2003). However, a recently reported study utilising liver-specific deletion of *Akt2* has demonstrated a critical role of this molecule in lipogenic gene expression and *de novo* lipogenesis in insulin resistant mouse models (Leavens et al. 2009).

Liver-specific inactivation of the PI3K p110β by means of adenoviral-delivery of Cre recombinase has also been shown to affect glucose homeostasis (Jia et al. 2008). These mice were glucose and insulin intolerant, although no defect in Akt phosphorylation was observed in the liver upon insulin stimulation. Interestingly, phosphoenolpyruvate carboxykinase (PEPCK) was the only one from a panel of gluconeogenic genes found up-regulated in these mutants and this correlated with increased glucose production in a pyruvate tolerance test.

Mice with liver-specific deletion of PTEN display an unusual phenotype. These mice have enhanced insulin sensitivity and glucose tolerance, but at the same time, they suffer from hepatomegaly and liver steatosis. The overall body fat content and circulating free fatty acids in these mutants are markedly reduced suggesting that deletion of PTEN in the liver promotes redistribution of fat from other tissues to the liver. In contrast, global deletion of SHIP2, the other PIP_3 degrading phosphatase, protects mice from high fat diet-induced obesity and obesity-associated diabetes (Sleeman et al. 2005).

Deletion of PDK1 in the liver results in glucose intolerance in the face of normal insulin sensitivity due to impaired regulation of hepatic glucose output (Mora et al. 2005). It appears that the underlying defect is loss of the normal effect of feeding on the expression of a set of gluconeogenesis regulatory genes. Liver-specific PDK1 KO mice died between 4 and 16 weeks of age due to liver failure. In these mice, feeding did not exert proper control on genes regulating gluconeogenesis such as *Pck1* (phosphoenolpyruvate carboxykinase), *G6pc* (glucose-6-phosphatase) and *Srebf1* (sterol-regulatory-element binding protein 1) and on insulin-responsive genes (insulin-like growth factor binding protein 1, *Irs2* and glucokinase).

Muscle. Muscle-specific insulin receptor knockout (MIRKO) has provided important insights with regard to the role of this tissue in body glucose homeostasis and pathogenesis of type-2 diabetes (Bruning et al. 1998). These mice displayed elevated adiposity and serum lipids, but normal glucose and insulin levels and tolerance. The study demonstrated that isolated insulin receptor-deficient muscle was insulin resistant and defective in glucose uptake and thus highlighted the role of other insulin-sensitive tissues, i.e. adipose tissue, in glucose clearance *in vivo*. These findings challenged the widely held assumption that a defect in insulin sensitivity and glucose uptake in muscle is the primary lesion in the pathogenesis of diabetes.

The role of muscle in glucose uptake has also been assessed by tissue-specific disruption of PKCλ, which as mentioned above, has been implicated in this process. Despite normal activation of AKT by insulin, glucose uptake was diminished in isolated PKCλ-deficient muscle (Farese et al. 2007). Similar to MIRKO mice, these mutants displayed increased abdominal adiposity and hyperlipidemia. However, in contrast with the MIRKO mice, muscle PKCλ KO mice were also systemically insulin resistant, glucose intolerant or diabetic and developed hepatosteatosis.

mTOR inactivation results in early embryonic lethality (Gangloff et al. 2004; Murakami et al. 2004). Information with regard to the role of mTOR complexes in cellular and organismal metabolism has only recently been obtained by means of tissue-specific inactivation of unique components of the two distinct mTOR complexes. The rapamycin sensitive mTORC1 has been targeted by inactivating Raptor in muscle. Although no overt direct metabolic phenotypes occurred, these mice exhibited muscle dystrophy and early lethality (Bentzinger et al. 2008). It was shown that deletion of Raptor downregulated molecules promoting mitochondrial biogenesis, such as PGC-1α. It should be noted though that this effect is different from that of global compound S6K1/2 deletion, which results in increased mitochondrial biogenesis in skeletal muscle (Aguilar et al. 2007). This suggests that disruption of mTORC1 has broader and/or different consequences than those resulting from targeting S6K. This fact would likely be reflected in the effects of rapamycin versus S6K-specific inhibitors. mTORC2 targeting by means of Rictor inactivation in muscle showed no substantial metabolic phenotypes (Bentzinger et al. 2008). However, double deletion of Raptor and Rictor in the muscle resulted in hyperphosphorylation of AKT presumably due to relief of the S6K negative feedback loop (see below) inhibiting PI3K (Bentzinger et al. 2008).

Adipose tissue. As realised from studies in MIRKO mice, adipose tissue has an important role in glucose homeostasis. Fat-specific *Insr* KO (FIRKO) display remarkable phenotypes (Bluher et al. 2002): These mice are protected against age-and hyperphagy-induced obesity and related glucose intolerance. Adipose tissue from these mice comprised two different types of adipocytes according to their size, big and small. Small adipocytes, being refractory to triglyceride accumulation and to the anti-lipolytic effect of insulin, likely account for the favourable metabolic profiles of these mutants. Interestingly, brown adipocyte-specific *Insr* KO (BATIRKO) mice display glucose intolerance, mainly due to impaired insulin secretion (Guerra et al. 2001). Concurrent deletion of INSR in the white adipose cells is sufficient to overcome this phenotype and improve the overall metabolic profile in these mutants.

PTEN deletion specifically in the adipose tissue results in improved glucose tolerance and insulin sensitivity. These mutants display decreased serum resistin levels, which likely underlie the enhanced insulin and AMP-kinase signalling observed in the liver (Kurlawalla-Martinez et al. 2005).

The role of mTORC1 signalling in this tissue has recently been studied by adipose tissue-specific deletion of Raptor. These mutants had improved glucose tolerance and insulin sensitivity and they were resistant to high-fat diet (Polak et al. 2008). They displayed increased thermogenesis owing to upregulation of the uncoupling protein UCP1.

3.2 Pancreatic Islets

In addition to regulating the function of the classical target tissues liver, muscle and fat, it is now well established that insulin receptor signalling plays key roles in the pancreatic beta cell and other islet endocrine cell-types. Despite being constantly exposed to insulin (and in contrast to the previously held view that insulin would have a negative feedback effect on beta cell function) insulin signalling positively regulates many aspects of beta cell physiology (reviewed in Leibiger et al. 2008). Furthermore, most components of the canonical PI3K signalling cassette are expressed in beta cells suggesting an important role for this pathway in this cell type (Leibiger et al. 2008). Although in vitro studies demonstrated that insulin activated its signalling cascade in islets and transformed beta cell lines, the role of PI3K signalling in peripheral tissues and the use of mouse gene targeting both global and tissue-specific have started to reveal the precise contribution of individual components of the PI3K signalling cascade in the pancreatic islet.

The first demonstration that the insulin signalling cascade regulates islet function *in vivo* was the finding as described above that mice globally lacking IRS2 developed type-2 diabetes due to a combination of peripheral insulin resistance and beta cell failure largely due to reduced beta cell mass (Withers et al. 1998). In contrast mice globally lacking IRS1 displayed an appropriate increase in beta cell mass in the face of insulin resistance (Withers et al. 1998). Subsequent studies with the *Irs2* null model revealed interplay between IGF1 receptor and IRS2 signalling in the maintenance of beta cell mass (Withers et al. 1999). Furthermore global deletion of *Irs2* combined with haploinsufficiency of *Irs1* caused rapidly progressive diabetes in young animals, while animals lacking *Irs1* but carrying a single allele of *Irs2* maintained beta cell mass in the face of marked insulin resistance. Together these studies indicated the dominant role of IRS2 signalling in beta cell function (Withers et al. 1999). Consistent with this finding re-expression of IRS2 back into *Irs2* null islets rescued the hypoplastic islet phenotype and restored downstream signalling events such as AKT activation (Hennige et al. 2003). FOXO1 has also been implicated in the effects of IRS2 signalling in beta cells as heterozygote insufficiency of this molecule rescued the defects in beta cell mass in *Irs2* null mice (Kitamura et al. 2002). Likewise haploinsufficiency of the AKT substrate GSK3β or PTEN also corrected the beta cell failure seen in these animals (Kushner et al. 2005; Tanabe et al. 2008). Taken together these studies indirectly implicate key components of the PI3K signalling network downstream of IRS2 in beta cell function.

As described above, no insights into the precise role of PI3K catalytic and regulatory subunits have been gained from global gene targeting approaches, but such approaches have started to reveal roles for key downstream effector molecules such as AKT and S6K1. Global deletion of *Akt2* leads to impaired glucose handling, but the effects on beta cell are dependent on genetic background with a reported loss of beta cell mass seen in DBA mice but in not C57BL/6 × 129 hybrids (Cho et al. 2001; Garofalo et al. 2003). In contrast reduction in beta cell AKT activity achieved through expression of a dominant negative mutant led to impaired insulin secretion but no alterations in beta cells mass (Bernal-Mizrachi et al. 2004). However, expression of a constitutively active AKT in beta cells resulted in a marked expansion of beta cell mass with increased cell proliferation and size (Bernal-Mizrachi et al. 2001). Furthermore, mice lacking S6K1 have glucose intolerance with reduced islet mass and beta cell size and hypoinsulinemia indicating a key role for this signalling molecule in beta cells (Pende et al. 2000). Together these studies demonstrate a key role for PI3K effector pathways in the maintenance of normal beta cell mass. However, this pathway is also likely to have additional effects in the endocrine pancreas, such as regulation of insulin synthesis and secretion, as studies, for example, on islets of global *Irs1* null mice revealed that these processes are defective in this model (reviewed in Leibiger et al. 2008).

The studies in mice with global gene deletion described above have given useful insights into the role of PI3K signalling pathways in islet function. However, the complex tissue interplay, that is key not only to normal metabolism but also seen in the development of pathological situations such as insulin resistance and type-2 diabetes, has meant that such approaches have not defined the role of beta cell intrinsic PI3K signalling. Conditional gene targeting utilising the Cre/loxP system has started to provide important insights to this question, but to date, the class IA PI3K catalytic and regulatory subunits have not been specifically deleted from these cells. However, many other signalling components have been studied with this approach and, with a couple of caveats, these indicate that beta cell autonomous PI3K signalling plays a key role in the function of this cell type.

Deletion of the insulin receptor in beta cells resulted in mice with abnormal first phase insulin secretion and progressive reduction in beta cell mass (Kulkarni et al. 1999). Deletion of the IGF1 receptor resulted in mice with normal beta cell mass but defects in glucose-stimulated insulin secretion (Kulkarni et al. 2002). Mice with combined deletion of both receptors had a progressive diabetic phenotype with reduced beta cell mass (Ueki et al. 2006). Furthermore studies on compound mutants of both alleles suggested that INSR has a dominant role in regulating beta cell mass (Ueki et al. 2006). Three groups generated mice with beta cell specific deletion of Irs2 (Choudhury et al. 2005; Kubota et al. 2004; Lin et al. 2004). All three models developed impaired glucose homeostasis and reduced beta cell mass, although not as profound as that seen in mice with global deletion of *Irs2*. Combined global heterozygote deletion of *Irs1* with beta cell-specific deletion of *Irs2* caused a more progressive diabetes phenotype (Lin et al. 2004). Furthermore in two of these studies, deletion of *Irs2* in beta cells was not maintained throughout the lives of the mice and so beta cell repopulation of the islets occurred with cells that had not undergone deletion of *Irs2* (Choudhury et al. 2005; Lin et al. 2004). However, all three models also developed obesity due to hypothalamic deletion of *Irs2* due to expression of the Cre recombinase transgene in a population of hypothalamic neurons (Choudhury et al. 2005; Kubota et al. 2004; Lin et al. 2004). Therefore the beta cell phenotypes reported could also reflect the obesity phenotype. Furthermore, the interpretation of the phenotype of other mouse models generated using the same Cre transgenic mouse has to be taken in the context that there is hypothalamic deletion in these strains too. To overcome the complication of CNS derived phenotype, one group generated mice lacking *Irs2* in the CNS and not the beta cell (Choudhury et al. 2005). This animal had the expected obesity phenotype but no reduction in beta cell mass. Subsequently, deletion of *Irs2* in all pancreatic endocrine cell types but not the hypothalamus achieved using a Cre transgenic mouse with the pancreatic duodenum homeobox gene 1 promoter demonstrated that beta cell intrinsic IRS2 signalling was required for the maintenance of beta cell mass (Cantley et al. 2007). In addition, these animals had defects in insulin secretion and alpha cell mass (Choudhury et al. 2005; Kubota et al. 2004; Lin et al. 2004). Deletion of the insulin receptor in alpha cells has also recently implicated insulin signalling in the modulation of glucagon secretion (Kawamori et al. 2009). Together these studies strongly implicate a key role for beta cell intrinsic insulin and IGF1 receptor and IRS signalling in the function of this cell type and suggest roles for INSR signalling in other islet cells.

Few studies that examine the function of more distal components of the PI3K signalling cascade in islet function have been undertaken. Deletion of PDK1 in beta cells resulted in a progressive diabetic phenotype with reduced beta cell mass (due to both a reduction in cell size and number) and insulin content (Hashimoto et al. 2006). Reduced AKT and S6K1 phosphorylation was seen in islets from these mice consistent with the role of PDK1 in activating members of the AGC family of kinases. The phenotype of these animals was partially rescued by concomitant heterozygote deletion of FOXO1 (Hashimoto et al. 2006). Mice with beta cell deletion of GSK3β rescued the diabetes phenotype in IRS2 global null mice indicating that GSK3β acts to negatively regulate the insulin signalling pathway in beta cells (Tanabe et al. 2008).

Taken together these studies in mice with either global or beta cell specific mutations demonstrate that the canonical insulin/IGF1 signalling pathway plays a key role in the regulation of beta cell function. Beta cell mass, insulin synthesis and secretion have all been shown to require intact insulin signalling. Although the catalytic and regulatory subunits of the class IA PI3Ks have not been directly manipulated in beta cells, many other models have implicated PI3K signalling as they mediated many of the effects of insulin. These findings suggest that beta cell resistance to the action of insulin, as seen in peripheral tissues, may be an important component of the pathophysiology of type-2 diabetes. Furthermore, PI3K signalling may also mediate the effects of other receptor signalling pathways that have been implicated in beta cell function and have been shown to activate PI3K in other tissues. These include leptin and vascular endothelial growth factor suggesting that

PI3K signalling may be a point of integration for multiple signals that regulate beta cell function.

3.3 Central Nervous System

The CNS is a second non-classical insulin target tissue. Indeed, it is increasingly recognised that the CNS is the key site for orchestrating metabolic regulation and, furthermore, that key neuronal circuits are targets for peripheral hormones that engage PI3K signalling mechanisms including insulin and the adipocyte hormone leptin (Barsh and Schwartz 2002; Konner et al. 2009). Much research effort in this area has focussed on the mediobasal hypothalamus, which is an important homeostatic regulatory region. The arcuate nucleus, which has access to the peripheral circulation and therefore to hormones and nutrients that signal energy status, has received particular attention. Within the arcuate nucleus, identification of anorexigenic proopiomelanocortin (POMC) expressing neurons and orexigenic neuropeptide Y/agouti-related protein (AgRP) expressing neurons, which are targets for hormones and nutrients that regulate energy homeostasis, has been a key observation (Barsh et al. 2000; Barsh and Schwartz 2002). Therefore, much work on understanding the signalling mechanisms regulating energy balance has focussed on these cell types. As for studies delineating the role of PI3K signalling in the pancreatic beta cell, initial pharmacological studies suggested a role for PI3K in the hypothalamic regulation of energy homeostasis. Administration of broad-spectrum PI3K inhibitors blocked the ability of insulin and leptin to inhibit food intake and inhibited the effects of these hormones on hypothalamic neuronal excitability (Niswender et al. 2001, 2003). Both leptin and insulin stimulated the accumulation of PIP_3 in hypothalamic neurons and increased IRS-associated PI3K activity.

Insights into the hypothalamic regulation of energy balance also accrued from studies in mice with global deletion of insulin signalling components. Mice with global deletion of *Irs2* displayed increased food intake, obesity, hyperleptinemia and defective hypothalamic leptin action; these studies further suggested that IRS2/PI3K signalling may act as a point of convergence for leptin and insulin action (Burks et al. 2000). The heterozygote p110α KI mice described above displayed hyperphagia and increased adiposity (Foukas et al. 2006). However, the systemic nature of both these mutants did not define the CNS role of PI3K signalling, and therefore, the greatest progress in this area has again derived from the use of conditional gene targeting.

Deletion of *Insr* in all neurons results in mild diet-sensitive obesity and reproductive abnormalities (Bruning et al. 2000). Subsequent studies using a combination of mouse models in which inducible deletion of *Insr* occurs in either/all tissues or is restricted to peripheral tissues and not to the CNS suggested that central insulin action positively regulated white adipose tissue mass and controlled glucose metabolism acting via a hepatic pathway that includes liver activation of STAT3

(Koch et al. 2008). Deletion of *Insr* in POMC neurons resulted in no discernable metabolic phenotype while INSR signalling in AgRP neurons regulated hepatic glucose production (Konner et al. 2007). These findings were consistent with studies either knocking down INSR expression in the hypothalamus or studying central insulin signalling in rat models of diabetes (Obici et al. 2002a,b). Together these results demonstrate that CNS insulin receptor signalling regulates energy homeostasis and metabolism.

Progress has been made using conditional gene targeting in understanding the role of key downstream insulin signalling elements in the CNS. As described above, three groups generated mice with beta cell deletion of *Irs2*, but in these animals, the deletion event also occurred in the hypothalamus resulting in the development of an obesity phenotype (Choudhury et al. 2005; Kubota et al. 2004; Lin et al. 2004). Studies from one of these groups revealed that the hypothalamic deletion was occurring in a population of neurons distinct from POMC or AgRP neurons but which played a key role in the melanocortin circuitry (Choudhury et al. 2005). Furthermore, deletion of *Irs2* in all neurons resulted in marked obesity but did not alter leptin action suggesting that IRS2 is not a point of convergence for leptin and insulin action in the CNS (Choudhury et al. 2005). In contrast, mice lacking *Irs2* in either POMC or AgRP neurons displayed no energy homeostasis phenotype (Choudhury et al. 2005).

Subsequent studies have revealed the role of PI3K signalling in the hypothalamus, although for global targeting of this pathway, the approaches have, in general, employed mutants that cannot categorically implicate PI3K in the reported phenotypes. Deletion of PTEN in POMC neurons increased PIP_3 levels in this cell type as would be anticipated if PI3K was active and yet the mice developed obesity in contrast to the expected lean phenotype (Plum et al. 2006). This most likely reflected the marked anatomical abnormalities in PTEN-deleted POMC neurons or because the protein phosphatase activity of PTEN as well as its lipid phosphatase action plays a role in leptin action (Ning et al. 2006). Deletion of PDK1 in POMC neurons resulted in hyperphagia, but this phenotype was complicated by the development of a marked reduction in cortisol levels which led to attenuation of the defect in energy homeostasis (Belgardt et al. 2008). Deletion of p85α in POMC neurons combined with global deletion of p85β resulted in mice with defects in the acute regulation of food intake by leptin and blockade of the effects of insulin and leptin on POMC neuron excitability, but no long-term alteration in energy homeostasis (Hill et al. 2008). However, as described above, mice with global deletion of p85β have improved insulin sensitivity and reduced body weight phenotypes that could again mask any energy phenotype resulting from loss of PI3K signalling in POMC neurons.

The most comprehensive examination of the role of PI3K signalling in POMC and AgRP neurons has been recently reported by Al-Qassab et al. who utilised floxed alleles of p110α and p110β, which preserve the signalling stoichiometry in the pathway upon deletion (Al-Qassab et al. 2009). These studies revealed that mice lacking p110β in POMC neurons displayed increased food intake, central leptin resistance, increased adiposity with absent electrophysiological responses to leptin and insulin in this cell type. In contrast, mice lacking p110α in POMC neurons displayed obesity only on a high fat diet and responded normally to leptin. While mice lacking p110α in AgRP neurons had no defect in energy homeostasis, deletion of p110β in these cells resulted in a lean phenotype with increased sensitivity to leptin. This phenotype might seem surprising as it has been shown that disruption of leptin receptor expression in AgRP neurons results in mild obesity, suggesting that loss of leptin action via PI3K signalling might result in a similar phenotype. However, leptin withdrawal rather than administration has been described to activate PI3K in AgRP neurons and therefore loss of PI3K signalling in this cell type might mimic the effects of leptin and therefore result in negative energy balance (Xu et al. 2005). These new findings suggest that at least in respect to POMC and AgRP neurons p110β plays the predominant role in mediating the effects of leptin and insulin upon the hypothalamic regulation of energy homeostasis, although recent studies have also implicated p110α in POMC neurons as playing a role in the regulation of hepatic glucose metabolism (Hill et al. 2009).

Downstream of PI3K, studies have implicated mTOR/S6K1 signalling and the forkhead protein FOXO1 in the central regulation of energy balance. The mTOR inhibitor rapamycin attenuates the anorectic effects of leptin when administered into the third ventricle and hypothalamic mTOR pathway activity is modulated by alterations in feeding status (Cota et al. 2006). Up-regulation of mTOR activity via deletion of *Tsc1* in POMC neurons resulted in not only hyperphagia but also a concomitant alteration in neuronal size and morphology, which is likely to influence the phenotype (Mori et al. 2009). The mTOR effector S6K1 has also been implicated in hypothalamic function (Ono et al. 2008). FOXO1 is a major downstream effector of PI3K signalling. FOXO1 has been shown to co-ordinate the expression of AgRP and POMC in the hypothalamus and most recently has been implicated in the processing of pro-opiomelanocortin peptides via regulation of carboxypeptidase E (Kitamura et al. 2006; Plum et al. 2009). It is therefore clear that PI3K signalling impacts upon a variety of mechanisms involved in the hypothalamic regulation of energy balance.

4 Role of the INSR/PI3K Pathway in the Integration of Nutrient Availability and Growth Factor/Hormonal Signals

A coordinated response of cells to growth factor/hormonal signals and availability of nutrients is a requisite for effective metabolic regulation. The INSR/PI3K pathway plays a role in this fundamental function, as well. Nutrients such as glucose and aminoacids are not mere substrates for energy producing or macromolecular biosynthetic reactions, but they also act as molecular signals in the hexosamine, mTOR and AMP-kinase (AMPK) signalling pathways (Marshall 2006). Nutrient overload results in development of pathologic conditions such as obesity, diabetes and cancer (Marshall 2006; Um et al. 2006).

The mTORC1 effector S6K has a key role as the central node of the network, where nutrient sensing and growth factor signals are integrated. Recent developments have increased our understanding of the mechanisms by which aminoacids activate mTOR (Kim et al. 2008; Sancak et al. 2008). Cellular energy status, reflected on cellular ATP/AMP ratio, controls the activity AMPK (Lage et al. 2008). AMPK has been shown to phosphorylate and activate TSC2, thus inhibiting mTOR (Inoki et al. 2003). Therefore, mTOR and AMPK have opposing actions with mTOR activation promoting and AMPK activation inhibiting cellular growth.

Class III PI3K, Vps34p has been shown to be an essential component of the nutrient sensing mechanism, which permits signals leading to activation of S6K by insulin (Byfield et al. 2005; Nobukuni et al. 2005). These two studies demonstrated that glucose and aminoacids activate S6K via activation of Vps34p instead of class IA PI3K. The enzymatic product of Vps34p, phosphatidylinositol-3-phosphate (PI3P) appears to be required for this effect, since overexpression of a PI3P lipid binding domain FYVE prevents activation of S6K. Therefore, inputs by both insulin stimulation via class IA PI3K and class III PI3K converge on S6K (Byfield et al. 2005). AMPK activation by glucose deprivation seems also to inhibit Vps34p activity as shown by treatment of cells with AICAR or oligomycin (Byfield et al. 2005). How exactly Vps34p activates S6K remains to be discovered. This is an important goal because as detailed in the following paragraph S6K activity impacts on insulin sensitivity by regulating IRS protein levels.

5 Molecular Basis of Insulin Resistance Development: Signal Termination Feedback Loops

A mechanism of insulin signal termination at the level of IRS protein expression has received great attention because of its implication in the development of insulin resistance (Zick 2004). Serine phosphorylation interferes with tyrosine phosphorylation by the insulin receptor and/or targets IRS for degradation thus reducing their cellular levels. This in turn results in diminished signalling output downstream the insulin receptor. Serine/threonine kinases activated in a phosphoinositide-regulated manner, such as mTOR and its downstream target p70S6K have been shown to phosphorylate IRS in the context of this feedback loop (Harrington et al. 2004; Shah et al. 2004). Ser/Thr kinases other than S6K have also been implicated in IRS phosphorylation, notably GSK3 (Liberman and Eldar-Finkelman 2005) and ERK (Bouzakri et al. 2003; De Fea and Roth 1997; Mothe and Van Obberghen 1996). Coupling of ERK signalling to insulin receptor activation seems to be mediated by the adaptor protein Gab1, as liver specific knock-out of Gab1 results in improved insulin sensitivity and glucose tolerance (Bard-Chapeau et al. 2005).

Obesity has been associated with activation of cellular stress and inflammation signalling leading to insulin resistance and development of type-2 diabetes. Serine phosphorylation of IRS by the stress activated kinase c-Jun-N-terminal kinase (JNK) has been shown to play a key role in this process (Hirosumi et al. 2002;

Ozcan et al. 2004). It is interesting to note that the p85 regulatory subunit of PI3K has been shown to play a role, not requiring PI3K activity, in the activation of JNK by endoplasmic reticulum stress or high fat feeding (Taniguchi et al. 2007).

In summary, IRS serine phosphorylation is thought to be a cause of insulin resistance induced by hyperinsulinemia, inflammation and dyslipidemia. Thus, targeting IRS phosphorylating Ser/Thr kinases, notably S6K or JNK, would be expected to have a therapeutic effect in pathologic conditions associated with insulin resistance. It should be mentioned though that a recent study employing ectopic expression of PDGF-receptors in insulin sensitive tissues, as a means of IRS-independent activation of the PI3K/AKT pathway, showed that exposure to such insults (persistent insulin stimulation, inflammatory cytokines and excessive free fatty acids) resulted in impaired glucose uptake, even in such an IRS-independent signalling system (Hoehn et al. 2008). These data suggest that in addition to IRS serine phosphorylation, other mechanisms likely contribute to the development of insulin resistance.

6 Discussion

Evidently, the composition and regulation of phosphoinositide-activated signalling pathways in metabolic control has been studied and described to a large extent. Nevertheless, there are still outstanding questions. These relate both to the precise modes of action of various constituent molecules of these signalling pathways and to their potential as therapeutic targets in metabolic disease.

The role of individual PI3K isoforms in insulin and leptin signalling requires further clarification given that disruption of distinct class I PI3K isoforms has resulted in metabolic phenotypes. Moreover, in vitro data have suggested promiscuity of class IA PI3K isoforms in insulin signalling in transformed cell lines (Chaussade et al. 2007). It has recently been shown though that the pattern of PI3K isoform coupling to receptors can change upon cellular transformation (Papakonstanti et al. 2008) and therefore transformed cell lines do not always accurately represent primary tissues. Furthermore, the regulation of glucose homeostasis has a complex physiology, and all the reported phenotypes of PI3K isoform deficiency, at the organismal level, do not necessarily relate to impaired insulin receptor signalling. In any case, a more detailed understanding of the involvement of the individual PI3K isoforms in metabolic signalling is necessary, not least because these molecules are currently being pursued as therapeutic targets in diseases such as cancer and inflammation and therefore the effect of putative drugs in metabolism needs to be addressed. To this end, additional tissue-specific mutagenesis studies in combination with the use of isoform-selective small molecule inhibitors, which are becoming increasingly available, will be instrumental.

Thus far, the main research focus has been on the roles of class I PI3K and their product, PIP_3. However, another group of PI3Ks, namely class II PI3Ks, and their enzymatic product phosphatidylinositol-3-phosphate has been shown to be activated by insulin and to regulate processes such as glucose transport (Brown et al. 1999; Maffucci et al. 2003). Studies on mouse mutants of these molecules are currently lacking, but are awaited with interest as they can shed light on the possible implication of these molecules in metabolic diseases (Falasca and Maffucci 2009).

Another important question relates to the extent these pathways can be therapeutically exploited in metabolic disease. A matter that should be rigorously assessed is the potential for tumourigenesis resulting from targeting of molecules with tumour suppressor functions. In this regard, GSK3 is considered a potential therapeutic target in diabetes, as the respective mouse mutants have shown beneficial effects on glucose homeostasis in the absence of increased incidence of tumourigenesis. Whether targeting of PTEN could have a therapeutic effect on metabolic disease is dubious. It is difficult to predict, which aspect of the PTEN inactivation phenotypes would prevail upon PTEN inhibition. Morbidity associated with hepatosteatosis, seen in liver-specific PTEN KO, could cancel any beneficial effect of enhanced insulin sensitivity. Furthermore, PTEN is a tumour suppressor whose loss-of-function mutations are a major cause of cancer. However, SHIP2, the other PIP_3 degrading phosphatase could potentially be a better therapeutic target. There is also an emerging theme that adipose-specific inactivation of molecules in the insulin and nutrient signalling pathways (e.g. INSR, mTORC1) results in improved glucose and lipid homeostasis. Therefore, targeting "druggable" molecules within these pathways e.g. S6K can have a therapeutic effect owing to their impact on such adipose tissue related metabolic processes.

Importantly, modern genomic technologies and large scale screening of human patient populations have started to provide clues about the identity of molecules involved in the development of metabolic diseases (O'Rahilly 2009). In some occasions, the findings of these studies corroborate and add weight to conclusions made by use of experimental models. For instance, genome wide association studies have recently identified *IRS1* as a gene associated with insulin resistance and type-2 diabetes (Rung et al. 2009). Added to the volume of experimental data implicating IRS1 in the development of insulin resistance in various models, these findings will likely influence therapeutic efforts towards modulation of IRS1 protein levels as a means for improving insulin sensitivity.

Undoubtedly, human genetics will have a decisive role guiding therapeutic efforts towards specific molecules in metabolic disease. Research exploiting animal models has generated the knowledge and a framework which will facilitate the identification of suitable molecular targets for rational design of therapeutic regimes.

References

Accili D, Drago J, Lee EJ, Johnson MD, Cool MH, Salvatore P, Asico LD, Jose PA, Taylor SI, Westphal H (1996) Early neonatal death in mice homozygous for a null allele of the insulin receptor gene. Nat Genet 12:106–109

Aguilar V, Alliouachene S, Sotiropoulos A, Sobering A, Athea Y, Djouadi F, Miraux S, Thiaudiere E, Foretz M, Viollet B, Diolez P, Bastin J, Benit P, Rustin P, Carling D, Sandri

M, Ventura-Clapier R, Pende M (2007) S6 kinase deletion suppresses muscle growth adaptations to nutrient availability by activating AMP kinase. Cell Metab 5:476–487

Al-Qassab H, Smith MA, Irvine EE, Guillermet-Guibert J, Claret M, Choudhury AI, Selman C, Piipari K, Clements M, Lingard S, Chandarana K, Bell JD, Barsh GS, Smith AJ, Batterham RL, Ashford ML, Vanhaesebroeck B, Withers DJ (2009) Dominant role of the p110beta isoform of PI3K over p110alpha in energy homeostasis regulation by POMC and AgRP neurons. Cell Metab 10:343–354

Araki E, Lipes MA, Patti ME, Bruning JC, Haag B 3rd, Johnson RS, Kahn CR (1994) Alternative pathway of insulin signalling in mice with targeted disruption of the IRS-1 gene. Nature 372:186–190

Bard-Chapeau EA, Hevener AL, Long S, Zhang EE, Olefsky JM, Feng GS (2005) Deletion of Gab1 in the liver leads to enhanced glucose tolerance and improved hepatic insulin action. Nat Med 11:567–571

Barsh GS, Schwartz MW (2002) Genetic approaches to studying energy balance: perception and integration. Nat Rev Genet 3:589–600

Barsh GS, Farooqi IS, O'Rahilly S (2000) Genetics of body-weight regulation. Nature 404:644–651

Bayascas JR, Wullschleger S, Sakamoto K, Garcia-Martinez JM, Clacher C, Komander D, van Aalten DM, Boini KM, Lang F, Lipina C, Logie L, Sutherland C, Chudek JA, van Diepen JA, Voshol PJ, Lucocq JM, Alessi DR (2008) Mutation of the PDK1 PH domain inhibits protein kinase B/Akt, leading to small size and insulin resistance. Mol Cell Biol 28:3258–3272

Belgardt BF, Husch A, Rother E, Ernst MB, Wunderlich FT, Hampel B, Klockener T, Alessi D, Kloppenburg P, Bruning JC (2008) PDK1 deficiency in POMC-expressing cells reveals FOXO1-dependent and -independent pathways in control of energy homeostasis and stress response. Cell Metab 7:291–301

Bentzinger CF, Romanino K, Cloetta D, Lin S, Mascarenhas JB, Oliveri F, Xia J, Casanova E, Costa CF, Brink M, Zorzato F, Hall MN, Ruegg MA (2008) Skeletal muscle-specific ablation of raptor, but not of rictor, causes metabolic changes and results in muscle dystrophy. Cell Metab 8:411–424

Bernal-Mizrachi E, Wen W, Stahlhut S, Welling CM, Permutt MA (2001) Islet beta cell expression of constitutively active Akt1/PKB alpha induces striking hypertrophy, hyperplasia, and hyperinsulinemia. J Clin Invest 108:1631–1638

Bernal-Mizrachi E, Fatrai S, Johnson JD, Ohsugi M, Otani K, Han Z, Polonsky KS, Permutt MA (2004) Defective insulin secretion and increased susceptibility to experimental diabetes are induced by reduced Akt activity in pancreatic islet beta cells. J Clin Invest 114:928–936

Bluher M, Michael MD, Peroni OD, Ueki K, Carter N, Kahn BB, Kahn CR (2002) Adipose tissue selective insulin receptor knockout protects against obesity and obesity-related glucose intolerance. Dev Cell 3:25–38

Bouzakri K, Roques M, Gual P, Espinosa S, Guebre-Egziabher F, Riou JP, Laville M, Le Marchand-Brustel Y, Tanti JF, Vidal H (2003) Reduced activation of phosphatidylinositol-3 kinase and increased serine 636 phosphorylation of insulin receptor substrate-1 in primary culture of skeletal muscle cells from patients with type 2 diabetes. Diabetes 52:1319–1325

Brown RA, Domin J, Arcaro A, Waterfield MD, Shepherd PR (1999) Insulin activates the alpha isoform of class II phosphoinositide 3-kinase. J Biol Chem 274:14529–14532

Bruning JC, Michael MD, Winnay JN, Hayashi T, Horsch D, Accili D, Goodyear LJ, Kahn CR (1998) A muscle-specific insulin receptor knockout exhibits features of the metabolic syndrome of NIDDM without altering glucose tolerance. Mol Cell 2:559–569

Bruning JC, Gautam D, Burks DJ, Gillette J, Schubert M, Orban PC, Klein R, Krone W, Muller-Wieland D, Kahn CR (2000) Role of brain insulin receptor in control of body weight and reproduction. Science 289:2122–2125

Burks DJ, Font de Mora J, Schubert M, Withers DJ, Myers MG, Towery HH, Altamuro SL, Flint CL, White MF (2000) IRS-2 pathways integrate female reproduction and energy homeostasis. Nature 407:377–382

Byfield MP, Murray JT, Backer JM (2005) hVps34 is a nutrient-regulated lipid kinase required for activation of p70 S6 kinase. J Biol Chem 280:33076–33082

Cantley J, Choudhury AI, Asare-Anane H, Selman C, Lingard S, Heffron H, Herrera P, Persaud SJ, Withers DJ (2007) Pancreatic deletion of insulin receptor substrate 2 reduces beta and alpha cell mass and impairs glucose homeostasis in mice. Diabetologia 50:1248–1256

Chaussade C, Rewcastle GW, Kendall JD, Denny WA, Cho K, Gronning LM, Chong ML, Anagnostou SH, Jackson SP, Daniele N, Shepherd PR (2007) Evidence for functional redundancy of class IA PI3K isoforms in insulin signalling. Biochem J 404:449–458

Chen D, Mauvais-Jarvis F, Bluher M, Fisher SJ, Jozsi A, Goodyear LJ, Ueki K, Kahn CR (2004) p50alpha/p55alpha phosphoinositide 3-kinase knockout mice exhibit enhanced insulin sensitivity. Mol Cell Biol 24:320–329

Cho H, Mu J, Kim JK, Thorvaldsen JL, Chu Q, Crenshaw EB 3rd, Kaestner KH, Bartolomei MS, Shulman GI, Birnbaum MJ (2001) Insulin resistance and a diabetes mellitus-like syndrome in mice lacking the protein kinase Akt2 (PKB beta). Science 292:1728–1731

Choudhury AI, Heffron H, Smith MA, Al-Qassab H, Xu AW, Selman C, Simmgen M, Clements M, Claret M, Maccoll G, Bedford DC, Hisadome K, Diakonov I, Moosajee V, Bell JD, Speakman JR, Batterham RL, Barsh GS, Ashford ML, Withers DJ (2005) The role of insulin receptor substrate 2 in hypothalamic and beta cell function. J Clin Invest 115:940–950

Ciraolo E, Iezzi M, Marone R, Marengo S, Curcio C, Costa C, Azzolino O, Gonella C, Rubinetto C, Wu H, Dastru W, Martin EL, Silengo L, Altruda F, Turco E, Lanzetti L, Musiani P, Ruckle T, Rommel C, Backer JM, Forni G, Wymann MP, Hirsch E (2008) Phosphoinositide 3-kinase p110beta activity: key role in metabolism and mammary gland cancer but not development. Sci Signal 1:ra3

Cota D, Proulx K, Smith KA, Kozma SC, Thomas G, Woods SC, Seeley RJ (2006) Hypothalamic mTOR signaling regulates food intake. Science 312:927–930

Dann SG, Selvaraj A, Thomas G (2007) mTOR Complex1-S6K1 signaling: at the crossroads of obesity, diabetes and cancer. Trends Mol Med 13:252–259

De Fea K, Roth RA (1997) Modulation of insulin receptor substrate-1 tyrosine phosphorylation and function by mitogen-activated protein kinase. J Biol Chem 272:31400–31406

Dong XC, Copps KD, Guo S, Li Y, Kollipara R, DePinho RA, White MF (2008) Inactivation of hepatic Foxo1 by insulin signaling is required for adaptive nutrient homeostasis and endocrine growth regulation. Cell Metab 8:65–76

Eldar-Finkelman H (2002) Glycogen synthase kinase 3: an emerging therapeutic target. Trends Mol Med 8:126–132

Falasca M, Maffucci T (2009) Rethinking phosphatidylinositol 3-monophosphate. Biochim Biophys Acta 1793:1795–1803

Farese RV, Sajan MP, Standaert ML (2005) Atypical protein kinase C in insulin action and insulin resistance. Biochem Soc Trans 33:350–353

Farese RV, Sajan MP, Yang H, Li P, Mastorides S, Gower WR Jr, Nimal S, Choi CS, Kim S, Shulman GI, Kahn CR, Braun U, Leitges M (2007) Muscle-specific knockout of PKC-lambda impairs glucose transport and induces metabolic and diabetic syndromes. J Clin Invest 117: 2289–2301

Foukas LC, Claret M, Pearce W, Okkenhaug K, Meek S, Peskett E, Sancho S, Smith AJ, Withers DJ, Vanhaesebroeck B (2006) Critical role for the p110alpha phosphoinositide-3-OH kinase in growth and metabolic regulation. Nature 441:366–370

Fruman DA, Mauvais-Jarvis F, Pollard DA, Yballe CM, Brazil D, Bronson RT, Kahn CR, Cantley LC (2000) Hypoglycaemia, liver necrosis and perinatal death in mice lacking all isoforms of phosphoinositide 3-kinase p85 alpha. Nat Genet 26:379–382

Gangloff YG, Mueller M, Dann SG, Svoboda P, Sticker M, Spetz JF, Um SH, Brown EJ, Cereghini S, Thomas G, Kozma SC (2004) Disruption of the mouse mTOR gene leads to early postimplantation lethality and prohibits embryonic stem cell development. Mol Cell Biol 24:9508–9516

Garofalo RS, Orena SJ, Rafidi K, Torchia AJ, Stock JL, Hildebrandt AL, Coskran T, Black SC, Brees DJ, Wicks JR, McNeish JD, Coleman KG (2003) Severe diabetes, age-dependent loss of adipose tissue, and mild growth deficiency in mice lacking Akt2/PKB beta. J Clin Invest 112: 197–208

George S, Rochford JJ, Wolfrum C, Gray SL, Schinner S, Wilson JC, Soos MA, Murgatroyd PR, Williams RM, Acerini CL, Dunger DB, Barford D, Umpleby AM, Wareham NJ, Davies HA, Schafer AJ, Stoffel M, O'Rahilly S, Barroso I (2004) A family with severe insulin resistance and diabetes due to a mutation in AKT2. Science 304:1325–1328

Guerra C, Navarro P, Valverde AM, Arribas M, Bruning J, Kozak LP, Kahn CR, Benito M (2001) Brown adipose tissue-specific insulin receptor knockout shows diabetic phenotype without insulin resistance. J Clin Invest 108:1205–1213

Harrington LS, Findlay GM, Gray A, Tolkacheva T, Wigfield S, Rebholz H, Barnett J, Leslie NR, Cheng S, Shepherd PR, Gout I, Downes CP, Lamb RF (2004) The TSC1-2 tumor suppressor controls insulin-PI3K signaling via regulation of IRS proteins. J Cell Biol 166:213–223

Hashimoto N, Kido Y, Uchida T, Asahara S, Shigeyama Y, Matsuda T, Takeda A, Tsuchihashi D, Nishizawa A, Ogawa W, Fujimoto Y, Okamura H, Arden KC, Herrera PL, Noda T, Kasuga M (2006) Ablation of PDK1 in pancreatic beta cells induces diabetes as a result of loss of beta cell mass. Nat Genet 38:589–593

Hennige AM, Burks DJ, Ozcan U, Kulkarni RN, Ye J, Park S, Schubert M, Fisher TL, Dow MA, Leshan R, Zakaria M, Mossa-Basha M, White MF (2003) Upregulation of insulin receptor substrate-2 in pancreatic beta cells prevents diabetes. J Clin Invest 112:1521–1532

Hill JW, Williams KW, Ye C, Luo J, Balthasar N, Coppari R, Cowley MA, Cantley LC, Lowell BB, Elmquist JK (2008) Acute effects of leptin require PI3K signaling in hypothalamic proopiomelanocortin neurons in mice. J Clin Invest 118:1796–1805

Hill JW, Xu Y, Preitner F, Fukuda M, Cho YR, Luo J, Balthasar N, Coppari R, Cantley LC, Kahn BB, Zhao JJ, Elmquist JK (2009) Phosphatidyl inositol 3-kinase signaling in hypothalamic proopiomelanocortin neurons contributes to the regulation of glucose homeostasis. Endocrinology 150:4874–4882

Hirosumi J, Tuncman G, Chang L, Gorgun CZ, Uysal KT, Maeda K, Karin M, Hotamisligil GS (2002) A central role for JNK in obesity and insulin resistance. Nature 420:333–336

Hoehn KL, Hohnen-Behrens C, Cederberg A, Wu LE, Turner N, Yuasa T, Ebina Y, James DE (2008) IRS1-independent defects define major nodes of insulin resistance. Cell Metab 7:421–433

Inoki K, Zhu T, Guan KL (2003) TSC2 mediates cellular energy response to control cell growth and survival. Cell 115:577–590

Jia S, Liu Z, Zhang S, Liu P, Zhang L, Lee SH, Zhang J, Signoretti S, Loda M, Roberts TM, Zhao JJ (2008) Essential roles of PI(3)K-p110beta in cell growth, metabolism and tumorigenesis. Nature 454:776–779

Joshi RL, Lamothe B, Cordonnier N, Mesbah K, Monthioux E, Jami J, Bucchini D (1996) Targeted disruption of the insulin receptor gene in the mouse results in neonatal lethality. EMBO J 15:1542–1547

Kawamori D, Kurpad AJ, Hu J, Liew CW, Shih JL, Ford EL, Herrera PL, Polonsky KS, McGuinness OP, Kulkarni RN (2009) Insulin signaling in alpha cells modulates glucagon secretion *in vivo*. Cell Metab 9:350–361

Kim E, Goraksha-Hicks P, Li L, Neufeld TP, Guan KL (2008) Regulation of TORC1 by Rag GTPases in nutrient response. Nat Cell Biol 10:935–945

Kitamura T, Nakae J, Kitamura Y, Kido Y, Biggs WH 3rd, Wright CV, White MF, Arden KC, Accili D (2002) The forkhead transcription factor Foxo1 links insulin signaling to Pdx1 regulation of pancreatic beta cell growth. J Clin Invest 110:1839–1847

Kitamura T, Feng Y, Kitamura YI, Chua SC Jr, Xu AW, Barsh GS, Rossetti L, Accili D (2006) Forkhead protein FoxO1 mediates Agrp-dependent effects of leptin on food intake. Nat Med 12:534–540

Knight ZA, Gonzalez B, Feldman ME, Zunder ER, Goldenberg DD, Williams O, Loewith R, Stokoe D, Balla A, Toth B, Balla T, Weiss WA, Williams RL, Shokat KM (2006) A pharmacological map of the PI3-K family defines a role for p110alpha in insulin signaling. Cell 125:733–747

Koch L, Wunderlich FT, Seibler J, Konner AC, Hampel B, Irlenbusch S, Brabant G, Kahn CR, Schwenk F, Bruning JC (2008) Central insulin action regulates peripheral glucose and fat metabolism in mice. J Clin Invest 118:2132–2147

Konner AC, Janoschek R, Plum L, Jordan SD, Rother E, Ma X, Xu C, Enriori P, Hampel B, Barsh GS, Kahn CR, Cowley MA, Ashcroft FM, Bruning JC (2007) Insulin action in AgRP-expressing neurons is required for suppression of hepatic glucose production. Cell Metab 5:438–449

Konner AC, Klockener T, Bruning JC (2009) Control of energy homeostasis by insulin and leptin: targeting the arcuate nucleus and beyond. Physiol Behav 97:632–638

Kubota N, Terauchi Y, Tobe K, Yano W, Suzuki R, Ueki K, Takamoto I, Satoh H, Maki T, Kubota T, Moroi M, Okada-Iwabu M, Ezaki O, Nagai R, Ueta Y, Kadowaki T, Noda T (2004) Insulin receptor substrate 2 plays a crucial role in beta cells and the hypothalamus. J Clin Invest 114: 917–927

Kulkarni RN, Bruning JC, Winnay JN, Postic C, Magnuson MA, Kahn CR (1999) Tissue-specific knockout of the insulin receptor in pancreatic beta cells creates an insulin secretory defect similar to that in type 2 diabetes. Cell 96:329–339

Kulkarni RN, Holzenberger M, Shih DQ, Ozcan U, Stoffel M, Magnuson MA, Kahn CR (2002) beta-cell-specific deletion of the Igf1 receptor leads to hyperinsulinemia and glucose intolerance but does not alter beta-cell mass. Nat Genet 31:111–115

Kurlawalla-Martinez C, Stiles B, Wang Y, Devaskar SU, Kahn BB, Wu H (2005) Insulin hypersensitivity and resistance to streptozotocin-induced diabetes in mice lacking PTEN in adipose tissue. Mol Cell Biol 25:2498–2510

Kushner JA, Simpson L, Wartschow LM, Guo S, Rankin MM, Parsons R, White MF (2005) Phosphatase and tensin homolog regulation of islet growth and glucose homeostasis. J Biol Chem 280:39388–39393

Lage R, Dieguez C, Vidal-Puig A, Lopez M (2008) AMPK: a metabolic gauge regulating whole-body energy homeostasis. Trends Mol Med 14:539–549

Leavens KF, Easton RM, Shulman GI, Previs SF, Birnbaum MJ (2009) Akt2 is required for hepatic lipid accumulation in models of insulin resistance. Cell Metab 10:405–418

Leibiger IB, Leibiger B, Berggren PO (2008) Insulin signaling in the pancreatic beta-cell. Annu Rev Nutr 28:233–251

Liberman Z, Eldar-Finkelman H (2005) Serine 332 phosphorylation of insulin receptor substrate-1 by glycogen synthase kinase-3 attenuates insulin signaling. J Biol Chem 280:4422–4428

Lin X, Taguchi A, Park S, Kushner JA, Li F, Li Y, White MF (2004) Dysregulation of insulin receptor substrate 2 in beta cells and brain causes obesity and diabetes. J Clin Invest 114:908–916

MacAulay K, Doble BW, Patel S, Hansotia T, Sinclair EM, Drucker DJ, Nagy A, Woodgett JR (2007) Glycogen synthase kinase 3alpha-specific regulation of murine hepatic glycogen metabolism. Cell Metab 6:329–337

MacDonald PE, Joseph JW, Yau D, Diao J, Asghar Z, Dai F, Oudit GY, Patel MM, Backx PH, Wheeler MB (2004) Impaired glucose-stimulated insulin secretion, enhanced intraperitoneal insulin tolerance, and increased beta-cell mass in mice lacking the p110gamma isoform of phosphoinositide 3-kinase. Endocrinology 145:4078–4083

Maffucci T, Brancaccio A, Piccolo E, Stein RC, Falasca M (2003) Insulin induces phosphatidylinositol-3-phosphate formation through TC10 activation. EMBO J 22:4178–4189

Marshall S (2006) Role of insulin, adipocyte hormones, and nutrient-sensing pathways in regulating fuel metabolism and energy homeostasis: a nutritional perspective of diabetes, obesity, and cancer. Sci STKE 2006:re7

Matsumoto M, Ogawa W, Akimoto K, Inoue H, Miyake K, Furukawa K, Hayashi Y, Iguchi H, Matsuki Y, Hiramatsu R, Shimano H, Yamada N, Ohno S, Kasuga M, Noda T (2003) PKClambda in liver mediates insulin-induced SREBP-1c expression and determines both hepatic lipid content and overall insulin sensitivity. J Clin Invest 112:935–944

Matsumoto M, Pocai A, Rossetti L, Depinho RA, Accili D (2007) Impaired regulation of hepatic glucose production in mice lacking the forkhead transcription factor Foxo1 in liver. Cell Metab 6:208–216

McManus EJ, Sakamoto K, Armit LJ, Ronaldson L, Shpiro N, Marquez R, Alessi DR (2005) Role that phosphorylation of GSK3 plays in insulin and Wnt signalling defined by knockin analysis. EMBO J 24:1571–1583

Michael MD, Kulkarni RN, Postic C, Previs SF, Shulman GI, Magnuson MA, Kahn CR (2000) Loss of insulin signaling in hepatocytes leads to severe insulin resistance and progressive hepatic dysfunction. Mol Cell 6:87–97

Miyake K, Ogawa W, Matsumoto M, Nakamura T, Sakaue H, Kasuga M (2002) Hyperinsulinemia, glucose intolerance, and dyslipidemia induced by acute inhibition of phosphoinositide 3-kinase signaling in the liver. J Clin Invest 110:1483–1491

Mora A, Komander D, van Aalten DM, Alessi DR (2004) PDK1, the master regulator of AGC kinase signal transduction. Semin Cell Dev Biol 15:161–170

Mora A, Lipina C, Tronche F, Sutherland C, Alessi DR (2005) Deficiency of PDK1 in liver results in glucose intolerance, impairment of insulin-regulated gene expression and liver failure. Biochem J 385:639–648

Mori H, Inoki K, Munzberg H, Opland D, Faouzi M, Villanueva EC, Ikenoue T, Kwiatkowski D, MacDougald OA, Myers MG Jr, Guan KL (2009) Critical role for hypothalamic mTOR activity in energy balance. Cell Metab 9:362–374

Mothe I, Van Obberghen E (1996) Phosphorylation of insulin receptor substrate-1 on multiple serine residues, 612, 632, 662, and 731, modulates insulin action. J Biol Chem 271: 11222–11227

Murakami M, Ichisaka T, Maeda M, Oshiro N, Hara K, Edenhofer F, Kiyama H, Yonezawa K, Yamanaka S (2004) mTOR is essential for growth and proliferation in early mouse embryos and embryonic stem cells. Mol Cell Biol 24:6710–6718

Nikoulina SE, Ciaraldi TP, Mudaliar S, Mohideen P, Carter L, Henry RR (2000) Potential role of glycogen synthase kinase-3 in skeletal muscle insulin resistance of type 2 diabetes. Diabetes 49:263–271

Ning K, Miller LC, Laidlaw HA, Burgess LA, Perera NM, Downes CP, Leslie NR, Ashford ML (2006) A novel leptin signalling pathway via PTEN inhibition in hypothalamic cell lines and pancreatic beta-cells. EMBO J 25:2377–2387

Niswender KD, Morton GJ, Stearns WH, Rhodes CJ, Myers MG Jr, Schwartz MW (2001) Intracellular signalling. Key enzyme in leptin-induced anorexia. Nature 413:794–795

Niswender KD, Morrison CD, Clegg DJ, Olson R, Baskin DG, Myers MG Jr, Seeley RJ, Schwartz MW (2003) Insulin activation of phosphatidylinositol 3-kinase in the hypothalamic arcuate nucleus: a key mediator of insulin-induced anorexia. Diabetes 52:227–231

Niswender KD, Baskin DG, Schwartz MW (2004) Insulin and its evolving partnership with leptin in the hypothalamic control of energy homeostasis. Trends Endocrinol Metab 15:362–369

Nobukuni T, Joaquin M, Roccio M, Dann SG, Kim SY, Gulati P, Byfield MP, Backer JM, Natt F, Bos JL, Zwartkruis FJ, Thomas G (2005) Amino acids mediate mTOR/raptor signaling through activation of class 3 phosphatidylinositol 3OH-kinase. Proc Natl Acad Sci USA 102:14238–14243

Obici S, Feng Z, Karkanias G, Baskin DG, Rossetti L (2002a) Decreasing hypothalamic insulin receptors causes hyperphagia and insulin resistance in rats. Nat Neurosci 5:566–572

Obici S, Zhang BB, Karkanias G, Rossetti L (2002b) Hypothalamic insulin signaling is required for inhibition of glucose production. Nat Med 8:1376–1382

Ono H, Pocai A, Wang Y, Sakoda H, Asano T, Backer JM, Schwartz GJ, Rossetti L (2008) Activation of hypothalamic S6 kinase mediates diet-induced hepatic insulin resistance in rats. J Clin Invest 118:2959–2968

O'Rahilly S (2009) Human genetics illuminates the paths to metabolic disease. Nature 462:307–314

Ozcan U, Cao Q, Yilmaz E, Lee AH, Iwakoshi NN, Ozdelen E, Tuncman G, Gorgun C, Glimcher LH, Hotamisligil GS (2004) Endoplasmic reticulum stress links obesity, insulin action, and type 2 diabetes. Science 306:457–461

Papakonstanti EA, Zwaenepoel O, Bilancio A, Burns E, Nock GE, Houseman B, Shokat K, Ridley AJ, Vanhaesebroeck B (2008) Distinct roles of class IA PI3K isoforms in primary and immortalised macrophages. J Cell Sci 121:4124–4133

Patel S, Doble BW, MacAulay K, Sinclair EM, Drucker DJ, Woodgett JR (2008) Tissue-specific role of glycogen synthase kinase 3beta in glucose homeostasis and insulin action. Mol Cell Biol 28:6314–6328

Pende M, Kozma SC, Jaquet M, Oorschot V, Burcelin R, Le Marchand-Brustel Y, Klumperman J, Thorens B, Thomas G (2000) Hypoinsulinaemia, glucose intolerance and diminished beta-cell size in S6K1-deficient mice. Nature 408:994–997

Plum L, Ma X, Hampel B, Balthasar N, Coppari R, Munzberg H, Shanabrough M, Burdakov D, Rother E, Janoschek R, Alber J, Belgardt BF, Koch L, Seibler J, Schwenk F, Fekete C, Suzuki A, Mak TW, Krone W, Horvath TL, Ashcroft FM, Bruning JC (2006) Enhanced PIP3 signaling in POMC neurons causes KATP channel activation and leads to diet-sensitive obesity. J Clin Invest 116:1886–1901

Plum L, Lin HV, Dutia R, Tanaka J, Aizawa KS, Matsumoto M, Kim AJ, Cawley NX, Paik JH, Loh YP, DePinho RA, Wardlaw SL, Accili D (2009) The obesity susceptibility gene Cpe links FoxO1 signaling in hypothalamic pro-opiomelanocortin neurons with regulation of food intake. Nat Med 15:1195–1201

Polak P, Cybulski N, Feige JN, Auwerx J, Ruegg MA, Hall MN (2008) Adipose-specific knockout of raptor results in lean mice with enhanced mitochondrial respiration. Cell Metab 8:399–410

Rung J, Cauchi S, Albrechtsen A, Shen L, Rocheleau G, Cavalcanti-Proenca C, Bacot F, Balkau B, Belisle A, Borch-Johnsen K, Charpentier G, Dina C, Durand E, Elliott P, Hadjadj S, Jarvelin MR, Laitinen J, Lauritzen T, Marre M, Mazur A, Meyre D, Montpetit A, Pisinger C, Posner B, Poulsen P, Pouta A, Prentki M, Ribel-Madsen R, Ruokonen A, Sandbaek A, Serre D, Tichet J, Vaxillaire M, Wojtaszewski JF, Vaag A, Hansen T, Polychronakos C, Pedersen O, Froguel P, Sladek R (2009) Genetic variant near IRS1 is associated with type 2 diabetes, insulin resistance and hyperinsulinemia. Nat Genet 41:1110–1115

Sancak Y, Peterson TR, Shaul YD, Lindquist RA, Thoreen CC, Bar-Peled L, Sabatini DM (2008) The Rag GTPases bind raptor and mediate amino acid signaling to mTORC1. Science 320: 1496–1501

Shah OJ, Wang Z, Hunter T (2004) Inappropriate activation of the TSC/Rheb/mTOR/S6K cassette induces IRS1/2 depletion, insulin resistance, and cell survival deficiencies. Curr Biol 14:1650–1656

Shepherd PR, Withers DJ, Siddle K (1998) Phosphoinositide 3-kinase: the key switch mechanism in insulin signalling. Biochem J 333(Pt 3):471–490

Simmgen M, Knauf C, Lopez M, Choudhury AI, Charalambous M, Cantley J, Bedford DC, Claret M, Iglesias MA, Heffron H, Cani PD, Vidal-Puig A, Burcelin R, Withers DJ (2006) Liver-specific deletion of insulin receptor substrate 2 does not impair hepatic glucose and lipid metabolism in mice. Diabetologia 49:552–561

Sleeman MW, Wortley KE, Lai KM, Gowen LC, Kintner J, Kline WO, Garcia K, Stitt TN, Yancopoulos GD, Wiegand SJ, Glass DJ (2005) Absence of the lipid phosphatase SHIP2 confers resistance to dietary obesity. Nat Med 11:199–205

Stiles B, Wang Y, Stahl A, Bassilian S, Lee WP, Kim YJ, Sherwin R, Devaskar S, Lesche R, Magnuson MA, Wu H (2004) Liver-specific deletion of negative regulator Pten results in fatty liver and insulin hypersensitivity [corrected]. Proc Natl Acad Sci USA 101: 2082–2087

Tamemoto H, Kadowaki T, Tobe K, Yagi T, Sakura H, Hayakawa T, Terauchi Y, Ueki K, Kaburagi Y, Satoh S et al (1994) Insulin resistance and growth retardation in mice lacking insulin receptor substrate-1. Nature 372:182–186

Tanabe K, Liu Z, Patel S, Doble BW, Li L, Cras-Meneur C, Martinez SC, Welling CM, White MF, Bernal-Mizrachi E, Woodgett JR, Permutt MA (2008) Genetic deficiency of glycogen synthase kinase-3beta corrects diabetes in mouse models of insulin resistance. PLoS Biol 6:e37

Taniguchi CM, Emanuelli B, Kahn CR (2006a) Critical nodes in signalling pathways: insights into insulin action. Nat Rev Mol Cell Biol 7:85–96

Taniguchi CM, Kondo T, Sajan M, Luo J, Bronson R, Asano T, Farese R, Cantley LC, Kahn CR (2006b) Divergent regulation of hepatic glucose and lipid metabolism by phosphoinositide 3-kinase via Akt and PKClambda/zeta. Cell Metab 3:343–353

Taniguchi CM, Aleman JO, Ueki K, Luo J, Asano T, Kaneto H, Stephanopoulos G, Cantley LC, Kahn CR (2007) The p85alpha regulatory subunit of phosphoinositide 3-kinase potentiates c-Jun N-terminal kinase-mediated insulin resistance. Mol Cell Biol 27:2830–2840

Terauchi Y, Tsuji Y, Satoh S, Minoura H, Murakami K, Okuno A, Inukai K, Asano T, Kaburagi Y, Ueki K, Nakajima H, Hanafusa T, Matsuzawa Y, Sekihara H, Yin Y, Barrett JC, Oda H, Ishikawa T, Akanuma Y, Komuro I, Suzuki M, Yamamura K, Kodama T, Suzuki H, Koyasu S, Aizawa S, Tobe K, Fukui Y, Yazaki Y, Kadowaki T (1999) Increased insulin sensitivity and hypoglycaemia in mice lacking the p85 alpha subunit of phosphoinositide 3-kinase. Nat Genet 21:230–235

Ueki K, Yballe CM, Brachmann SM, Vicent D, Watt JM, Kahn CR, Cantley LC (2002) Increased insulin sensitivity in mice lacking p85beta subunit of phosphoinositide 3-kinase. Proc Natl Acad Sci USA 99:419–424

Ueki K, Okada T, Hu J, Liew CW, Assmann A, Dahlgren GM, Peters JL, Shackman JG, Zhang M, Artner I, Satin LS, Stein R, Holzenberger M, Kennedy RT, Kahn CR, Kulkarni RN (2006) Total insulin and IGF-I resistance in pancreatic beta cells causes overt diabetes. Nat Genet 38:583–588

Um SH, D'Alessio D, Thomas G (2006) Nutrient overload, insulin resistance, and ribosomal protein S6 kinase 1, S6K1. Cell Metab 3:393–402

Vanhaesebroeck B, Ali K, Bilancio A, Geering B, Foukas LC (2005) Signalling by PI3K isoforms: insights from gene-targeted mice. Trends Biochem Sci 30:194–204

Wijesekara N, Konrad D, Eweida M, Jefferies C, Liadis N, Giacca A, Crackower M, Suzuki A, Mak TW, Kahn CR, Klip A, Woo M (2005) Muscle-specific Pten deletion protects against insulin resistance and diabetes. Mol Cell Biol 25:1135–1145

Withers DJ, Gutierrez JS, Towery H, Burks DJ, Ren JM, Previs S, Zhang Y, Bernal D, Pons S, Shulman GI, Bonner-Weir S, White MF (1998) Disruption of IRS-2 causes type 2 diabetes in mice. Nature 391:900–904

Withers DJ, Burks DJ, Towery HH, Altamuro SL, Flint CL, White MF (1999) Irs-2 coordinates Igf-1 receptor-mediated beta-cell development and peripheral insulin signalling. Nat Genet 23:32–40

Xu AW, Kaelin CB, Takeda K, Akira S, Schwartz MW, Barsh GS (2005) PI3K integrates the action of insulin and leptin on hypothalamic neurons. J Clin Invest 115:951–958

Zick Y (2004) Uncoupling insulin signalling by serine/threonine phosphorylation: a molecular basis for insulin resistance. Biochem Soc Trans 32:812–816

Role of RAS in the Regulation of PI 3-Kinase

Esther Castellano and Julian Downward

Contents

Abstract Ras proteins are key regulators of signalling cascades, controlling many processes such as proliferation, differentiation and apoptosis. Mutations in these proteins or in their effectors, activators and regulators are associated with pathological conditions, particularly the development of various forms of human cancer. RAS proteins signal through direct interaction with a number of effector enzymes, one of the best characterized being type I phosphatidylinositol (PI) 3-kinases. Although the ability of RAS to control PI 3-kinase has long been well established in cultured cells, evidence for a role of the interaction of endogenous RAS with PI 3-kinase in normal and malignant cell growth *in vivo* has only been obtained recently. Mice with mutations in the PI 3-kinase catalytic p110a isoform that block its ability to interact with RAS are highly resistant to endogenous KRAS oncogene induced lung tumourigenesis and HRAS oncogene induced skin carcinogenesis. Cells from these mice show proliferative defects and selective disruption of signalling from certain growth factors to PI 3-kinase, while the mice also display delayed development of the lymphatic vasculature. The interaction of RAS with

E. Castellano and J. Downward (✉)
Signal Transduction Laboratory, Cancer Research UK London Research Institute, 44 Lincoln's Inn Fields, London WC2A 3PX, UK
e-mail: julian.downward@cancer.org.uk

C. Rommel et al. (eds.). *Phosphoinositide 3-kinase in Health and Disease*, Volume 1
Current Topics in Microbiology and Immunology 346, DOI 10.1007/82_2010_56

p110a is thus required *in vivo* for some normal growth factor signalling and also for RAS-driven tumour formation.

RAS family members were among the first oncogenes identified over 40 years ago. In the late 1960s, the rat-derived Harvey and Kirsten murine sarcoma retroviruses were discovered and subsequently shown to promote cancer formation through related oncogenes, termed *RAS* (from rat sarcoma virus). The central role of RAS proteins in human cancer is highlighted by the large number of tumours in which they are activated by mutation: approximately 20% of human cancers carry a mutation in RAS proteins. Because of the complex signalling network in which RAS operates, with multiple activators and effectors, each with a different pattern of tissue-specific expression and a distinct set of intracellular functions, one of the critical issues concerns the specific role of each effector in RAS-driven oncogenesis. In this chapter, we summarize current knowledge about how RAS regulates one of its best-known effectors, phosphoinositide 3-kinase (PI3K).

1 RAS Proteins

The canonical members of the RAS family of GTPases, the 21-kDa protein products of the human *HRAS*, *NRAS*, and *KRAS* genes, are highly conserved proteins across species and have established roles in numerous basic cellular functions, including control of proliferation, differentiation and apoptosis.

RAS proteins are part of a large and diverse family of small GTPases. Expression of RAS proteins is nearly ubiquitous and broadly conserved across species, although there are differences in the expression level depending on the tissue and the developmental stage (Leon et al. 1987; Su et al. 2004). Mammalian cells express four different RAS proteins, HRAS, NRAS and KRAS. Alternative splicing of exon 4 in *KRAS* leads to two different protein isoforms: KRAS 4A and KRAS 4B, the latter being the most highly expressed isoform (Bar-Sagi 2001; Barbacid 1987; Lowy and Willumsen 1993; Plowman and Hancock 2005). Comparison of the sequences of RAS proteins shows that they share more that 80% homology, but the different isoforms differ in their last 25 amino acids except for the last 4 amino acids of their carboxy-terminus where the sequence Cys^{186}-A-A-X (or CAAX motif, where C is cysteine, A is any aliphatic amino acid and X represent any amino acid) is present in all the RAS isoforms. Due to the heterogeneity in the sequence of these 25 last amino acids this region is known as hypervariable region (HVR).

RAS proteins need to be attached to the plasma membrane to become completely functional. Following translation on cytosolic ribosomes, RAS proteins undergo a number of post-translational modifications (Lowy and Willumsen 1989). The first of these modifications is the addition of a farnesyl isoprenoid lipid group on the cysteine of the CAAX motif by the enzyme farnesyl transferase. This signal targets RAS proteins to the endoplasmic reticulum (ER), where they undergo proteolytic cleavage of their -AAX motif by RAS-converting enzyme-1 (RCE1), followed by carboxymethylation of the new carboxy terminal cysteine residue (Choy et al. 1999; Dai et al. 1998; Hancock et al. 1989; Kim et al. 1999). Further processing takes place as the proteins transit the ER. Specifically, a second set of processing events occurs at other cysteine residues in the protein's C-terminus. In the case of HRAS and NRAS, cysteines at 184 and 186, respectively, are palmitoylated. These modifications serve as a second signal that directs further trafficking along the exocytic pathway. KRAS 4B does not undergo palmitoylation, but rather relies on a polybasic region in its HVR, consisting of a stretch of lysines, for proper trafficking and membrane anchorage (Hancock 2003). Fully processed H-RAS and N-RAS traffic via vesicles budded from the Golgi body to the plasma membrane. Thus, the differences found in the HVR of RAS proteins determine differences in the post-translational modifications and trafficking and localization in the plasma membrane that will result in differences in the biological activity determined by their co-localization with upstream regulators (such as tyrosine kinase receptors (RTKs) and G-protein coupled receptors (GPCRs)) or downstream effectors such as Raf or PI3K (Plowman and Hancock 2005; Hancock 2003; Ehrhardt et al. 2004; Hamilton and Wolfman 1998; Matallanas et al. 2006; Rocks et al. 2005; Voice et al. 1999).

As a result of the differences in the post-translational modification, palmitoylated and polybasic-targeted RAS proteins are directed to different subdomains of the plasma membrane. Thus, HRAS, but not KRAS 4B, is preferentially localized to cholesterol-rich microdomains (lipid rafts and caveolae), with signalling through palmitoylated HRAS, but not through KRAS, being sensitive to perturbations of plasma membrane cholesterol (Roy et al. 1999). The localization of RAS proteins with the different membrane subdomains is dynamic and depends on the activation state. Thus, for example HRAS in its active form redistributes from rafts into bulk plasma membrane by a mechanism requiring the adjacent HVR (Prior et al. 2001). This change in membrane domain localization is necessary for efficient activation of effectors. By contrast, KRAS is located outside rafts irrespective of bound nucleotide (Prior et al. 2001). Finally, NRAS, like HRAS, localizes into the lipid rafts in the plasma membrane, but it is never associated to caveolae (Matallanas et al. 2003).

RAS proteins cycle between two conformational states, one when they are in their active form – GTP bound – and another when they are in their inactive form – GDP bound (Bourne et al. 1990; Field et al. 1987; Satoh et al. 1987; Wittinghofer and Pai 1991). The structural changes are restricted to two motile regions named as switch I and switch II. Thus, RAS proteins alternate between an active state in which they can interact with effectors or GAPs through switch I and an inactive state in which switch I is inaccessible, but GEFs can interact with switch II (Hall et al. 2001; Herrmann 2003; Ma and Karplus 1997). This binary aspect enables RAS proteins to function as molecular switches in a broad range of signalling processes, often in the transduction of extracellular signals to the interior of cells.

Intrinsic rates of GTP hydrolysis and nucleotide exchange by RAS proteins are too slow to facilitate efficient GDP-GTP cycling for physiological reactions and accessory proteins exist that enhance and regulate GDP-GTP cycling. Guanine nucleotide exchange factors, or GEFs, promote formation of active, GTP-bound

RAS (Wolfman and Macara 1990), whereas GTPase activating proteins or GAPs stimulates the hydrolysis of GTP and return RAS proteins to their inactive state (Trahey and McCormick 1987). There are three families of GEFs, Sos, RASGRF and RASGRP, each one with multiple members and a number of GAPs, including p120GAP and NF1.

Though wild type RAS proteins play an important role in normal developmental processes, it was the oncogenic potential of constitutively activated mutant RAS proteins that led to their initial discovery. The study of the oncogenic contribution of RAS signalling has played a critical role in establishing a molecular basis for the pathogenesis of human cancer (Hirakawa and Ruley 1988; Land et al. 1983a, b; Ruley 1983), with activating mutations in members of the *RAS* family of genes among the most common genetic lesions in human tumours (Bos 1989). These mutations lock RAS proteins into a constitutively activated state in which they signal to downstream effectors even in the absence of extracellular stimuli. Although mutations in a number of codons have been found in *RAS* genes in human tumours, the greatest impairment of GTPase activity has been linked to mutations in residues 12, 13 and 61 (Barbacid 1987; Broach and Deschenes 1990) and mutations in these sites account for more than 99% of the changes seen in human cancer. The importance of RAS signalling in tumour initiation and maintenance is emphasized by the prevalence not only of *RAS* mutations, but also the deregulation of many of its activators or effector pathways, thus affecting RAS pathway activity. For example, loss of the tumour suppressor NF1, one of the RAS GAP proteins, results in elevated levels of RAS-GTP and is associated with syndromes characterized by benign and malignant neoplasias (Cichowski and Jacks 2001; Shannon et al. 1994; Side et al. 1997). Also, while the frequency of *KRAS* mutations in colorectal cancer is approximately 50%, oncogenic mutations of BRAF are also observed in approximately 12% of colorectal adenocarcinomas (Davies et al. 2002); activating mutations affecting the PI3K pathway are also observed in approximately 40% of colorectal cancers (Parsons et al. 2005).

2 RAS Effectors

Activated RAS stimulates a multitude of downstream signalling pathways (Fig. 1). Many of these effector pathways are essential for oncogenic cellular transformation (Repasky et al. 2004; Rojas and Santos 2002). During the past two decades a wide spectrum of proteins have been shown to specifically interact with RAS-GTP: RAF proteins, PI3K, RalGDS family, p120GAP, NF1, MEKK1, Rin1, AF-6, PKC-ς, Nore1, Canoe, etc. The interaction of these proteins with RAS is GTP-dependent through the effector loop region (Wittinghofer and Nassar 1996).

RAS family members, as well as their effectors, have unique intracellular localizations and overlap of these localizations likely restricts many of the interactions. In addition, many of the best-studied RAS effector molecules are members of large families. For example, the class I family of PI3K has p110α, p110β, p110δ,

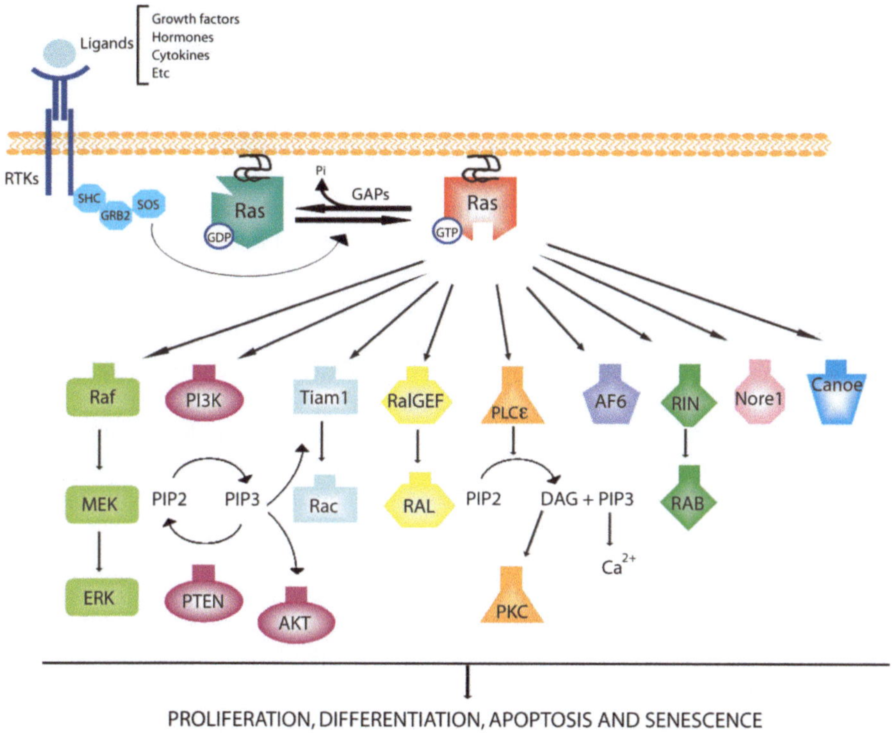

Fig. 1 *Ras effectors*. Once activated, Ras proteins signal through multiple effector pathways, thus activating many different signal transduction pathways. Most of these pathways are involved in proliferation, differentiation, apoptosis and senescence, although it is still unclear the contribution of each Ras-effector pathway to these functions. Among all the pathways that are known to be activated by Ras the best characterized are the MAPK (Raf-MEK-ERK) pathway, the PI3K pathway and the Ral pathway

and p110γ isoforms; the Raf kinase family comprises Raf1, A-Raf and B-Raf; and RalGEFs include RalGDS, RGL, RGL2/Rlf, and RGL3. RAS molecules might be able to regulate different members of the same family, and these selective interactions may have important biological consequences. It should be mentioned that the different RAS isoforms exhibit quantitative and qualitative differences in their ability to activate a particular effector, and that some RAS effectors also serve as effectors for other RAS-family proteins, and that not all isoforms within a class of effectors have been verified as *bona fide* RAS effectors (Rodriguez-Viciana et al. 2004).

All RAS effectors contain a RAS-binding domain (RBD). Three different classes of sequences have been identified, all of them containing ~100 amino acids: the RBD of Raf or Tiam1, the RBDs found in class I phosphoinositide 3-kinases (PI3K-RBD), and the RAS association (RA) domains found in the majority of other RAS effectors (Repasky et al. 2004). Although they do have only limited primary sequence identity, all three domains form the same topology of an ubiquitin superfold, characterized by a βαβαβ tertiary structure (Herrmann 2003). This common topology for otherwise divergent primary sequences accounts for the similar mode of interaction of RAS with effectors with different RBDs.

The first RAS-effector pathway to be identified was the RAF-MEK-ERK pathway (Moodie et al. 1993; Warne et al. 1993; Vojtek et al. 1993; Zhang et al. 1993). It appears to be an essential, shared element of mitogenic signalling involving tyrosine kinase receptors. The RAF family of proteins (Raf-1, A-Raf and B-Raf) are serine/threonine kinases that bind to the effector region of RAS-GTP, thus inducing translocation of the protein to the plasma membrane. Once there, RAF proteins are phosphorylated by protein kinases such as protein kinase C (PKC) and bind to 14-3-3 proteins (Fabian et al. 1994; Marais et al. 1995, 1998; Morrison 1994; Avruch et al. 2001). Active RAF phosphorylates MEK that, in turn, phosphorylates and activates extracellular signal regulated kinases 1 and 2 (ERK1/2) (Crews and Erikson 1993). The ERKs have numerous substrates including Rsk, Mnk, cPLA2 and PHAS-1 (Sturgill et al. 1988; Waskiewicz et al. 1997; Wang et al. 1998). ERK phosphorylation promotes its homodimerization and translocation to the nucleus, where they stimulate the activity of different transcription factors, such as p62/Elk-1 and Ets-2, by direct phosphorylation (Marais et al. 1993; Khokhlatchev et al. 1998; Treisman 1996). The importance of this effector pathway in normal and oncogenic RAS function is supported by (1) the ability of Raf and MEK to transform rodent fibroblasts (Rapp et al. 1983), (2) the largely mutually exclusive occurrence of BRAF and RAS mutations in tumours (Rojas and Santos 2002) and (3) the ability of RAF inhibitors to revert aspects of the RAS-driven transformed phenotype, such as growth in soft agar (Lyons et al. 2001).

PI3Ks are the next-best understood effectors of RAS and play important roles as mediators of RAS-mediated cell survival and proliferation (Vivanco and Sawyers 2002). When active, PI3K converts phosphatidylinositol (4, 5)-bisphosphate (PIP_2) into phosphatidylinositol (3, 4, 5)-trisphosphate (PIP_3). PIP_3, in turn, phosphorylates Akt/PKB, stimulating the catalytic activity of Akt, resulting in the phosphorylation of a host of other proteins that affect cell growth, cell cycle entry, and cell survival. PI3K can also activate Rac and this activation is involved in cytoskeleton reorganization (Cantley 2002). The functional importance of the RAS-PI3K pathway will be more extensively discussed in the next sections.

Another well-characterized RAS effector pathway involves RalGEF proteins. RalGEFs have been systematically found in yeast two-hybrid screens as RAS-binding proteins and they link RAS proteins to activation of the RalA and RalB small GTPases (D'Adamo et al. 1997). Four RalGEFs have been identified as effectors of RAS signalling: RalGDS, RGL, RGL2 (also known as Rlf) and RGL3. Although the biological function of these proteins is not yet fully understood, there is evidence that play an important role of this pathway in RAS-mediated transformation and tumorigenesis *in vivo* (Rodriguez-Viciana and McCormick 2005; Gonzalez-Garcia et al. 2005). Expression of RalGDS cooperates with constitutively activated RAF to induce focus formation, suggesting a cooperation of RalGEF in RAS-mediated transformation *in vitro* (White et al. 1996). The importance of RalA, but not RalB, in the contribution to transformation mediated by RAS in cancerous human cell lines with oncogenic RAS mutations has been recently reinforced by the fact that RalA silencing results in decrease of the tumoral behaviour of cancerous cell lines in xenografts and soft agar studies (Lim et al. 2005).

Apart from the above-mentioned effectors, in recent years an increasing number of molecules that specifically interact with RAS have been described, including Tiam1, p120GAP, NF1, MEKK1, Rin1, AF-6, PLC-ε, Nore1, Canoe, etc. These proteins bind to RAS exclusively when it is in an activated state (Ehrhardt et al. 2002). Aberrant RAS signalling causes deregulation of the Rho GTPases family, including Rac1 and RhoA, though no direct association has been found yet. It is known that abnormal Rho activation promotes tumoral invasion and metastasis (Sahai and Marshall 2002), so those effectors that facilitate RAS-mediated Rho activation might also facilitate the role of oncogenic RAS in tumour progression. An effector presenting this ability is Tiam1, a specific Rac GTPase that contains a RAS binding domain similar to that of Raf. The finding that knockout mice for Tiam1 have a normal development but are resistant to H-RAS-induced skin carcinogenesis demonstrate its importance in RAS-mediated tumour formation (Malliri et al. 2002). MEKK1, Rin1, AF-6, PLC-ε, Nore1 and Canoe have also been identified as RAS effectors, but their physiological function has not been yet elucidated; it seems unlikely that all of these proteins can be playing physiologically meaningful roles as direct downstream effectors of RAS. There are also some proteins that cooperate with RAS in the activation of effectors, such as KSR (Cacace et al. 1999), 14-3-3 (Roy et al. 1998), CNK (Therrien et al. 1999) and Sur8 (Selfors et al. 1998; Sieburth et al. 1998; Li et al. 2000).

RAS proteins mediate a great diversity of cellular responses, some requiring the activity of more that one effector pathway. For example, although oncogenic RAS promotes transformation and differentiation in many circumstances, in some others RAS inhibits cell growth or blocks differentiation (Shields et al. 2000). Depending on the biological system, RAS activity can be anti-apoptotic or promote apoptosis. In fact inhibition exerted by RAS in anoikis is dependent of PI3K but not Raf activation (Frisch et al. 1996; Khwaja et al. 1997). On the other hand, RAS induced senescence in primary fibroblasts seems to be dependent principally on RAF activation (Lin et al. 1998; Zhu and Parada 2001). In some other instances, synergistic cooperation of different RAS effectors is needed. Thus, inhibition of differentiation in myoblasts is dependent on RAF, RalGDS and some other effectors not yet identified. Also, regulation of G1 progression by RAS is partially mediated by RAF, RalGDS and PI3K (Takuwa and Takuwa 1997; Taylor and Shalloway 1996) through regulation of cyclin D1 expression and the transcriptional activity of E2F (Gille and Downward 1999).

Numerous studies have described the biological differences among the different members of RAS family in the regulation of the same effectors or that ascribe certain functions to one of the members of RAS family but not the others, pointing out the complexity of RAS-effector interaction. For example, HRAS and KRAS differentially interact with and activate Raf-1 and PI3K. In COS cells overexpression of oncogenic KRAS increases Raf-1 kinase activity and membrane translocation in a stronger manner than HRAS does. On the other hand, overexpression of activated HRAS is linked to a greater increase in the kinase activity of the catalytic subunit of PI3K (Yan et al. 1998). Similarly, in REF52 cells oncogenic KRAS is a more potent activator of Rac1 than HRAS (Walsh and Bar-Sagi 2001).

RAS signals from different locations in the cell, and this spatiotemporal regulation is important in the diversity and specificity of RAS signalling. Recently, it has been shown that, among different RAS effectors, PI3K was specifically implicated in the signalling in the endosome and that PI3K that is recruited to endosomal membranes depends for its activation on EGF and RAS activation (Tsutsumi et al. 2009).

3 RAS-PI3K Interaction

Activation by RAS is just one of the pathways contributing to PI3K activation. The initial step in the activation is the binding of ligand to receptor tyrosine kinases (RTKs). This causes dimerization of the receptor and autophosphorylation at tyrosine residue, which allows them to interact with Src homology 2 (SH2) domain-containing molecules (Pawson and Nash 2003; Schlessinger 2002), such as GRB2. GRB2 activates RAS through the activation of SOS (Son of Sevenless). RAS, in turn, activates p110 independently of p85. GRB2 also exists in a complex with GAB or other scaffolding proteins which interact with p85, bringing these activators into close proximity with p110 PI3K (Ong et al. 2001). There is evidence that RAS has to function in concert with phosphotyrosine-bound p85 to activate p110. For example, HRAS promotes the catalytic activity of PI3K only when p85 is bound to phosphorylated tyrosine residues (Chan et al. 2002). So, RAS-mediated PI3K activation in response to growth factors might require two steps: (1) phosphorylation of the RTK and, in some cases, adaptor proteins and (2) activation of RAS small GTPases. It is still unclear the contribution of each of the different PI3K activation pathways in different physiological situations.

The first indication that RAS proteins interact with PI3K came from the observation that some phosphatidylinositol (3) kinase activity could be found in RAS immunoprecipitates from RAS transformed cells (Sjolander et al. 1991). Soon after that, Pablo Rodriguez-Viciana described that RAS immobilized on agarose beads could bind to p110α/p85 when RAS was in its active GTP-bound form and demonstrated that the level of PI3K products generated *in vivo* could be increased or decreased depending on whether activated or dominant negative mutant RAS proteins are expressed. He also described that the interaction was direct and did not involve other proteins, suggesting a role for PI3K as an effector of RAS (Rodriguez-Viciana et al. 1994). Similar results were subsequently found using the fission yeast *Schizosaccharomyces pombe* as a model (Kodaki et al. 1994). Later on it was shown that for the interaction of RAS with PI3K, lysine residue 227 is essential and that RAS activates the Rho family GTPase Rac in a PI3K-dependent manner that is required for efficient transformation of fibroblasts by RAS (Rodriguez-Viciana et al. 1996).

The interaction of RAS with the different isoforms of class I PI3Ks has been described (Rodriguez-Viciana et al. 1994, 1996; Rubio et al. 1997; Vanhaesebroeck et al. 1997). For example it was found that active RAS also binds to and activates p110γ catalytic subunit provoking an approximately 8- to 20-fold increase in PI3Kγ activity. This activation is strictly dependent of RAS-GTP and is at least as great as the stimulation of the p110α/p85 heterodimer by RAS (Rodriguez-Viciana et al. 1996; Pacold et al. 2000). The activating interaction of PI3Kγ with RAS is limited by the dissociation of the RAS effector complex and not by the catalytic cycle of the effector enzyme. The transience of the complex ensures that PI3Kγ is not constitutively activated.

Initial crystallization trials of RAS-PI3K interaction were unsuccessful but in 2000 Roger William's group described the crystal structure of the RAS-PI3Kγ complex (Pacold et al. 2000; Walker et al. 1999). Interaction between RAS and PI3K is transient and reversible. The structure of the RBD in PI3K comprises a five-stranded mixed β-sheets (Rβ1-Rβ5), flanked by two α-helices (Rα1–Rα2). All RAS-RBD complexes use a similar model of RAS-effector interaction in which a β sheet in the RAS and a β sheet in the RBD are aligned to form a single β sheet connecting the two proteins. Contacts between the switch I region of RAS and the RBD stabilize the interaction and ensure its dependence on RAS-GTP. PI3K, RAF and RalGDS interact with many of the same switch I residues of RAS (Nassar et al. 1995; Huang et al. 1998). In the structure of the RAS-PI3Kγ complex contacts are primarily made via the switch I region of RAS and the RBD of PI3Kγ. However, a unique feature of the RAS-PI3Kγ complex is that the helix Rα1 and the Rα1–Rβ3 loop are significantly longer and, as a consequence, it causes a significant rotation of RAS relative to the RBD in PI3Kγ resulting in essential intermolecular contacts with the RBD, thereby revealing the structural specificity of the RAS-PI3Kγ interaction (Pacold et al. 2000; Djordjevic and Driscoll 2002). Uniquely, for RAS interactions with effectors, there is contact between RAS and the catalytic domain of PI3Kγ, with R73 in the switch II region of RAS contacting E919 in the C-terminal lobe of the catalytic domain of PI3Kγ. This may contribute to the allosteric regulation of PI3K lipid kinase activity by RAS.

The nature of the catalytic activation of PI3K by RAS has been a theme of debate, as it is difficult to distinguish between the contribution of membrane localization and RAS-induced allosteric lipid-kinase activation.

4 RAS-PI3K in Normal Signalling

PI3Ks are recognized as one of the principal effector families of RAS signalling. A major activity of these lipids kinases is the conversion of phosphatidylinositol (4, 5)-bisphosphate (PI(4, 5)P_2) and phosphatidylinositol (4)-phosphate (PI(4)P) to phosphatidylinositol (3, 4, 5)-trisphosphate (PIP_3) and phosphatidylinositol (3, 4)-bisphosphate (PI(3, 4)P_2), respectively. Signalling proteins with pleckstrin-homology (PH) domains accumulate at sites of PI3K activation by directly binding to PIP_3. The lipid products act in different pathways controlling mitogenic signalling, inhibition of apoptosis, intracellular vesicle trafficking and secretion, regulation of actin and integrin functions and metabolic changes, often through two different protein kinases: Akt (also called protein kinase B or PKB) and p70 ribosomal protein S6 kinases (p70S6K). The lipid second-messenger molecules produced (PIP_3 and PI (3, 4)P_2) activate the phosphoinositide-dependent kinases PDK1 and PDK2, which then activate Akt/PKB and isoforms of PKC (Rojas and Santos 2002; Cantley 2002; Shields et al. 2000). The p70S6K route participates in protein biosynthesis, whereas Akt controls signalling pathways regulating different cellular processes such as cell survival, glucose uptake and glycogen metabolism (Rojas and Santos 2002). PI3K can also activate the Rac GTPase, and this Rho family protein is an important mediator of oncogenic RAS transformation. Indeed, RAS and PI3K activities are necessary for Rac activation through the exchangers Sos and Vav (Nimnual et al. 1998; Han et al. 1998). The tumour suppressor PTEN antagonizes PI3K activity via its intrinsic lipid phosphatase activity, which reduces the cellular pool of PIP_3 by converting it back to PI(4,5)P_2 (Maehama and Dixon 1998; Maehama et al. 2001; Wishart and Dixon 2002).

It is well established that PI3Ks are one of the principal effectors of RAS signalling to the cell-cycle control machinery (Jones et al. 1999; Stacey and Kazlauskas 2002). RAS activity is needed throughout all the different phases of cell cycle. In quiescent cells stimulated with growth factors, PI3K is activated twice as cells transition from G0 into G1 phase (Jones et al. 1999; Auger et al. 1989; Whiteford et al. 1996), and then later in G1 phase. But the use of PI3K inhibitors has demonstrated that it is only during the later stages of G1 phase that PI3K activity promotes entry into S-phase (Jones et al. 1999).

RAS-PI3K activation is linked to pro-survival signalling due to the activation of Akt. The importance of PI3K/Akt signalling to survival has been shown repeatedly by both the ability of activated PI3K or Akt to abrogate apoptosis and the ability of a dominant-negative Akt to enhance it (Vivanco and Sawyers 2002; Cox and Der 2003; Downward 1998, 2004). The mechanism by which AKT protects cells from death could be considered as multifactorial because Akt phosphorylates a number of substrates important for the regulation of apoptosis. Akt phosphorylation of Bad, a pro-apoptotic member of the Bcl-2 family of apoptotic regulators, causes Bad to bind preferentially to 14-3-3 in an inactive complex, thereby preventing it from sequestering and inactivating the anti-apoptotic proteins Bcl-2 and Bcl-X_L (Cox and Der 2003; Gire et al. 2000; Datta et al. 1997). Also, Akt and Rac facilitate RAS activation of the NF-κB transcription factor (Irani et al. 1997; Romashkova and Makarov 1999; Ozes et al. 1999), which serves an anti-apoptotic role in RAS function (Mayo et al. 1997). Moreover, PI3K is absolutely required for the proliferative response to RAS in human thyroid epithelial cells, acting via suppression of RAS induced apoptosis (Gire et al. 2000). Akt also phosphorylates the FOXO family of transcription factors (Brunet et al. 1999). Thus, by increasing the ability for growth and by decreasing the capacity for apoptosis, PI3K signalling supports tumorigenesis.

Most of the initial knowledge about the interaction between mammalian PI3K and RAS has been acquired using purified or overexpressed proteins *in vitro* and in cultured cells. Prober and Edgar first demonstrated a genetic link between RAS and PI3K in *Drosophila* (Prober and Edgar 2002). Orme et al. subsequently showed that direct interaction of PI3K with RAS was critical in this setting (Orme et al. 2006). They created transgenic flies carrying a Dp110 (the only *Drosophila* p110) with mutations in the RBD that did not interact with RAS but was otherwise biochemically normal. They found that RAS-mediated Dp110 regulation is dispensable for viability, although the survivor flies were smaller and females exhibited reduced fecundity, laying 60% fewer eggs. Dp110 mutant flies also had decreased activation of Akt in response to insulin in the brain and imaginal discs and decreased basal Akt activation in the ovaries. This work suggested that different developmental contexts might require different levels of signalling and, more importantly, it provided the first demonstration that the interaction of a PI3K with RAS is essential in specific developmental contexts requiring maximal PI3K signalling.

In mammals, the first report indicating the *in vivo* function of the RAS-PI3K interaction was done using p110γ in neutrophils (Suire et al. 2006). p110γ is stimulated by G protein-coupled receptors and by RAS. Suire et al. generated a mouse model in which they mutated the RBD of p110γ ($p110\gamma^{DASAA}$); the resulting mice were viable, fertile, of normal size and with generally normal blood counts. In isolated neutrophils, production of $PI(3, 4)P_2$ and PIP_3, activation of Akt and chemotaxis were decreased compared with wild-type cells or cells from p110γ heterozygous mice. The production of reactive oxygen species in response to agonist stimulation was decreased in p110γ-RBD mutant mice, revealing an important role for the interaction in the normal function of neutrophils.

Further evidence of the importance of the RAS-PI3K interaction *in vivo* was provided by the generation in our lab of a knock-in mouse having two point mutations in the RBD of p110α (T208D and K227A) that completely disrupt the interaction between RAS and the catalytic subunit p110α, but do not affect the enzymatic activity of p110α (Gupta et al. 2007). In cultured mouse embryo fibroblasts, loss of p110α binding to RAS strongly reduces Akt activation in response to certain growth factors such as EGF or FGF2. The number of homozygous mice born was significantly reduced, and they were smaller in size that their wild-type counterparts. Furthermore, they exhibited defective branching and development of the lymphatic vasculature system. This defect was linked with the development of chylous ascites in the newborn mice and is similar to developmental anomalies seen in mice with defective VEGF-C signalling (Karkkainen et al. 2004).

5 RAS-PI3K in Oncogenesis

Though the importance of RAS proteins in developmental processes is generally appreciated, it is the oncogenic potential of constitutively activated RAS proteins that has attracted most attention. The implication of PI3K as an important target in RAS-dependent transformation was established soon after the discovery of the interaction of these two molecules (Rodriguez-Viciana et al. 1996, 1997; Sheng et al. 2001; Li et al. 2004).

RAS signalling pathways are activated in tumorigenesis by a number of different mechanisms, such as mutations in RAS, loss of GAP proteins, overexpression of RTKs (EGFR, ERBB2), and also by mutations or amplifications in any of their effector molecules (Vivanco and Sawyers 2002; Hennessy et al. 2005; Shaw and Cantley 2006; Jia et al. 2009). In recent years it has become evident that many human tumours harbour somatic missense mutations in *PIK3CA* at high frequency, including tumours of the brain, breast, colon, liver, stomach, lung and ovary (Cancer Genome Atlas Research Network 2008; Parsons et al. 2008; Samuels et al. 2004; Simi et al. 2008; Stemke-Hale et al. 2008; Thomas et al. 2007; Yamamoto et al. 2008). These are point mutations that are usually clustered within three hotspots in helical and kinase domains: E542K, E545K and H1047R. It has been shown in different systems that these are oncogenic gain-of-functions mutations (Isakoff et al. 2005; Zhao et al. 2005; Ikenoue et al. 2005; Kang et al. 2005; Samuels et al. 2005). Recently, it has been shown that helical and kinase domain mutations trigger gain of function through different mechanisms, showing a differential requirement for interaction with the regulatory subunit p85 and with RAS-GTP. Gain of function by E542K and E545K mutations is highly dependent on RAS binding, whereas H1047R mutations are dependent on allosteric change mediated by p85 and this allosteric change mimic RAS-GTP binding, making this mutation independent of interaction with active RAS (Zhao and Vogt 2008a, b).

Data from different studies suggest that RAS needs p110α and, in some cases, p110β interaction to exert its oncogenic role. Thus, MEFs lacking p110α are resistant to oncogenic RAS transformation (Zhao et al. 2006) and in some human breast cancer cell lines that express an oncogenic HRAS but carry wild-type PIK3CA and PTEN, inhibitors of p110α, but not p110β, block PI3K signalling, without affecting cell proliferation (Torbett et al. 2008). In some other cases, RAS-p110β interaction has been shown to be essential for RAS driven tumorigenesis, since ablation of this isoform completely abrogates focus formation induced by oncogenic HRAS (Jia et al. 2008).

One of the clearest examples showing the importance of the RAS-PI3K interaction in RAS-driven tumorigenesis was provided by an RBD-p110α mutant mouse model generated in our lab. By using two well-characterized mouse models of endogenous RAS-driven tumourigenesis, it was shown that p110α is a critical effector for RAS-driven tumorigenesis, at least in two tissues. First, the K-RAS LA2 mice (Johnson et al. 2001) were crossed with mice homozygous for mutant *PIK3CA* and in this model an impressive 95% reduction in lung tumour formation was found. Second, when *PIK3CA* mutant mice were subjected to DMBA-induced carcinogenesis, a near-complete reduction in H-RAS driven tumour formation was seen (Gupta et al. 2007).

Recently, it has been shown that tumours with mutant p110α often coexist with mutations in RAS in colorectal and endometrial cancer (Parsons et al. 2005; Samuels et al. 2005; Oda et al. 2008; Velho et al. 2005). A possible explanation for this might be that additional mutations strengthen PI3K pathway signalling caused by oncogenic RAS and might activate other pathways, thus enhancing the oncogenic transformation. The use of inhibitors and shRNA studies have shown that signalling to downstream molecules such as AKT is mediated by p110α in breast, colon and endometrial cancers with PIK3CA and coexisting RAS mutations (Oda et al. 2008). In agreement, in HCT116 and DLD1 colon cancer cell lines harbouring mutant p110α and K-RAS, somatic cell knockout of p110α inhibits downstream PI3K signalling, cell growth and transformation, and maintenance of tumour xenografts (Samuels et al. 2005).

Tumour cells become addicted to the expression of initiating oncogenes, such that loss of oncogene expression in established tumours leads to tumour regression (Giuriato et al. 2004). It is well accepted that oncogenic RAS promotes both initiation of tumorigenesis and maintenance of tumour growth (Chin et al. 1999). Given the sizable array of RAS effectors, an important subject to address is which of these effectors is essential for RAS tumour maintenance. In this regard, work from Counter's group showed that PI3K signalling is indispensable to maintain transformed growth in RAS mutant cell lines both *in vitro* and in xenografts in mice. They showed that, although multiple RAS effectors are essential to initiate tumour formation, only signalling through the PI3K/Akt pathway is necessary to maintain tumour growth (Lim and Counter 2005). The reduction of RAS oncogene dependence to activation of PI3K/Akt pathway appears to be a consequence of redundant signalling provided by the established tumour microenvironment. It has also been described that the requirement for PI3K signalling during tumour maintenance in RAS-driven tumour growth might be due, at least in part, to the Akt-dependent activation of eNOS (by phosphorylation in S1177). This, in turn, leads to *S*-nitrosylation of normal RAS proteins, activating them perhaps as a means to diversify the signal beyond that of oncogenic RAS (Lim et al. 2008).

While the data above suggest that PI3K signalling is essential for RAS-driven tumour maintenance, other recent evidence has suggested that while PI3K may be required for KRAS induced tumorigenesis, inhibition of PI3K alone is not sufficient to cause regression of these tumours once established. When mice with genetic deletion of the p85 PI3K regulatory subunits were crossed with a tetracycline-inducible *K-rasG12D* transgenic model (Fisher et al. 2001) or with the *LSL K-RAS* model (Jackson et al. 2001), the expected induction of lung tumour formation was reduced relative to a PI3K wild type background. However, the lung tumours that developed in response to KRAS activation in mice with normal PI3K signalling failed to shrink when these animals were treated with NVP-BEZ235, a dual pan-PI3K/mTOR inhibitor, even though tumours driven by expression of activated PI3K did regress with this drug (Engelman et al. 2008). Interestingly, combined treatment of these tumours with the PI3K inhibitor and a MEK inhibitor led to impressive regression, suggesting that a combination of PI3K and MEK inhibitors may have potential for the treatment of RAS mutant tumours in the clinic, provided, of course, that toxicities are not limiting.

Much of our knowledge of human cell transformation derives from studies undertaken in rodent cells. However, it has been described that there exist fundamental differences in the behaviour of rodent and human cells, especially in certain biological features relevant to transformation (Rangarajan and Weinberg 2003; Rangarajan et al. 2004; Hamad et al. 2002). In this regard, RAS-dependent transformation presents differences in the effector pathway needed for effective transformation not only between mouse and human cell lines but also when comparing human cell lines originating from different tissue types (Rangarajan et al. 2004). While the *ab initio* creation of tumour cell lines from normal cells *in vitro* that is used in these studies has provided many insights into the process of tumorigenesis, it is still unclear exactly how closely this correlates with the processes occurring during the normal pathological process of tumour formation *in vivo*.

An important feature in human cancer is formation of metastasis since it is the main cause of death in cancer patients. Metastasis establishment is a sequential and interrelated multistep process including dissociation of tumoral cells in the primary tumour, invasion of extracellular matrix, angiogenesis, intravasation into the vasculature or lymphatic systems, survival in these places, extravasation and proliferation at a distant site (Steeg 2006; Chambers et al. 2002; Duffy et al. 2008; Nguyen et al. 2009). Oncogenic activation of RAS has been implicated in facilitating almost all aspects of malignant phenotype (Giehl 2005; Campbell and Der 2004). However, although there is no doubt about the involvement of RAS in human cancer, much less is known regarding the specific role it plays in tumour invasion and metastasis or the main effector pathway through RAS contribute to metastasis formation.

One of the first requirements in metastasis formation is the acquisition of an increased migratory phenotype, which is accompanied by extensive remodelling of the actin cytoskeleton. This is in part regulated by Rho family GTPases (including RhoA, Rac1 and Cdc42) which have also been implicated as important components of RAS transformation (Zohn et al. 1998). It has been suggested that PI3K is involved in tumour cell motility and invasion mainly through the regulation of RhoGTPases (Sahai and Marshall 2002; Etienne-Manneville and Hall 2002). RAS can cause Rac activation via PI3K and actin rearrangement correlates with the ability of RAS mutants to activate PI3K. Inhibition of PI3K activity blocks RAS induction of membrane ruffling, while activated PI3K is sufficient to induce membrane ruffling, acting through Rac. The ability of activated RAS to stimulate PI3K in addition to RAF is therefore important in RAS transformation of mammalian cells and essential in RAS-induced cytoskeletal reorganization (Rodriguez-Viciana et al. 1997). How RAS regulates RhoA and Cdc42 function is still not clearly understood. Rho GTPases can also regulate epithelial cell morphogenesis (Van Aelst and Symons 2002). It has been suggested that in cancer cells there is a feedback loop between Rho proteins and PI3K such that when epithelial-mesenchymal transition occurs, PI3K can induce the activation of Rho, Rac and Cdc42, which in turn lead to the generation of PIP_3 (Benard et al. 1999; Genot et al. 2000; Weiner et al. 2002; Cozzolino et al. 2003). Whether RAS might mediate this process remains unclear. PI3K can also activate Rac GEFs such as Sos or Vav to promote activation of Rac (Nimnual et al. 1998; Han et al. 1998). Rac regulation of actin reorganization and membrane ruffling can promote increased cell motility and contribute to tumour cell invasion and metastasis (Etienne-Manneville and Hall 2002). In addition to work in mammalian cells, many insights into the role of RAS and PI3K in cell motility, and especially chemotaxis, have also come from the study of the slime mould *Dictyostelium discoideum* (Sasaki and Firtel 2006).

An interesting connexion between PI3K signalling and RAS-induced malignant transformation was given by the fact that small GTPase RhoB is a potent suppressor of transformation, migration and invasion of different cancer cells and NIH3T3 fibroblasts (Jiang et al. 2004). Expression of RhoB is down-regulated by oncogenic HRAS dependent on PI3K but not RAF signalling pathways. In contrast, ectopic expression of RhoB antagonizes RAS-PI3K-dependent malignant transformation and apoptosis.

PI3K signalling controls changes in the stability of cell-cell adhesion and cell-ECM interaction with important consequences for the regulation of cell motility. For example, in MDA-MB-435 breast carcinoma cells, upregulation of PI3K via $\alpha 6\beta 4$ integrin signalling leads to increased invasion that is Rac-dependent but MAPK and p70S6K independent (Shaw et al. 1997; Burgering and Coffer 1995). Since $\beta 4$ integrin contains Shc motifs that can recruit Grb2 and Sos, it can be speculated that this increase in migration might occur via RAS activation. Also, in uveal melanoma, the PI3K pathway is an important mediator of increased motility by downregulation of the cell adhesion molecules E-cadherin and β-catenin, thus attenuating cell-cell adhesion and promoting the enhanced motility and migration typically seen in these cancer cells (Ye et al. 2008). It has been demonstrated that PI3K activation can enhance invasion by regulating expression of different matrix metalloproteases such as MMP-9 in HT1080 cells (Kim et al. 2001) or MMP-2 in mouse mammary epithelial cells (Park et al. 2001). However, whether these mechanisms are directly related to RAS regulation of PI3K is not clear.

The capacity of tumour cells to form metastases requires their ability to escape matrix deprivation-induced apoptosis or anoikis (Chiarugi and Giannoni 2008; Zhan et al. 2004; Simpson et al. 2008). Oncogenic RAS and PI3K can promote the loss of anchorage-dependent growth. In MDCK epithelial cells the PI3K pathway, but not RAF pathway, is both necessary and sufficient for the protection provided by RAS from anoikis (Frisch et al. 1996; Khwaja et al. 1997). This inhibition of anoikis gives the detached tumour cell the possibility to migrate to a conduit by which it can reach distal tissues.

Although much work has been accomplished in discovering the contribution of oncogenic RAS to increased motility, invasiveness and metastatic potential, mechanisms downstream of RAS are much more complex that originally thought. There are still many facets of the contribution of PI3K to oncogenic RAS invasiveness and metastatic potential that remain unanswered. Appropriate knowledge of this contribution might be critical for the development of proper therapy against metastasis.

6 RAS-PI3K-RAF Pathway Interconnections

It has been shown that PI3K and RAF pathways can interact in multiple ways, both in normal and transforming conditions (Rodriguez-Viciana et al. 1997; Chang et al. 2003) (Fig. 2). Cross-talk regulation between RAF-MAPK and PI3K-AKT

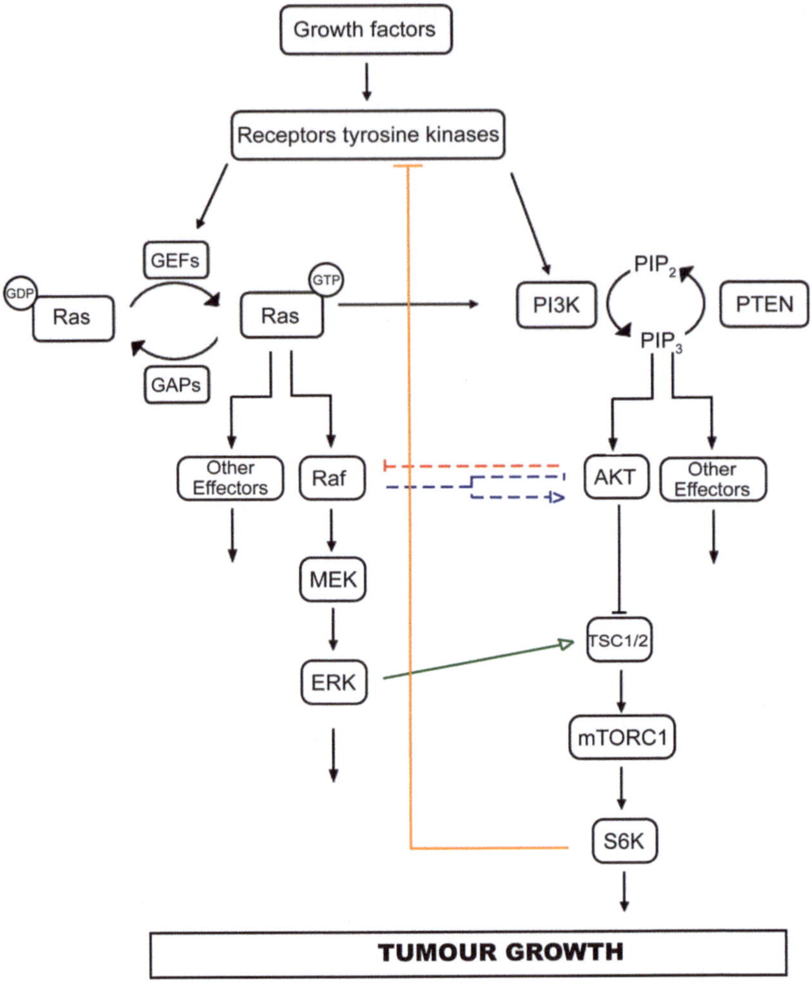

Fig. 2 *PI3K and MAPK signalling pathways are interconnected.* PI3K pathway becomes activated by RAS and also directly by RTKs. Once activated the signal is transduced to different downstream effectors. There are several connexions between the MAPK and the PI3K pathways, leading to either activation or inhibition at different levels. Raf can be inhibited by PI3K in different situations such as muscle differentiation, in case of some breast tumour cell lines and in vascular smooth muscle and intestinal epithelial cell differentiation. PI3K pathway can also be activated or inhibited by Raf under certain conditions. Another level of connexion of these two pathways involves mTORC1. When mTORC1 is inhibited a feedback signal through RTKs is initiated that activates MAPK in a RAS-dependent manner, thus provoking ERK and AKT phosphorylation and activation of both pathways. Abnormal activation of both pathways leads to tumour growth

pathways was first observed during muscle cell differentiation (Rommel et al. 1999; Zimmermann and Moelling 1999). During this process both pathways exert opposing effects. PI3K-Akt inhibits the RAF pathway in muscle cell hypertrophy and this cross-regulation depends on the differentiation state of the cell: Akt activation inhibits the RAF-MEK-Erk pathway in differentiated myotubes, but not in their myoblast precursors (Rommel et al. 1999). Interestingly, Akt interacts with, and phosphorylates, RAF protein at a highly conserved serine residue (Ser259) of its regulatory domain. This phosphorylation of RAF by Akt inhibits activation of the RAF signalling pathway and shifts the cellular response in a human breast cancer cell line from cell cycle arrest to proliferation, thus providing a molecular basis for the cross-talk between these two signalling pathways at the level of RAF and Akt (Zimmermann and Moelling 1999). PI3K inhibition of Raf-MAPK signalling is also required for vascular smooth muscle and intestinal epithelial cell differentiation (Reusch et al. 2001; Laprise et al. 2004). The PI3K inhibitor LY294002 and the downstream p70S6K inhibitor rapamycin are potent inhibitors of Raf-mediated proliferation, suggesting that the PI3K/Akt/p70S6K pathway mediates some effects of RAF on cell growth (Chang et al. 2003). It has also been established a cross-talk regulation where RAF-MAPK signalling can inhibit PI3K activation and such inhibition provokes cellular arrest. Constitutive RAF-MEK1 signalling leads to negative feedback inhibition of RAS and PI3K through the Eprhin receptor $EphA_2$, and this event is required for cellular arrest (Menges and McCance 2008). However, RAF-MAPK signalling can be a potent inducer of both RAS and PI3K activation, especially in epithelial cells, through the autocrine production of growth factors such as HB-EGF (Schulze et al. 2001). Overall, the cross talk between the PI3K and RAF pathways is made up of a complex network of events, some of which are likely to be more significant than others under physiological signalling conditions.

Recently, it has been shown that mTOR (mammalian target of rapamycin) is also connecting these two pathways in a RAS-dependent manner (Fig. 2) (Carracedo et al. 2008). mTOR signals through two different complexes, mTORC1 (containing RAPTOR) and mTORC2 (containing RICTOR). Both PI3K and MAPK signalling pathways regulate mTORC1 activity through phosphorylation of the tuberous sclerosis complex 2 (TSC2) (Ballif et al. 2005; Inoki et al. 2002; Ma et al. 2005; Manning et al. 2002). A negative feedback loop has been described in which mTORC1 acting via p70S6K phosphorylates the adaptor protein IRS1, leading to inhibition of insulin signalling to PI3K and Akt (Zhang et al. 2007; Shah et al. 2004; O'Reilly et al. 2006). Pandolfi's group has recently described that inhibition of mTORC1 by RAD001, a derivative of rapamycin, in patients with metastatic disease leads to an activation of the MAPK pathway. This feedback signalling is also present in several cancer cell lines and primary cultures after rapamycin treatment as well as in a mouse model of prostate cancer driven by PTEN inactivation treated with RAD001. They proposed that in all these cases mTORC1 inhibition increases RTKs activity toward RAS/MAPK, thus promoting AKT and ERK activation in a dual feedback mechanism (Carracedo et al. 2008). On the other hand, it has also been described that in human pancreatic duct epithelial cells inactivation of the MAPK pathway using different inhibitors has as a consequence a decrease in Akt phosphorylation (Campbell et al. 2007), suggesting that the interaction between Raf/MEK/ERK and PI3K/Akt pathways is important in the RAS-mediated transformation in human cells.

7 Inhibitors

The fact that RAS is one of the most mutated oncogenes in human cancer makes RAS proteins and the signalling pathways that they regulate attractive targets for drug development. By far the greatest pharmaceutical effort made to abolish RAS oncogenic signalling was the development of drugs targeting farnesyltransferases, which were thought to be essential for RAS's binding to membranes and hence its biological activity. Although many drugs were successfully developed that inhibited the farnesylation process, they failed in human patients studies principally due to the ability of KRAS and NRAS to use an alternative modification enzyme (geranylgeranyltransferease) to localize to membranes (Downward 2003).

A more effective approach for targeting RAS signalling pathways has proved to be the inhibition of RTK function (Garcia-Echeverria 2009). However the use of these compounds is restricted to patients presenting with oncogenic RTKs signalling through normal RAS proteins, while patients with tumours expressing an oncogenic mutant form of RAS do not benefit from such compounds since oncogenic RAS acts downstream to circumvent the need for an oncogenic RTK to induce cell proliferation and survival (Ramjaun and Downward 2007).

More recent therapeutic attempts have focused on the inhibition of PI3Ks. The fact that PI3K activation might have an important role during tumour maintenance highlights the importance of this pathway as an anticancer target (Lim and Counter 2005; Lim et al. 2008). Since cancers are treated at the tumour maintenance stage, targeting this aspect of RAS oncogenesis may hold promise as an approach to the management of RAS-driven-human cancers. In this way, the mouse model generated in our lab in which the RAS-PI3K interaction has been disrupted by mutations in the RBD of p110α (Gupta et al. 2007), provides helpful insight into the benefits that might be achieved from therapeutic compounds developed to target the RAS-PI3K interaction in human tumours. The development of such a drug might improve the condition of cancer patients in which the oncogenicity of RAS is exerted via PI3K pathway and, as a consequence of the specificity of the interaction targeted, might minimize the side effects in the patients, one of the main prerequisite for all novel therapeutic compounds. It would also overcome the feedback loops that exist between MAPK and PI3K pathways.

It might also be considered that p110α and p110β are mutated in different types of cancer. Although the role of oncogenic RAS signalling through p110β is not very well established, both p110α and p110β might be useful drug targets in particular tumour types bearing mutant RAS proteins. Hence, compounds targeting the specific isoform might be a good option in tumours featuring these lesions (Jia et al. 2009).

Given the coordinate regulation of PI3K and RAF/MEK/MAPK signalling pathways by RAS, the approach of using a combination of MAPK pathway (either MEK, pan-RAF or BRAF) and PI3K inhibitors in the treatment of cancer is an attractive one. As well as recent data discussed above favouring the use of a combination of PI3K and MEK inhibition to hit established RAS-induced tumours (Engelman et al. 2008), it has also been shown that activation of the PI3K pathway, either by *PIK3CA* mutations or PTEN loss, is a major resistance mechanism that impairs the efficiency of MEK inhibitors in KRAS mutated cancers. Downregulation of p110α resensitizes tumours with *KRAS* and *PIK3CA* mutations to MEK pathway inhibition both *in vitro* and *in vivo* (Wee et al. 2009). This might be an explanation of why RAS mutant cancers exhibit a variable response to MEK inhibitors (Solit et al. 2006). Similar results regarding to the clinical benefits of downregulation of both pathways have been shown in a mouse model of prostate cancer (Kinkade et al. 2008) and in thyroid tumours cells (Miller et al. 2009), and in haematopoietic cell lines it has been proven that the combined inhibition of MEK and PI3K pathways provokes a stronger effect in apoptosis induction and growth inhibition than single pathway inhibition (Shelton et al. 2003). Taken together these data provide a strong rationale for the combination of PI3K and MEK inhibitors in cancers having an oncogenic *KRAS* mutation.

8 Conclusions

The direct link between RAS and PI3K has now been investigated in great detail in a variety of systems and found to play important roles in both normal and oncogenic signalling. The ability of RAS to control several enzyme systems undoubtedly contributes to the significant difficulty of effectively targeting tumour cells with activating RAS mutations. While the impact of activated RAS on cells is strengthened by its ability to regulate PI3K, the simultaneous occurrence of *PIK3CA* mutations in some tumour types, such as colon, suggests that further advantage can be gained by the tumour cell by the acquisition of lesions that activate the PI3K pathway more strongly. In the immediate future, the most attractive prospect for the successful targeting of RAS mutant tumours in the clinic must lie in the exploitation of combinations of PI3K and MEK or RAF inhibitors. However, given the complexity of RAS signalling, which clearly involves much more than just activation of RAF and PI3K, in the long term a direct pharmacological assault on RAS itself is likely to be unavoidable if we are to make significant progress against the major killer tumour types where RAS mutations are prevalent, in particular pancreas, lung and colon. Developing novel strategies for this must be one of the biggest challenges for cancer research in the second decade of the twenty-first century.

References

Auger KR et al (1989) PDGF-dependent tyrosine phosphorylation stimulates production of novel polyphosphoinositides in intact cells. Cell 57:167–175

Avruch J et al (2001) Ras activation of the Raf kinase: tyrosine kinase recruitment of the MAP kinase cascade. Recent Prog Horm Res 56:127–155

Ballif BA et al (2005) Quantitative phosphorylation profiling of the ERK/p90 ribosomal S6 kinase-signaling cassette and its targets, the tuberous sclerosis tumor suppressors. Proc Natl Acad Sci USA 102:667–672

Barbacid M (1987) ras genes. Annu Rev Biochem 56:779–827

Bar-Sagi D (2001) A Ras by any other name. Mol Cell Biol 21:1441–1443

Benard V et al (1999) Characterization of rac and cdc42 activation in chemoattractant-stimulated human neutrophils using a novel assay for active GTPases. J Biol Chem 274:13198–13204

Bos JL (1989) ras oncogenes in human cancer: a review. Cancer Res 49:4682–4689

Bourne HR et al (1990) The GTPase superfamily: a conserved switch for diverse cell functions. Nature 348:125–132

Broach JR, Deschenes RJ (1990) The function of ras genes in Saccharomyces cerevisiae. Adv Cancer Res 54:79–139

Brunet A et al (1999) Akt promotes cell survival by phosphorylating and inhibiting a Forkhead transcription factor. Cell 96:857–868

Burgering BM, Coffer PJ (1995) Protein kinase B (c-Akt) in phosphatidylinositol-3-OH kinase signal transduction. Nature 376:599–602

Cacace AM et al (1999) Identification of constitutive and ras-inducible phosphorylation sites of KSR: implications for 14-3-3 binding, mitogen-activated protein kinase binding, and KSR overexpression. Mol Cell Biol 19:229–240

Campbell PM, Der CJ (2004) Oncogenic Ras and its role in tumor cell invasion and metastasis. Semin Cancer Biol 14:105–114

Campbell PM et al (2007) K-Ras promotes growth transformation and invasion of immortalized human pancreatic cells by Raf and phosphatidylinositol 3-kinase signaling. Cancer Res 67:2098–2106

Cancer Genome Atlas Research Network (2008) Comprehensive genomic characterization defines human glioblastoma genes and core pathways. Nature 455:1061–1068

Cantley LC (2002) The phosphoinositide 3-kinase pathway. Science 296:1655–1657

Carracedo A et al (2008) Inhibition of mTORC1 leads to MAPK pathway activation through a PI3K-dependent feedback loop in human cancer. J Clin Invest 118:3065–3074

Chambers AF et al (2002) Dissemination and growth of cancer cells in metastatic sites. Nat Rev Cancer 2:563–572

Chan TO et al (2002) Small GTPases and tyrosine kinases coregulate a molecular switch in the phosphoinositide 3-kinase regulatory subunit. Cancer Cell 1:181–191

Chang F et al (2003) Involvement of PI3K/Akt pathway in cell cycle progression, apoptosis, and neoplastic transformation: a target for cancer chemotherapy. Leukemia 17:590–603

Chiarugi P, Giannoni E (2008) Anoikis: a necessary death program for anchorage-dependent cells. Biochem Pharmacol 76:1352–1364

Chin L et al (1999) Essential role for oncogenic Ras in tumour maintenance. Nature 400: 468–472

Choy E et al (1999) Endomembrane trafficking of ras: the CAAX motif targets proteins to the ER and Golgi. Cell 98:69–80

Cichowski K, Jacks T (2001) NF1 tumor suppressor gene function: narrowing the GAP. Cell 104:593–604

Cox AD, Der CJ (2003) The dark side of Ras: regulation of apoptosis. Oncogene 22:8999–9006

Cozzolino M et al (2003) p120 Catenin is required for growth factor-dependent cell motility and scattering in epithelial cells. Mol Biol Cell 14:1964–1977

Crews CM, Erikson RL (1993) Extracellular signals and reversible protein phosphorylation: what to Mek of it all. Cell 74:215–217

D'Adamo DR et al (1997) rsc: a novel oncogene with structural and functional homology with the gene family of exchange factors for Ral. Oncogene 14:1295–1305

Dai Q et al (1998) Mammalian prenylcysteine carboxyl methyltransferase is in the endoplasmic reticulum. J Biol Chem 273:15030–15034

Datta SR et al (1997) Akt phosphorylation of BAD couples survival signals to the cell-intrinsic death machinery. Cell 91:231–241

Davies H et al (2002) Mutations of the BRAF gene in human cancer. Nature 417:949–954

Djordjevic S, Driscoll PC (2002) Structural insight into substrate specificity and regulatory mechanisms of phosphoinositide 3-kinases. Trends Biochem Sci 27:426–432

Downward J (1998) Ras signalling and apoptosis. Curr Opin Genet Dev 8:49–54

Downward J (2003) Targeting RAS signalling pathways in cancer therapy. Nat Rev Cancer 3:11–22

Downward J (2004) PI 3-kinase, Akt and cell survival. Semin Cell Dev Biol 15:177–182

Duffy MJ et al (2008) Cancer invasion and metastasis: changing views. J Pathol 214:283–293

Ehrhardt A et al (2002) Ras and relatives–job sharing and networking keep an old family together. Exp Hematol 30:1089–1106

Ehrhardt A et al (2004) Distinct mechanisms determine the patterns of differential activation of H-Ras, N-Ras, K-Ras 4B, and M-Ras by receptors for growth factors or antigen. Mol Cell Biol 24:6311–6323

Engelman JA et al (2008) Effective use of PI3K and MEK inhibitors to treat mutant Kras G12D and PIK3CA H1047R murine lung cancers. Nat Med 14:1351–1356

Etienne-Manneville S, Hall A (2002) Rho GTPases in cell biology. Nature 420:629–635

Fabian JR et al (1994) A single amino acid change in Raf-1 inhibits Ras binding and alters Raf-1 function. Proc Natl Acad Sci USA 91:5982–5986

Field J et al (1987) Guanine nucleotide activation of, and competition between, RAS proteins from Saccharomyces cerevisiae. Mol Cell Biol 7:2128–2133

Fisher GH et al (2001) Induction and apoptotic regression of lung adenocarcinomas by regulation of a K-Ras transgene in the presence and absence of tumor suppressor genes. Genes Dev 15:3249–3262

Frisch SM et al (1996) Control of adhesion-dependent cell survival by focal adhesion kinase. J Cell Biol 134:793–799

Garcia-Echeverria C (2009) Protein and lipid kinase inhibitors as targeted anticancer agents of the Ras/Raf/MEK and PI3K/PKB pathways. Purinergic Signal 5:117–125

Genot EM et al (2000) The T-cell receptor regulates Akt (protein kinase B) via a pathway involving Rac1 and phosphatidylinositide 3-kinase. Mol Cell Biol 20:5469–5478

Giehl K (2005) Oncogenic Ras in tumour progression and metastasis. Biol Chem 386:193–205

Gille H, Downward J (1999) Multiple ras effector pathways contribute to G(1) cell cycle progression. J Biol Chem 274:22033–22040

Gire V et al (2000) PI-3-kinase is an essential anti-apoptotic effector in the proliferative response of primary human epithelial cells to mutant RAS. Oncogene 19:2269–2276

Giuriato S et al (2004) Conditional animal models: a strategy to define when oncogenes will be effective targets to treat cancer. Semin Cancer Biol 14:3–11

Gonzalez-Garcia A et al (2005) RalGDS is required for tumor formation in a model of skin carcinogenesis. Cancer Cell 7:219–226

Gupta S et al (2007) Binding of ras to phosphoinositide 3-kinase p110alpha is required for ras-driven tumorigenesis in mice. Cell 129:957–968

Hall BE et al (2001) Structure-based mutagenesis reveals distinct functions for Ras switch 1 and switch 2 in Sos-catalyzed guanine nucleotide exchange. J Biol Chem 276:27629–27637

Hamad NM et al (2002) Distinct requirements for Ras oncogenesis in human versus mouse cells. Genes Dev 16:2045–2057

Hamilton M, Wolfman A (1998) Oncogenic Ha-Ras-dependent mitogen-activated protein kinase activity requires signaling through the epidermal growth factor receptor. J Biol Chem 273:28155–28162
Han J et al (1998) Role of substrates and products of PI 3-kinase in regulating activation of Rac-related guanosine triphosphatases by Vav. Science 279:558–560
Hancock JF (2003) Ras proteins: different signals from different locations. Nat Rev Mol Cell Biol 4:373–384
Hancock JF et al (1989) All ras proteins are polyisoprenylated but only some are palmitoylated. Cell 57:1167–1177
Hennessy BT et al (2005) Exploiting the PI3K/AKT pathway for cancer drug discovery. Nat Rev Drug Discov 4:988–1004
Herrmann C (2003) Ras-effector interactions: after one decade. Curr Opin Struct Biol 13:122–129
Hirakawa T, Ruley HE (1988) Rescue of cells from ras oncogene-induced growth arrest by a second, complementing, oncogene. Proc Natl Acad Sci USA 85:1519–1523
Huang L et al (1998) Structural basis for the interaction of Ras with RalGDS. Nat Struct Biol 5:422–426
Ikenoue T et al (2005) Functional analysis of PIK3CA gene mutations in human colorectal cancer. Cancer Res 65:4562–4567
Inoki K et al (2002) TSC2 is phosphorylated and inhibited by Akt and suppresses mTOR signalling. Nat Cell Biol 4:648–657
Irani K et al (1997) Mitogenic signaling mediated by oxidants in Ras-transformed fibroblasts. Science 275:1649–1652
Isakoff SJ et al (2005) Breast cancer-associated PIK3CA mutations are oncogenic in mammary epithelial cells. Cancer Res 65:10992–11000
Jackson EL et al (2001) Analysis of lung tumor initiation and progression using conditional expression of oncogenic K-ras. Genes Dev 15:3243–3248
Jia S et al (2008) Essential roles of PI(3)K-p110beta in cell growth, metabolism and tumorigenesis. Nature 454:776–779
Jia S et al (2009) Should individual PI3 kinase isoforms be targeted in cancer? Curr Opin Cell Biol 21:199–208
Jiang K et al (2004) Akt mediates Ras downregulation of RhoB, a suppressor of transformation, invasion, and metastasis. Mol Cell Biol 24:5565–5576
Johnson L et al (2001) Somatic activation of the K-ras oncogene causes early onset lung cancer in mice. Nature 410:1111–1116
Jones SM et al (1999) PDGF induces an early and a late wave of PI 3-kinase activity, and only the late wave is required for progression through G1. Curr Biol 9:512–521
Kang S et al (2005) Phosphatidylinositol 3-kinase mutations identified in human cancer are oncogenic. Proc Natl Acad Sci USA 102:802–807
Karkkainen MJ et al (2004) Vascular endothelial growth factor C is required for sprouting of the first lymphatic vessels from embryonic veins. Nat Immunol 5:74–80
Khokhlatchev AV et al (1998) Phosphorylation of the MAP kinase ERK2 promotes its homodimerization and nuclear translocation. Cell 93:605–615
Khwaja A et al (1997) Matrix adhesion and Ras transformation both activate a phosphoinositide 3-OH kinase and protein kinase B/Akt cellular survival pathway. EMBO J 16:2783–2793
Kim E et al (1999) Disruption of the mouse Rce1 gene results in defective Ras processing and mislocalization of Ras within cells. J Biol Chem 274:8383–8390
Kim D et al (2001) Akt/PKB promotes cancer cell invasion via increased motility and metalloproteinase production. FASEB J 15:1953–1962
Kinkade CW et al (2008) Targeting AKT/mTOR and ERK MAPK signaling inhibits hormone-refractory prostate cancer in a preclinical mouse model. J Clin Invest 118:3051–3064
Kodaki T et al (1994) The activation of phosphatidylinositol 3-kinase by Ras. Curr Biol 4:798–806
Land H et al (1983a) Tumorigenic conversion of primary embryo fibroblasts requires at least two cooperating oncogenes. Nature 304:596–602

Land H et al (1983b) Cellular oncogenes and multistep carcinogenesis. Science 222:771–778

Laprise P et al (2004) Down-regulation of MEK/ERK signaling by E-cadherin-dependent PI3K/Akt pathway in differentiating intestinal epithelial cells. J Cell Physiol 199:32–39

Leon J et al (1987) Differential expression of the ras gene family in mice. Mol Cell Biol 7: 1535–1540

Li W et al (2000) The leucine-rich repeat protein SUR-8 enhances MAP kinase activation and forms a complex with Ras and Raf. Genes Dev 14:895–900

Li W et al (2004) Transformation potential of Ras isoforms correlates with activation of phosphatidylinositol 3-kinase but not ERK. J Biol Chem 279:37398–37406

Lim KH, Counter CM (2005) Reduction in the requirement of oncogenic Ras signaling to activation of PI3K/AKT pathway during tumor maintenance. Cancer Cell 8:381–392

Lim KH et al (2005) Activation of RalA is critical for Ras-induced tumorigenesis of human cells. Cancer Cell 7:533–545

Lim KH et al (2008) Tumour maintenance is mediated by eNOS. Nature 452:646–649

Lin AW et al (1998) Premature senescence involving p53 and p16 is activated in response to constitutive MEK/MAPK mitogenic signaling. Genes Dev 12:3008–3019

Lowy DR, Willumsen BM (1989) Protein modification: new clue to Ras lipid glue. Nature 341:384–385

Lowy DR, Willumsen BM (1993) Function and regulation of ras. Annu Rev Biochem 62:851–891

Lyons JF et al (2001) Discovery of a novel Raf kinase inhibitor. Endocr Relat Cancer 8:219–225

Ma J, Karplus M (1997) Molecular switch in signal transduction: reaction paths of the conformational changes in ras p21. Proc Natl Acad Sci USA 94:11905–11910

Ma L et al (2005) Phosphorylation and functional inactivation of TSC2 by Erk implications for tuberous sclerosis and cancer pathogenesis. Cell 121:179–193

Maehama T, Dixon JE (1998) The tumor suppressor, PTEN/MMAC1, dephosphorylates the lipid second messenger, phosphatidylinositol 3, 4, 5-trisphosphate. J Biol Chem 273:13375–13378

Maehama T et al (2001) PTEN and myotubularin: novel phosphoinositide phosphatases. Annu Rev Biochem 70:247–279

Malliri A et al (2002) Mice deficient in the Rac activator Tiam1 are resistant to Ras-induced skin tumours. Nature 417:867–871

Manning BD et al (2002) Identification of the tuberous sclerosis complex-2 tumor suppressor gene product tuberin as a target of the phosphoinositide 3-kinase/akt pathway. Mol Cell 10:151–162

Marais R et al (1993) The SRF accessory protein Elk-1 contains a growth factor-regulated transcriptional activation domain. Cell 73:381–393

Marais R et al (1995) Ras recruits Raf-1 to the plasma membrane for activation by tyrosine phosphorylation. EMBO J 14:3136–3145

Marais R et al (1998) Requirement of Ras-GTP-Raf complexes for activation of Raf-1 by protein kinase C. Science 280:109–112

Matallanas D et al (2003) Differences on the inhibitory specificities of H-Ras, K-Ras, and N-Ras (N17) dominant negative mutants are related to their membrane microlocalization. J Biol Chem 278:4572–4581

Matallanas D et al (2006) Distinct utilization of effectors and biological outcomes resulting from site-specific Ras activation: Ras functions in lipid rafts and Golgi complex are dispensable for proliferation and transformation. Mol Cell Biol 26:100–116

Mayo MW et al (1997) Requirement of NF-kappaB activation to suppress p53-independent apoptosis induced by oncogenic Ras. Science 278:1812–1815

Menges CW, McCance DJ (2008) Constitutive activation of the Raf-MAPK pathway causes negative feedback inhibition of Ras-PI3K-AKT and cellular arrest through the EphA2 receptor. Oncogene 27:2934–2940

Miller KA et al (2009) Oncogenic Kras requires simultaneous PI3K signaling to induce ERK activation and transform thyroid epithelial cells *in vivo*. Cancer Res 69:3689–3694

Moodie SA et al (1993) Complexes of Ras.GTP with Raf-1 and mitogen-activated protein kinase kinase. Science 260:1658–1661

Morrison D (1994) 14-3-3: modulators of signaling proteins? Science 266:56–57

Nassar N et al (1995) The 2.2 A crystal structure of the Ras-binding domain of the serine/threonine kinase c-Raf1 in complex with Rap1A and a GTP analogue. Nature 375:554–560

Nguyen DX et al (2009) Metastasis: from dissemination to organ-specific colonization. Nat Rev Cancer 9:274–284

Nimnual AS et al (1998) Coupling of Ras and Rac guanosine triphosphatases through the Ras exchanger Sos. Science 279:560–563

Oda K et al (2008) PIK3CA cooperates with other phosphatidylinositol 3′-kinase pathway mutations to effect oncogenic transformation. Cancer Res 68:8127–8136

Ong SH et al (2001) Stimulation of phosphatidylinositol 3-kinase by fibroblast growth factor receptors is mediated by coordinated recruitment of multiple docking proteins. Proc Natl Acad Sci USA 98:6074–6079

O'Reilly KE et al (2006) mTOR inhibition induces upstream receptor tyrosine kinase signaling and activates Akt. Cancer Res 66:1500–1508

Orme MH et al (2006) Input from Ras is required for maximal PI(3)K signalling in Drosophila. Nat Cell Biol 8:1298–1302

Ozes ON et al (1999) NF-kappaB activation by tumour necrosis factor requires the Akt serine-threonine kinase. Nature 401:82–85

Pacold ME et al (2000) Crystal structure and functional analysis of Ras binding to its effector phosphoinositide 3-kinase gamma. Cell 103:931–943

Park BK et al (2001) Akt1 induces extracellular matrix invasion and matrix metalloproteinase-2 activity in mouse mammary epithelial cells. Cancer Res 61:7647–7653

Parsons DW et al (2005) Colorectal cancer: mutations in a signalling pathway. Nature 436:792

Parsons DW et al (2008) An integrated genomic analysis of human glioblastoma multiforme. Science 321:1807–1812

Pawson T, Nash P (2003) Assembly of cell regulatory systems through protein interaction domains. Science 300:445–452

Plowman SJ, Hancock JF (2005) Ras signaling from plasma membrane and endomembrane microdomains. Biochim Biophys Acta 1746:274–283

Prior IA et al (2001) GTP-dependent segregation of H-ras from lipid rafts is required for biological activity. Nat Cell Biol 3:368–375

Prober DA, Edgar BA (2002) Interactions between Ras1, dMyc, and dPI3K signaling in the developing Drosophila wing. Genes Dev 16:2286–2299

Ramjaun AR, Downward J (2007) Ras and phosphoinositide 3-kinase: partners in development and tumorigenesis. Cell Cycle 6:2902–2905

Rangarajan A, Weinberg RA (2003) Opinion: Comparative biology of mouse versus human cells: modelling human cancer in mice. Nat Rev Cancer 3:952–959

Rangarajan A et al (2004) Species- and cell type-specific requirements for cellular transformation. Cancer Cell 6:171–183

Rapp UR et al (1983) Structure and biological activity of v-raf, a unique oncogene transduced by a retrovirus. Proc Natl Acad Sci USA 80:4218–4222

Repasky GA et al (2004) Renewing the conspiracy theory debate: does Raf function alone to mediate Ras oncogenesis? Trends Cell Biol 14:639–647

Reusch HP et al (2001) Regulation of Raf by Akt controls growth and differentiation in vascular smooth muscle cells. J Biol Chem 276:33630–33637

Rocks O et al (2005) An acylation cycle regulates localization and activity of palmitoylated Ras isoforms. Science 307:1746–1752

Rodriguez-Viciana P, McCormick F (2005) RalGDS comes of age. Cancer Cell 7:205–206

Rodriguez-Viciana P et al (1994) Phosphatidylinositol-3-OH kinase as a direct target of Ras. Nature 370:527–532

Rodriguez-Viciana P et al (1996) Activation of phosphoinositide 3-kinase by interaction with Ras and by point mutation. EMBO J 15:2442–2451

Rodriguez-Viciana P et al (1997) Role of phosphoinositide 3-OH kinase in cell transformation and control of the actin cytoskeleton by Ras. Cell 89:457–467
Rodriguez-Viciana P et al (2004) Signaling specificity by Ras family GTPases is determined by the full spectrum of effectors they regulate. Mol Cell Biol 24:4943–4954
Rojas JM, Santos E (2002) ras genes and human cancer: different implications and different roles. Curr Genom 3:295–311
Romashkova JA, Makarov SS (1999) NF-kappaB is a target of AKT in anti-apoptotic PDGF signalling. Nature 401:86–90
Rommel C et al (1999) Differentiation stage-specific inhibition of the Raf-MEK-ERK pathway by Akt. Science 286:1738–1741
Roy S et al (1998) 14-3-3 facilitates Ras-dependent Raf-1 activation *in vitro* and *in vivo*. Mol Cell Biol 18:3947–3955
Roy S et al (1999) Dominant-negative caveolin inhibits H-Ras function by disrupting cholesterol-rich plasma membrane domains. Nat Cell Biol 1:98–105
Rubio I et al (1997) Interaction of Ras with phosphoinositide 3-kinase gamma. Biochem J 326 (Pt 3):891–895
Ruley HE (1983) Adenovirus early region 1A enables viral and cellular transforming genes to transform primary cells in culture. Nature 304:602–606
Sahai E, Marshall CJ (2002) RHO-GTPases and cancer. Nat Rev Cancer 2:133–142
Samuels Y et al (2004) High frequency of mutations of the PIK3CA gene in human cancers. Science 304:554
Samuels Y et al (2005) Mutant PIK3CA promotes cell growth and invasion of human cancer cells. Cancer Cell 7:561–573
Sasaki AT, Firtel RA (2006) Regulation of chemotaxis by the orchestrated activation of Ras, PI3K, and TOR. Eur J Cell Biol 85:873–895
Satoh T et al (1987) Induction of neurite formation in PC12 cells by microinjection of proto-oncogenic Ha-ras protein preincubated with guanosine-5′-O-(3-thiotriphosphate). Mol Cell Biol 7:4553–4556
Schlessinger J (2002) Ligand-induced, receptor-mediated dimerization and activation of EGF receptor. Cell 110:669–672
Schulze A et al (2001) Analysis of the transcriptional program induced by Raf in epithelial cells. Genes Dev 15:981–994
Selfors LM et al (1998) soc-2 encodes a leucine-rich repeat protein implicated in fibroblast growth factor receptor signaling. Proc Natl Acad Sci USA 95:6903–6908
Shah OJ et al (2004) Inappropriate activation of the TSC/Rheb/mTOR/S6K cassette induces IRS1/2 depletion, insulin resistance, and cell survival deficiencies. Curr Biol 14:1650–1656
Shannon KM et al (1994) Loss of the normal NF1 allele from the bone marrow of children with type 1 neurofibromatosis and malignant myeloid disorders. N Engl J Med 330:597–601
Shaw RJ, Cantley LC (2006) Ras, PI(3)K and mTOR signalling controls tumour cell growth. Nature 441:424–430
Shaw LM et al (1997) Activation of phosphoinositide 3-OH kinase by the alpha6beta4 integrin promotes carcinoma invasion. Cell 91:949–960
Shelton JG et al (2003) Effects of the RAF/MEK/ERK and PI3K/AKT signal transduction pathways on the abrogation of cytokine-dependence and prevention of apoptosis in hematopoietic cells. Oncogene 22:2478–2492
Sheng H et al (2001) Akt/PKB activity is required for Ha-Ras-mediated transformation of intestinal epithelial cells. J Biol Chem 276:14498–14504
Shields JM et al (2000) Understanding Ras: 'it ain't over 'til it's over'. Trends Cell Biol 10:147–154
Side L et al (1997) Homozygous inactivation of the NF1 gene in bone marrow cells from children with neurofibromatosis type 1 and malignant myeloid disorders. N Engl J Med 336:1713–1720
Sieburth DS et al (1998) SUR-8, a conserved Ras-binding protein with leucine-rich repeats, positively regulates Ras-mediated signaling in C. elegans. Cell 94:119–130

Simi L et al (2008) High-resolution melting analysis for rapid detection of KRAS, BRAF, and PIK3CA gene mutations in colorectal cancer. Am J Clin Pathol 130:247–253
Simpson CD et al (2008) Anoikis resistance and tumor metastasis. Cancer Lett 272:177–185
Sjolander A et al (1991) Association of p21ras with phosphatidylinositol 3-kinase. Proc Natl Acad Sci USA 88:7908–7912
Solit DB et al (2006) BRAF mutation predicts sensitivity to MEK inhibition. Nature 439: 358–362
Stacey D, Kazlauskas A (2002) Regulation of Ras signaling by the cell cycle. Curr Opin Genet Dev 12:44–46
Steeg PS (2006) Tumor metastasis: mechanistic insights and clinical challenges. Nat Med 12:895–904
Stemke-Hale K et al (2008) An integrative genomic and proteomic analysis of PIK3CA, PTEN, and AKT mutations in breast cancer. Cancer Res 68:6084–6091
Sturgill TW et al (1988) Insulin-stimulated MAP-2 kinase phosphorylates and activates ribosomal protein S6 kinase II. Nature 334:715–718
Su AI et al (2004) A gene atlas of the mouse and human protein-encoding transcriptomes. Proc Natl Acad Sci USA 101:6062–6067
Suire S et al (2006) Gbetagammas and the Ras binding domain of p110gamma are both important regulators of PI(3)Kgamma signalling in neutrophils. Nat Cell Biol 8:1303–1309
Takuwa N, Takuwa Y (1997) Ras activity late in G1 phase required for p27kip1 downregulation, passage through the restriction point, and entry into S phase in growth factor-stimulated NIH 3T3 fibroblasts. Mol Cell Biol 17:5348–5358
Taylor SJ, Shalloway D (1996) Cell cycle-dependent activation of Ras. Curr Biol 6:1621–1627
Therrien M et al (1999) Functional analysis of CNK in RAS signaling. Proc Natl Acad Sci USA 96:13259–13263
Thomas RK et al (2007) High-throughput oncogene mutation profiling in human cancer. Nat Genet 39:347–351
Torbett NE et al (2008) A chemical screen in diverse breast cancer cell lines reveals genetic enhancers and suppressors of sensitivity to PI3K isoform-selective inhibition. Biochem J 415:97–110
Trahey M, McCormick F (1987) A cytoplasmic protein stimulates normal N-ras p21 GTPase, but does not affect oncogenic mutants. Science 238:542–545
Treisman R (1996) Regulation of transcription by MAP kinase cascades. Curr Opin Cell Biol 8:205–215
Tsutsumi K et al (2009) Visualization of Ras-PI3K interaction in the endosome using BiFC. Cell Signal 21:1672–1679
Van Aelst L, Symons M (2002) Role of Rho family GTPases in epithelial morphogenesis. Genes Dev 16:1032–1054
Vanhaesebroeck B et al (1997) P110delta, a novel phosphoinositide 3-kinase in leukocytes. Proc Natl Acad Sci USA 94:4330–4335
Velho S et al (2005) The prevalence of PIK3CA mutations in gastric and colon cancer. Eur J Cancer 41:1649–1654
Vivanco I, Sawyers CL (2002) The phosphatidylinositol 3-Kinase AKT pathway in human cancer. Nat Rev Cancer 2:489–501
Voice JK et al (1999) Four human ras homologs differ in their abilities to activate Raf-1, induce transformation, and stimulate cell motility. J Biol Chem 274:17164–17170
Vojtek AB et al (1993) Mammalian Ras interacts directly with the serine/threonine kinase Raf. Cell 74:205–214
Walker EH et al (1999) Structural insights into phosphoinositide 3-kinase catalysis and signalling. Nature 402:313–320
Walsh AB, Bar-Sagi D (2001) Differential activation of the Rac pathway by Ha-Ras and K-Ras. J Biol Chem 276:15609–15615

Wang X et al (1998) The phosphorylation of eukaryotic initiation factor eIF4E in response to phorbol esters, cell stresses, and cytokines is mediated by distinct MAP kinase pathways. J Biol Chem 273:9373–9377
Warne PH et al (1993) Direct interaction of Ras and the amino-terminal region of Raf-1 *in vitro*. Nature 364:352–355
Waskiewicz AJ et al (1997) Mitogen-activated protein kinases activate the serine/threonine kinases Mnk1 and Mnk2. EMBO J 16:1909–1920
Wee S et al (2009) PI3K pathway activation mediates resistance to MEK inhibitors in KRAS mutant cancers. Cancer Res 69:4286–4293
Weiner OD et al (2002) A PtdInsP(3)- and Rho GTPase-mediated positive feedback loop regulates neutrophil polarity. Nat Cell Biol 4:509–513
White MA et al (1996) A role for the Ral guanine nucleotide dissociation stimulator in mediating Ras-induced transformation. J Biol Chem 271:16439–16442
Whiteford CC et al (1996) D-3 phosphoinositide metabolism in cells treated with platelet-derived growth factor. Biochem J 319(Pt 3):851–860
Wishart MJ, Dixon JE (2002) PTEN and myotubularin phosphatases: from 3-phosphoinositide dephosphorylation to disease. Trends Cell Biol 12:579–585
Wittinghofer A, Nassar N (1996) How Ras-related proteins talk to their effectors. Trends Biochem Sci 21:488–491
Wittinghofer A, Pai EF (1991) The structure of Ras protein: a model for a universal molecular switch. Trends Biochem Sci 16:382–387
Wolfman A, Macara IG (1990) A cytosolic protein catalyzes the release of GDP from p21ras. Science 248:67–69
Yamamoto H et al (2008) PIK3CA mutations and copy number gains in human lung cancers. Cancer Res 68:6913–6921
Yan J et al (1998) Ras isoforms vary in their ability to activate Raf-1 and phosphoinositide 3-kinase. J Biol Chem 273:24052–24056
Ye M et al (2008) Involvement of PI3K/Akt signaling pathway in hepatocyte growth factor-induced migration of uveal melanoma cells. Invest Ophthalmol Vis Sci 49:497–504
Zhan M et al (2004) Signalling mechanisms of anoikis. Histol Histopathol 19:973–983
Zhang XF et al (1993) Normal and oncogenic p21ras proteins bind to the amino-terminal regulatory domain of c-Raf-1. Nature 364:308–313
Zhang H et al (2007) PDGFRs are critical for PI3K/Akt activation and negatively regulated by mTOR. J Clin Invest 117:730–738
Zhao L, Vogt PK (2008a) Class I PI3K in oncogenic cellular transformation. Oncogene 27:5486–5496
Zhao L, Vogt PK (2008b) Helical domain and kinase domain mutations in p110alpha of phosphatidylinositol 3-kinase induce gain of function by different mechanisms. Proc Natl Acad Sci USA 105:2652–2657
Zhao JJ et al (2005) The oncogenic properties of mutant p110alpha and p110beta phosphatidylinositol 3-kinases in human mammary epithelial cells. Proc Natl Acad Sci USA 102: 18443–18448
Zhao JJ et al (2006) The p110alpha isoform of PI3K is essential for proper growth factor signaling and oncogenic transformation. Proc Natl Acad Sci USA 103:16296–16300
Zhu Y, Parada LF (2001) A particular GAP in mind. Nat Genet 27:354–355
Zimmermann S, Moelling K (1999) Phosphorylation and regulation of Raf by Akt (protein kinase B). Science 286:1741–1744
Zohn IM et al (1998) Rho family proteins and Ras transformation: the RHOad less traveled gets congested. Oncogene 17:1415–1438

More Than Just Kinases: The Scaffolding Function of PI3K

Carlotta Costa and Emilio Hirsch

Contents

Abstract Recently, it has been reported that some members of the PI3K family might have a "double identity"; in other words, PI3K have been found to act not only as classical kinases, but also as scaffolding proteins. Until now, the use of knockout mice has been considered sufficient to model the effects of PI3K inhibition and to predict the outcome of anti-PI3K pharmacological treatments by observing the resulting phenotypes. These studies supported the view that PI3K may represent promising pharmacological targets for cancer and inflammation. However, in selected cases, different experimental strategies of gene targeting of the same locus have resulted in distinct phenotypes. This demonstrates that "knocking-out" a gene is not necessarily equivalent to "knocking-in" an inactivating point mutation (Vanhaesebroeck et al. in Cell 118:274–276, 2004). Specifically, knockout and kinase-dead models have led to the discovery that PI3Kγ and β may act independently of their kinase activity, likely as adaptor proteins.

C. Costa and E. Hirsch (✉)
Molecular Biotechnology Center, University of Torino, Via Nizza 52, 10126 Torino, Italy

C. Rommel et al. (eds.). *Phosphoinositide 3-kinase in Health and Disease*, Volume 1

Current Topics in Microbiology and Immunology 346, DOI 10.1007/82_2010_57

1 The Double Identity of PI3Kγ

Studies in knockout and kinase-dead mice have demonstrated that PI3Kγ plays a key role as a kinase in controlling different aspects of inflammatory processes, such as leukocyte recruitment and ROS production (Hirsch et al. 2000; Li et al. 2000; Sasaki et al. 2000; Patrucco et al. 2004; Hirsch et al. 2008). In addition, PI3Kγ is implicated in heart function and in the cardiac β-adrenergic receptor (βAR) signaling pathway (Alloatti et al. 2004). Specifically, PI3Kγ regulates the termination of βAR signaling, controlling βAR internalization that occurs after ligand stimulation (Backer 2005). Termination of βAR signaling is initiated by G protein-coupled receptor kinase 2 (GRK2), via phosphorylation of the agonist-occupied receptor tail, followed by recruitment of β-arrestin and arrestin-mediated interaction with adaptins and clathrin (Shenoy and Lefkowitz 2003). The mechanism in which GRK2 acts is as follows: upon agonist stimulation, GRK2 mediates translocation of PI3Kγ to βARs, interacting directly with the PIK domain of PI3Kγ (Gaidarov and Keen 1999; Naga Prasad et al. 2002). A local pool of PIP3 is then generated by PI3Kγ in proximity to βAR, and this pool enhances recruitment of a number of phosphoinositide-binding endocytic proteins essential for βAR internalization, such as β-arrestin and AP-2 (Naga Prasad et al. 2002). Although the lipid kinase activity of PI3Kγ is sufficient to recruit the AP-2 adaptor, receptor internalization also requires protein kinase-mediated phosphorylation of non-muscle tropomyosin, a protein that binds actin filaments (Naga Prasad et al. 2005) (Fig. 1). These observations have led to hypothesize that the loss of PI3Kγ would cause increased βAR levels as well as potentiated βAR-dependent signaling. Specifically, βAR activation, upon catecholamine binding, triggers heterodimeric G proteins dissociation into G_α and

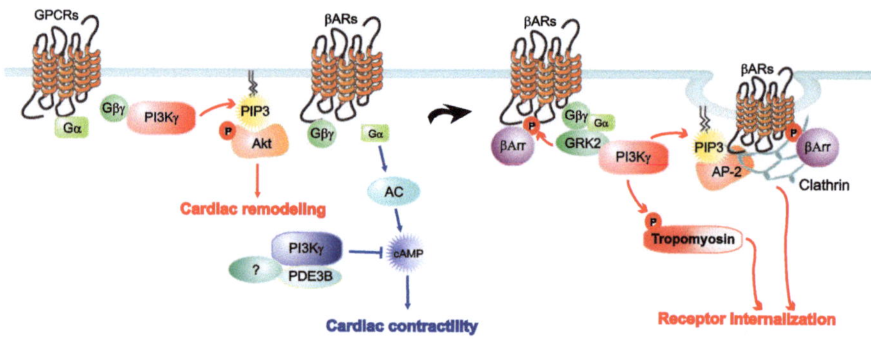

Fig. 1 Kinase-dependent and kinase-independent roles of PI3Kγ in the heart. GPCR-dependent activation of PI3Kγ catalytic function triggers PIP3 production, Akt activation as well as βAR internalization. In addition, PI3Kγ functions as a scaffold element for the cAMP digesting enzyme PDE3B and for proteins yet to be identified (represented by a *question mark*). The organization of this complex activates PDE3B to reduce cAMP production triggered by βAR activation. Kinase-dependent and -independent pathways are represented in *red* and *blue*, respectively

$G_{\beta\gamma}$ subunits. G_{α} subunits subsequently activate the effector adenylyl cyclase to generate the second messenger cAMP, leading to an increase in contractility. This is achieved by cAMP-dependent activation of PKA that, in turn, phosphorylates multiple substrates, including L-type Ca^{2+} channels and phospholamban that modulate Ca^{2+} fluxes and hence contraction and relaxation. In accordance, loss of PI3Kγ in knockout mice results in increased cardiac contractility (Crackower et al. 2002), although apparently without a significant rise in the βAR plasma membrane density (Nienaber et al. 2003). Overall, such observations have suggested that, in mice lacking PI3Kγ, increasing PIP3 production by cardiac loss of the PIP3 phosphatase PTEN would rescue the contractile phenotype. However, mice lacking both PTEN and PI3Kγ in the heart only show partial normalization of the contractile phenotype (Crackower et al. 2002). This indicates that, in controlling cardiac function, PI3Kγ is not only acting as a kinase but also plays roles that do not involve its enzymatic activity. Further studies support this view and detect kinase-dependent and -independent functions of PI3Kγ in cardiomyocytes, circulating endothelial progenitor cells (EPCs) and platelets.

1.1 Cardiomyocytes

PI3Kγ knockout and kinase-dead mice show distinct cardiac phenotypes, supporting the view that PI3Kγ has a "double identity". Indeed, in PI3Kγ knockout mice, the increased contractility correlates with augmented baseline concentration levels of cAMP, while PI3Kγ kinase-dead mice show normal cAMP as well as unaltered heart function (Patrucco et al. 2004; Vanhaesebroeck et al. 2004). These data suggest that PI3Kγ is involved in signaling functions unrelated to its kinase activity and further studies demonstrate that this process occurs via protein–protein interactions. Indeed, PI3Kγ negatively regulates heart contractility by forming a complex with phosphodiesterase 3B (PDE3B), a key enzyme involved in cAMP degradation (Patrucco et al. 2004). Supposedly, PI3Kγ acts by organizing a macromolecular complex that brings PDE3B in closer proximity to its potential activator/activators (Fig. 1). While all players of this complex are yet to be identified, the PI3Kγ–adaptor protein p84/p87 physically interacts with PDE3B, suggesting its involvement in the PI3Kγ/PDE3B complex (Voigt et al. 2005).

1.2 Circulating Endothelial Progenitor Cells

Bone marrow-derived EPCs complement the regenerative potential of resident vascular cells in the site of neovascularization (Urbich and Dimmeler 2004). PI3Kγ is expressed by EPCs and appears to modulate several aspects of the reparative angiogenesis supported by these cells. Indeed, PI3Kγ-deficient EPCs show defects in proliferation, survival and migration. In agreement, mice lacking PI3Kγ show reduced capillarization and arteriogenesis of hind limbs following ligation of the femoral artery as well as impaired incorporation of EPCs in the reforming vessels. However, this reparative neovascularization process is preserved in PI3Kγ kinase-dead mice, and PI3Kγ kinase-dead EPCs do not show defects in proliferation, survival and migration. Overall, this suggests that a kinase-independent function of PI3Kγ is required for correct vascular repair (Madeddu et al. 2008). However, the kinase-independent functions of PI3Kγ in EPCs as well as in other bone marrow-derived cells have not been completely elucidated, and further mechanistic explanations are still needed.

1.3 Platelets

In platelets, PI3Kβ and γ orchestrate the activation of adhesive function, required for efficient aggregation and thrombus formation, through affinity modulation of the $\alpha_{IIb}\beta_3$ integrin. While PI3Kβ predominately regulates G_i-dependent $\alpha_{IIb}\beta_3$-mediated platelet aggregation through a classical lipid kinase-dependent mechanism that involves Rap1 and Akt activation, PI3Kγ appears to regulate this process principally through a non-catalytic signaling mechanism. Indeed, combination of the lack of PI3Kγ with wortmannin-mediated inhibition of all PI3K, leads to a much greater defect in platelet aggregation than inhibition of the PI3K catalytic function alone. These data demonstrate that in platelets a cooperative PI3K signaling mechanism exists that involves the catalytic activity of PI3Kβ as well as the non-catalytic function of PI3Kγ (Schoenwaelder et al. 2007).

2 The Double Identity of PI3Kβ

The role of PI3Kβ has remained elusive for a long time because its genetic ablation leads to an embryonic lethal phenotype (Bi et al. 2002) that prevented an accurate characterization of its *in vivo* function. Nonetheless, studies with kinase-dead and conditional knockout mice are helping to better clarify the role of PI3Kβ and, unexpectedly, are contributing to the identification of distinct PI3Kβ kinase-dependent and -independent functions in cell growth, metabolism and oncogenesis (Jia et al. 2008; Ciraolo et al. 2008) (Fig. 2).

2.1 Cell Growth

Mice homozygous for a kinase-dead PI3Kβ mutant allele are able to survive to adulthood, provided a high expression of the catalytically inactive PI3Kβ. By contrast, low expresser siblings die in utero. This observation documents that, in

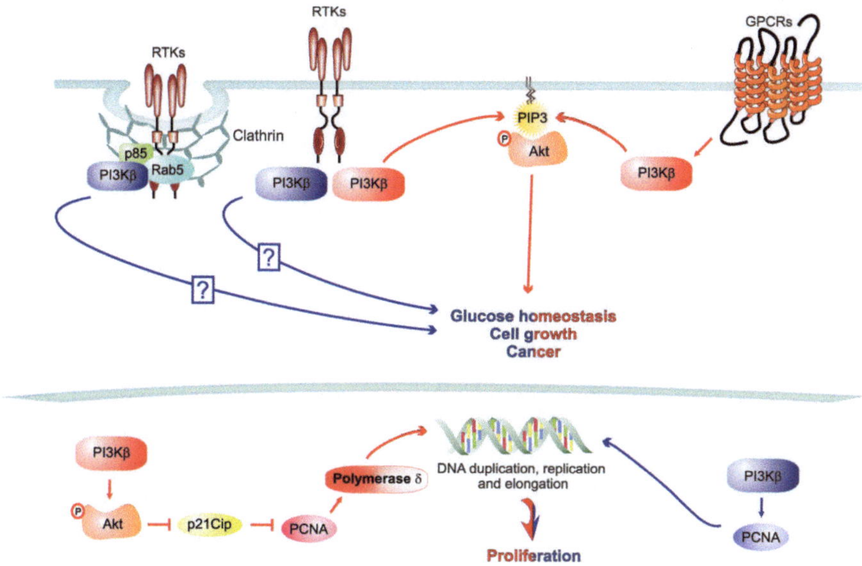

Fig. 2 Kinase-dependent and kinase-independent roles of PI3Kβ. Upon activation of RTKs and GPCRs, PI3Kβ is recruited to the plasma membrane in the proximity of its substrate, thus leading to PIP3 production. PIP3 acts as a second messenger that, through Akt activation, mediates many cellular responses, such as glucose homeostasis, cell growth and proliferation. These processes are also controlled in a kinase-independent manner. One proposed mechanism for the kinase-independent function of PI3Kβ is through the regulation of the receptor endocytic pathway, though the exact molecular mechanisms are yet to be defined (*question marks*). A further complex PI3Kβ function takes place in the nucleus, where a specific pool of PI3Kβ controls DNA duplication by regulating PCNA activity through its kinase activity as well as its scaffolding function. Kinase-dependent and -independent pathways are represented in *red* and *blue*, respectively

development and growth, the expression of PI3Kβ is critically required, while its kinase function is dispensable. PI3Kβ low expresser mouse embryonic fibroblasts (MEFs) show inhibition of growth, whereas MEFs expressing kinase-dead PI3Kβ replicate normally (Ciraolo et al. 2008). In agreement with these data, complete deletion of PI3Kβ in MEFs decreases cell proliferation (Jia et al. 2008) but reconstitution of these knockout MEFs with a kinase-dead PI3Kβ restores normal cell proliferation. This suggests that, similarly to PI3Kγ, PI3Kβ has an additional kinase-independent function, possibly linked to its ability to enucleate critical protein–protein interactions. Indeed, PI3Kβ can associate to the small GTPse Rab5, a crucial element controlling vesicular trafficking, and appears to be involved in receptor endocytosis (Joly et al. 1994; Christoforidis et al. 1999; Kurosu and Katada 2001; Shin et al. 2005). Consistent with such a role of PI3Kβ, epidermal growth factor receptor (EGFR) and transferrin uptake are defective when PI3Kβ is lost (Jia et al. 2008; Ciraolo et al. 2008). Since overexpression of a dominant-active Rab5 mutant does not rescue EGFR internalization, PI3Kβ is likely required in the early stages of endocytosis. In agreement, the loss of PI3Kβ correlates with significant reduction of clathrin-coated pits and vesicles. Both defects in EGFR and transferrin receptor recycling are rescued by the kinase-dead PI3Kβ, clearly demonstrating that PI3Kβ controls these processes without an involvement of its catalytic activity (Jia et al. 2008; Ciraolo et al. 2008). Interestingly, since endocytosis is thought to be involved in the modulation of cell proliferation, this kinase-independent function can, in principle, account for the reduced proliferation detected in fibroblasts lacking PI3Kβ expression. However, PI3Kβ can also control cell growth by other kinase-independent mechanisms. Indeed, PI3Kβ associates in the nucleus with the proliferating cell nuclear antigen (PCNA) and favors its loading onto chromatin, a process essential for DNA duplication (Marques et al. 2009). Interestingly, this process is tightly coupled with modulation of PI3Kβ catalytic function that, in the nucleus, promotes Akt activation and the subsequent nuclear Akt-mediated p21Cip phosphorylation (a PCNA-negative regulator), PCNA release and its binding to Polymerase δ (Marques et al. 2009). In agreement with these data showing that nuclear PI3Kβ triggers Akt phosphorylation after serum stimulation, previous studies have implicated a preferential involvement of the PI3Kβ catalytic function in G-protein-coupled receptor (GPCR)-dependent signaling (Roche et al. 1998; Murga et al. 2000; Guillermet-Guibert et al. 2008). For example, the absence of PI3Kβ abrogates Akt phosphorylation triggered by GPCR agonists such as LPA or S1P. Of note, the decrease in Akt phosphorylation in response to stimulation with LPA observed in PI3Kβ knockout cells is restored by reintroducing wild-type but not the kinase-dead allele of PI3Kβ, thus demonstrating that, besides its kinase-independent function, PI3Kβ plays a crucial role as a kinase involved in the proliferative response to selected agonists (Ciraolo et al. 2008; Jia et al. 2008).

2.2 Metabolism

Consistent with reports concluding that the kinase activity of PI3Kβ has only a minor function in insulin signaling (Foukas et al. 2006; Knight et al. 2006), deletion of PI3Kβ does not have negative effects on the peak of Akt phosphorylation in primary MEFs in response to stimulation by insulin, epidermal growth factor (EGF), platelet-derived growth factor (PDGF) or IGF-1 (Jia et al. 2008; Ciraolo et al. 2008). However, mice expressing a kinase-dead PI3Kβ show insulin resistance and reduced glucose tolerance, with higher levels of fasting plasma glucose and insulin. Similar observations are reported in mice with a liver-specific ablation of PI3Kβ expression. Although these results might be compatible with PI3Kβ contributing to metabolic regulation through a kinase-independent mechanism, other observations still suggest a role of PI3Kβ catalytic activity in insulin responses. Indeed, in livers of wild-type mice treated with a PI3Kβ inhibitor or kinase-dead mice, insulin-evoked Akt activation declines significantly faster than in untreated wild-type controls. Comparable results are observed in insulin-stimulated HepG2 hepatoma cells treated with a PI3Kβ inhibitor, thus further supporting the view that PI3Kβ catalytic activity is not necessary for peak signaling but is required, in the long run, to sustain the insulin-evoked response (Ciraolo et al. 2008). Nonetheless, taken together, these data suggest that insulin signaling requires a possible combination of the kinase-dependent and kinase-independent functions of PI3Kβ.

2.3 *Oncogenesis*

Alterations that lead to increased PI3K signaling confer a survival and growth advantage, and are therefore frequent in human tumors (Salmena et al. 2008; Vogt et al. 2009). Although activating mutations of PI3Kβ have not been observed in human cancers, PI3Kβ appears to play a key role in oncogenesis. Indeed, loss of PI3Kβ abrogates transformation of MEFs by mutant Ras and mutant EGFR, while loss of PI3Kα has a less pronounced effect. Cell cultures also regain partial susceptibility to transformation if, instead of the wild-type PI3Kβ, the kinase-dead PI3Kβ mutant is reintroduced to the PI3Kβ knockout background, thus indicating that PI3Kβ also has a partial kinase-independent role in this process. Nonetheless, *in vivo* studies show that, in oncogenic transformation, the PI3Kβ catalytic function is apparently more critical. Indeed, in a mouse model of prostate cancer induced by PTEN loss, concomitant ablation of PI3Kβ, but not PI3Kα, leads to decreased Akt phosphorylation in the prostate, and prevents development of high-grade prostatic intraepithelial neoplasia (PIN) (Jia et al. 2008). This supports the view that the loss of PTEN function uncovers a dominance of PI3Kβ catalytic function in the signal transduction pathways controlling cell proliferation and growth. Consistently, in PTEN-null human cancer cells, knockdown of PI3Kβ, but not of PI3Kα, interferes with signaling and cell replication (Wee et al. 2008). These findings are consistent with the model that PI3Kβ generates a basal pool of PIP3 that defines a threshold for PI3Kα activation necessary for signaling (Knight et al. 2006). Inactivation of PTEN increases basal PIP3 levels generated by PI3Kβ, thereby decreasing the threshold for Akt activation and transformation. In accordance with a kinase-dependent role of PI3Kβ in oncogenesis, mice homozygous for the kinase-dead mutation of PI3Kβ show protection from tumorigenesis in a model of mammary gland cancer triggered by an activated ERBB2 transgene (neuT) (Ciraolo et al. 2008). In agreement, treatment with a PI3Kβ inhibitor does not affect the growth of tumor cells derived from kinase-dead PI3Kβ/neuT mice, but causes a significant reduction in proliferation in wild-type tumor cells, showing that oncogenic ERBB2 drives tumor growth largely through PI3Kβ catalytic activity. These findings have potentially important implications for the rationale design of novel cancer therapies targeting PI3K signaling pathways.

3 The Double Identity of Adaptor Subunits

Several studies demonstrate that not only p110 catalytic subunits possess kinase-independent functions but also PI3K-adaptor subunits show different functions in addition to their ability of modulating p110-PI3K activity. For example, p85α contains a BH (breakpoint cluster region-homology) domain with sequence homology to GTPase-activating proteins (GAPs). Indeed, p85α shows GAP activity toward Rab5 and Rab4, and purified recombinant Rab5 and p85α show a direct interaction. p85α binds to the PDGF receptor (PDGFR) during receptor endocytosis in the same early endosomal compartment of Rab5 and Rab4. Physical colocalization and the ability of p85α to preferentially stimulate downregulation of Rab5 and Rab4 GTPases suggest that p85α regulates the length of time that Rab5 and Rab4 remain in their GTP-bound active state. Consistently, expression of BH domain mutants of p85α result in increased PDGFR activation and downstream signaling (Akt and MAPK phosphorylation) and in decreased PDGFR degradation (Chamberlain et al. 2004). Furthermore, disruption of RabGAP function of p85α because of a single point mutation (R274A) is sufficient to cause cellular transformation via a PI3K-independent mechanism, partially reversed by dominant negative Rab5-S34N expression. This suggests a novel role for p85α in controlling receptor signaling and trafficking through its effects on Rab GTPases (Chamberlain et al. 2008). Because p110β can directly bind Rab5 (Kurosu and Katada 2001), further studies are needed to elucidate p110-dependendent and -independent functions associated to p85α.

Furthermore, p85α controls mammalian cytokinesis in a PI3K-independent manner. Indeed, deletion of p85α adaptor subunit induces cell accumulation in telophase and appearance of binucleated cells, whereas inhibition of PI3K activity does not affect cytokinesis. In this case, p85α acts as a scaffolding protein, binding Cdc42 and septin 2 simultaneously. p85α is thus involved in the spatial control of cytosolic division by regulating the local activation of Cdc42 in the cleavage furrow and in turn septin 2 localization (Garcia et al. 2006).

4 Conclusions

The human kinome contains over 500 proteins that orchestrate much of the biology of the cell (Manning et al. 2002). Remarkably, about 10% of proteins included in the kinome are designated as "pseudokinases", because they lack one or more essential conserved catalytic residues in the catalytic pocket (Boudeau et al. 2006). Nonetheless, although most of their function are still unclear, several appear to act as key scaffolding proteins. It is thus possible to speculate that the protein kinase fold has evolved not only to position many key residues for ATP binding and phosphoryl transfer, but also to expose chemically diverse surfaces to solvents. These surfaces provide docking sites for many other proteins, including substrates, inhibitors, regulatory proteins and other signaling molecules (Kornev and Taylor 2009). It is thus not surprising to find that kinome members like PI3K might exert part of their functions independently from their catalytic activity, via protein–protein interactions. These kinase-independent activities may operate in concert with the kinase activity, like PI3Kβ in insulin signaling, or act in complete autonomy from the catalytic function, as shown by PI3Kγ in cardiac contractility.

Whether other members of class I (PI3Kδ and α), class II or III PI3K play similar kinase-independent roles has not yet been determined, and further studies should be aimed at addressing this issue. It is conceivable that these studies of PI3K scaffolding activity could extend the common knowledge of targeting the catalytic site to block the enzymatic function and define other PI3K surfaces as drug targets for the modulation of scaffolding activities. Indeed, extensive evidence shows that PI3K stand out as therapeutic targets and selective inhibitors are now available. However, these "classical" inhibitors abrogate catalytic activity of PI3K but, likely, still leave their kinase-independent functions intact (Vogt et al. 2009). Nonetheless, a future challenging task can be the generation of selective peptide-mimetic small molecules that, inhibiting protein–protein interactions, modulate kinase-independent functions of PI3K. These small molecules could act together with classical ATP competitors to synergistically ablate kinase-dependent and -independent functions and achieve a more thorough efficacy, for example, in anticancer activity.

References

Alloatti G, Montrucchio G, Lembo G, Hirsch E (2004) Phosphoinositide 3-kinase gamma: kinase-dependent and -independent activities in cardiovascular function and disease. Biochem Soc Trans 32:383–386

Backer JM (2005) Substrate specificity: PI(3)Kgamma has it both ways. Nat Cell Biol 7:773–774

Bi L, Okabe I, Bernard DJ, Nussbaum RL (2002) Early embryonic lethality in mice deficient in the p110beta catalytic subunit of PI 3-kinase. Mamm Genome 13:169–172

Boudeau J, Miranda-Saavedra D, Barton GJ, Alessi DR (2006) Emerging roles of pseudokinases. Trends Cell Biol 16:443–452

Chamberlain MD, Berry TR, Pastor MC, Anderson DH (2004) The p85alpha subunit of phosphatidylinositol 3′-kinase binds to and stimulates the GTPase activity of Rab proteins. J Biol Chem 279:48607–48614

Chamberlain MD, Chan T, Oberg JC, Hawrysh AD, James KM, Saxena A, Xiang J, Anderson DH (2008) Disrupted RabGAP function of the p85 subunit of phosphatidylinositol 3-kinase results in cell transformation. J Biol Chem 283:15861–15868

Christoforidis S, Miaczynska M, Ashman K, Wilm M, Zhao L, Yip SC, Waterfield MD, Backer JM, Zerial M (1999) Phosphatidylinositol-3-OH kinases are Rab5 effectors. Nat Cell Biol 1:249–252

Ciraolo E, Iezzi M, Marone R, Marengo S, Curcio C, Costa C, Azzolino O, Gonella C, Rubinetto C, Wu H et al (2008) Phosphoinositide 3-kinase p110beta activity: key role in metabolism and mammary gland cancer but not development. Sci Signal 1:ra3

Crackower MA, Oudit GY, Kozieradzki I, Sarao R, Sun H, Sasaki T, Hirsch E, Suzuki A, Shioi T, Irie-Sasaki J et al (2002) Regulation of myocardial contractility and cell size by distinct PI3K-PTEN signaling pathways. Cell 110:737–749

Foukas LC, Claret M, Pearce W, Okkenhaug K, Meek S, Peskett E, Sancho S, Smith AJ, Withers DJ, Vanhaesebroeck B (2006) Critical role for the p110alpha phosphoinositide-3-OH kinase in growth and metabolic regulation. Nature 441:366–370

Gaidarov I, Keen JH (1999) Phosphoinositide-AP-2 interactions required for targeting to plasma membrane clathrin-coated pits. J Cell Biol 146:755–764

Garcia Z, Silio V, Marques M, Cortes I, Kumar A, Hernandez C, Checa AI, Serrano A, Carrera AC (2006) A PI3K activity-independent function of p85 regulatory subunit in control of mammalian cytokinesis. EMBO J 25:4740–4751

Guillermet-Guibert J, Bjorklof K, Salpekar A, Gonella C, Ramadani F, Bilancio A, Meek S, Smith AJ, Okkenhaug K, Vanhaesebroeck B (2008) The p110beta isoform of phosphoinositide 3-kinase signals downstream of G protein-coupled receptors and is functionally redundant with p110gamma. Proc Natl Acad Sci USA 105:8292–8297

Hirsch E, Katanaev VL, Garlanda C, Azzolino O, Pirola L, Silengo L, Sozzani S, Mantovani A, Altruda F, Wymann MP (2000) Central role for G protein-coupled phosphoinositide 3-kinase gamma in inflammation. Science 287:1049–1053

Hirsch E, Ciraolo E, Ghigo A, Costa C (2008) Taming the PI3K team to hold inflammation and cancer at bay. Pharmacol Ther 118:192–205

Jia S, Liu Z, Zhang S, Liu P, Zhang L, Lee SH, Zhang J, Signoretti S, Loda M, Roberts TM et al (2008) Essential roles of PI(3)K-p110beta in cell growth, metabolism and tumorigenesis. Nature 454:776–779

Joly M, Kazlauskas A, Fay FS, Corvera S (1994) Disruption of PDGF receptor trafficking by mutation of its PI-3 kinase binding sites. Science 263:684–687

Knight ZA, Gonzalez B, Feldman ME, Zunder ER, Goldenberg DD, Williams O, Loewith R, Stokoe D, Balla A, Toth B et al (2006) A pharmacological map of the PI3-K family defines a role for p110alpha in insulin signaling. Cell 125:733–747

Kornev AP, Taylor SS (2009) Pseudokinases: functional insights gleaned from structure. Structure 17:5–7

Kurosu H, Katada T (2001) Association of phosphatidylinositol 3-kinase composed of p110beta-catalytic and p85-regulatory subunits with the small GTPase Rab5. J Biochem 130:73–78

Li Z, Jiang H, Xie W, Zhang Z, Smrcka AV, Wu D (2000) Roles of PLC-beta2 and -beta3 and PI3Kgamma in chemoattractant-mediated signal transduction. Science 287:1046–1049

Madeddu P, Kraenkel N, Barcelos LS, Siragusa M, Campagnolo P, Oikawa A, Caporali A, Herman A, Azzolino O, Barberis L et al (2008) Phosphoinositide 3-kinase gamma gene knockout impairs postischemic neovascularization and endothelial progenitor cell functions. Arterioscler Thromb Vasc Biol 28:68–76

Manning G, Plowman GD, Hunter T, Sudarsanam S (2002) Evolution of protein kinase signaling from yeast to man. Trends Biochem Sci 27:514–520

Marques M, Kumar A, Poveda AM, Zuluaga S, Hernandez C, Jackson S, Pasero P, Carrera AC (2009) Specific function of phosphoinositide 3-kinase beta in the control of DNA replication. Proc Natl Acad Sci USA 106:7525–7530

Murga C, Fukuhara S, Gutkind JS (2000) A novel role for phosphatidylinositol 3-kinase beta in signaling from G protein-coupled receptors to Akt. J Biol Chem 275:12069–12073

Naga Prasad SV, Laporte SA, Chamberlain D, Caron MG, Barak L, Rockman HA (2002) Phosphoinositide 3-kinase regulates beta2-adrenergic receptor endocytosis by AP-2 recruitment to the receptor/beta-arrestin complex. J Cell Biol 158:563–575

Naga Prasad SV, Jayatilleke A, Madamanchi A, Rockman HA (2005) Protein kinase activity of phosphoinositide 3-kinase regulates beta-adrenergic receptor endocytosis. Nat Cell Biol 7:785–796

Nienaber JJ, Tachibana H, Naga Prasad SV, Esposito G, Wu D, Mao L, Rockman HA (2003) Inhibition of receptor-localized PI3K preserves cardiac beta-adrenergic receptor function and ameliorates pressure overload heart failure. J Clin Invest 112:1067–1079

Patrucco E, Notte A, Barberis L, Selvetella G, Maffei A, Brancaccio M, Marengo S, Russo G, Azzolino O, Rybalkin SD et al (2004) PI3Kgamma modulates the cardiac response to chronic pressure overload by distinct kinase-dependent and -independent effects. Cell 118:375–387

Roche S, Downward J, Raynal P, Courtneidge SA (1998) A function for phosphatidylinositol 3-kinase beta (p85alpha-p110beta) in fibroblasts during mitogenesis: requirement for insulin- and lysophosphatidic acid-mediated signal transduction. Mol Cell Biol 18:7119–7129

Salmena L, Carracedo A, Pandolfi PP (2008) Tenets of PTEN tumor suppression. Cell 133:403–414

Sasaki T, Irie-Sasaki J, Jones RG, Oliveira-dos-Santos AJ, Stanford WL, Bolon B, Wakeham A, Itie A, Bouchard D, Kozieradzki I et al (2000) Function of PI3Kgamma in thymocyte development, T cell activation, and neutrophil migration. Science 287:1040–1046

Schoenwaelder SM, Ono A, Sturgeon S, Chan SM, Mangin P, Maxwell MJ, Turnbull S, Mulchandani M, Anderson K, Kauffenstein G et al (2007) Identification of a unique co-operative phosphoinositide 3-kinase signaling mechanism regulating integrin alpha IIb beta 3 adhesive function in platelets. J Biol Chem 282:28648–28658

Shenoy SK, Lefkowitz RJ (2003) Multifaceted roles of beta-arrestins in the regulation of seven-membrane-spanning receptor trafficking and signaling. Biochem J 375:503–515

Shin HW, Hayashi M, Christoforidis S, Lacas-Gervais S, Hoepfner S, Wenk MR, Modregger J, Uttenweiler-Joseph S, Wilm M, Nystuen A et al (2005) An enzymatic cascade of Rab5 effectors regulates phosphoinositide turnover in the endocytic pathway. J Cell Biol 170:607–618

Urbich C, Dimmeler S (2004) Endothelial progenitor cells functional characterization. Trends Cardiovasc Med 14:318–322

Vanhaesebroeck B, Rohn JL, Waterfield MD (2004) Gene targeting: attention to detail. Cell 118:274–276

Vogt PK, Gymnopoulos M, Hart JR (2009) PI 3-kinase and cancer: changing accents. Current Opin Genet Dev 19:12–17

Voigt P, Brock C, Nurnberg B, Schaefer M (2005) Assigning functional domains within the p101 regulatory subunit of phosphoinositide 3-kinase gamma. J Biol Chem 280:5121–5127

Wee S, Wiederschain D, Maira SM, Loo A, Miller C, deBeaumont R, Stegmeier F, Yao YM, Lengauer C (2008) PTEN-deficient cancers depend on PIK3CB. Proc Natl Acad Sci USA 105:13057–13062

PI3K Signaling in Neutrophils

Phillip T. Hawkins, Len R. Stephens, Sabine Suire, and Michael Wilson

Contents

Abstract PI3Ks play important roles in the signaling pathways used by a wide variety of cell surface receptors on neutrophils. Class IB PI3K plays a major role in the initial generation of PtdIns(3,4,5)P_3 by Gi-coupled G-protein coupled receptors (GPCRs) (e.g., receptors for fMLP, C5a, LTB_4). Class IA PI3Ks generate PtdIns (3,4,5)P_3 downstream of receptors which directly or indirectly couple to protein tyrosine kinases such as integrins, FcγRs, cytokine receptors, and GPCRs. The PtdIns(3,4,5)P_3 made by Class I PI3Ks regulates the activity of several different effector proteins, many of which are plasma membrane GEFs or GAPs for small GTPases. Class III PI3K generates PtdIns(3)P in the phagosome membrane and plays an important role in efficient assembly of the NADPH oxidase at this location. Much still remains to be discovered about the molecular details that govern activation of PI3Ks and the mechanisms by which these enzymes regulate complex cellular processes, such as neutrophil spreading, chemotaxis, phagocytosis, and killing of pathogens. However, it is clear from recent use of transgenic mouse models and isoform-selective PI3K inhibitors that these pathways are important in

P.T. Hawkins (✉), L.R. Stephens, S. Suire, and M. Wilson
The Babraham Institute, Babraham Research Campus, Cambridge CB22 3AT, UK
e-mail: Philip.hawkins@bbsrc.ac.uk

C. Rommel et al. (eds.). *Phosphoinositide 3-kinase in Health and Disease*, Volume 1
Current Topics in Microbiology and Immunology 346, DOI 10.1007/82_2010_40

regulating neutrophil recruitment to sites of infection and damage in vivo. Thus, PI3K pathways may present novel opportunities for selective inhibition in some inflammatory pathologies.

1 Neutrophil Biology

Neutrophils are the most abundantly circulating phagocytes in the bloodstream and the principal component of our innate immune system. They are usually the first immune cells to be delivered in numbers to the site of infection and play a vital role in the destruction of invading bacterial and fungal pathogens (Nauseef 2007). An overview of the central processes that are thought to regulate the passage of neutrophils out of inflamed venules (extravasation) and towards sources of damage and/or infection is given in Fig. 1.

Substances released from damaged cells or pathogens activate both immune and nonimmune cells to secrete cytokines, which in turn initiate the production of a complex mixture of proinflammatory agents, including further cytokines and chemoattractants (chemokines and lipid derivatives). These proinflammatory agents remodel the endothelial cell lining of neighboring capillary walls to express increased numbers of adhesion molecules which, together with entrapped chemokines, halt the flow of circulating neutrophils and stimulate their adherence, spread, and diapedesis across the endothelial wall (Ley et al. 2007). The precise combination of cells (e.g., macrophages, mast cells, platelets, smooth muscle cells, epithelial cells) and mediators which are most important in these inflammatory cascades depends on the context of the insult and where it occurs in the body. However, TNF-α and IL-1β appear to be generally important as initiating cytokines in many examples of inflammation and IL-8 (or the analogous KC in the mouse) and MIP-2 appear to play particularly important roles in the initial activation and recruitment of neutrophils.

Having exited the blood vessel, neutrophils sense a gradient of chemoattractants and use it to home in on the source of inflammation. Again, the precise molecules involved are very context specific, but IL-8 and LTB_4 are generally accepted to be important "intermediate" chemoattractants, leading the neutrophil through the layer of pericytes and extracellular matrix. fMLP and C5a are generally considered "end-point attractants", guiding the neutrophils to the immediate vicinity of the inflammation (Heit et al. 2002). The neutrophil that arrives at the site of inflammation is a substantially different cell from that originally circulating in the blood stream, having been "primed" by prior exposure to activating substances (low concentrations of chemoattractants and bacterial products) to express different levels of surface receptors and associated signal transduction elements. This priming generally allows enhanced neutrophil response to subsequent stimuli, which may be higher concentrations of the priming agents themselves, other inflammatory agents, or pathogens (Condliffe et al. 1998). In the case of pathogens, this will involve mounting an appropriate microbicidal attack against them. Where the pathogen is

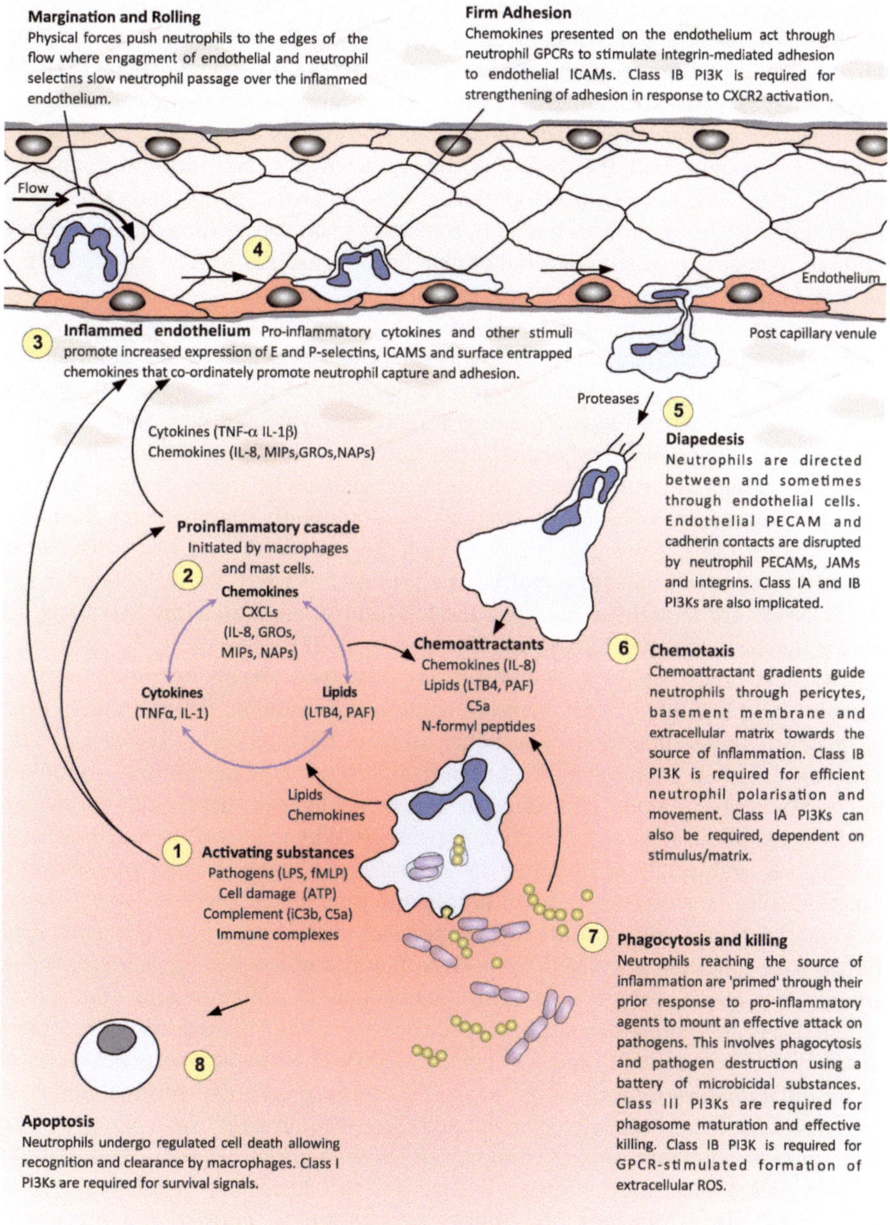

Fig. 1 Neutrophil recruitment to a site of inflammation. This figure is a simplified overview describing the major stages of neutrophil recruitment to a site of infection, showing some examples of the key molecules involved and the points at which PI3Ks are thought to act

small enough to be engulfed by phagocytosis (e.g., bacteria), this involves their uptake into a specialized form of endocytic organelle called a phagosome, into which a battery of destructive peptides, proteases, reactive oxygen species (ROS), and peroxidases are secreted from specialized granules (Steinberg and Grinstein 2008; Hampton et al. 1998). The rate of phagocytosis is usually increased by the binding of opsonins onto the surface of the pathogen (e.g., complement fragments) which can be recognized by neutrophil receptors. Where the pathogen is too large to be phagocytosed (e.g., fungal hyphae), secretory granules are directed to an extracellular "synapse" type structure rather than the phagosome, with an associated risk of host damage (Moraes et al. 2006).

Circulating neutrophils have a short half-life (approximately 6 h for human neutrophils) and die via a default pathway of apoptosis. Once at a site of inflammation, neutrophil apoptosis can be temporarily delayed by the action of survival factors (e.g., cytokines and complement fragments), which allow the neutrophil an extended time to combat pathogens. After the neutrophil has performed its duties, apoptosis ensues to ensure that it, and its contents, can be safely cleared from the site of inflammation by other phagocytes, predominantly macrophages (Kennedy and DeLeo 2009). Recently, another form of cell death has been discovered whereby neutrophils secrete a matrix of chromatin to form "extracellular NETs". These NETs are thought to entrap pathogens and provide an additional focus for killing mechanisms (Brinkmann and Zychlinsky 2007).

The neutrophil behaviors and functions described above are regulated by a complex series of interactions between soluble and insoluble ligands and an array of receptors on the neutrophil plasma membrane. Each receptor creates a cellular response based on the magnitude, duration, and location of receptor activation and the combination of signal transduction pathways engaged therein. PI3Ks are key components of the signal transduction network used by many different receptors on neutrophils and are thus involved in several important elements of neutrophil function (Fig. 1). However, a major problem in pinpointing precisely which receptors and neutrophil responses are most dependent on PI3Ks in vivo is that neutrophil function is generally regulated by a large number of coexisting, interdependent stimuli and events. Thus, neutrophil numbers at an inflammatory site can be affected by individual blockade of neutrophil capture, diapedesis, chemotaxis, phagocytosis/activation (because of positive feedback loops involving the secretion of proinflammatory molecules by neutrophils themselves), and survival/apoptosis (because of effects on neutrophil clearance by macrophages). Advances in intra vital microscopy have recently improved in vivo measurements of neutrophil capture and diapedesis, but these studies are still limited to a relatively small number of tissue locations and inflammatory stimuli. Further, the process of chemotaxis is still very difficult to measure in vivo, although careful histological examination can give crude measures of neutrophil numbers, activation (e.g., myeloperoxidase staining), and apoptosis. Thus, to a large extent, the role of PI3Ks in neutrophil biology is inferred from the effects of genetic and/or pharmacological blockade of PI3Ks in in vitro assays of neutrophil function and the consistent effects of these interventions in a smaller number of in vivo models of inflammation. It is generally accepted, however, that PI3Ks play vital roles in neutrophil spreading on inflamed endothelium, migration to sites of inflammation, and killing of pathogens (Ellson et al. 2006; Liu et al. 2007; Puri et al. 2005; Pinho et al. 2007; Smith et al. 2006).

2 Regulation of PI3Ks

Neutrophils express Class IB PI3K (a p101 or p84 regulatory subunit and a p110γ catalytic subunit), Class IA PI3Ks (a p85 family regulatory subunit and either a p110α, β or δ catalytic subunit), Class II PI3Ks (α and β isoforms), and Class III PI3K (Stephens et al. 2002). Class I PI3Ks predominantly phosphorylate the membrane lipid PtdIns(4,5)P_2 to form PtdIns(3,4,5)P_3; these enzymes are acutely activated by many different types of cell surface receptor and the PtdIns (3,4,5)P_3 thus formed is now accepted to play a major role in receptor-proximal signal transduction pathways (Fig. 2; Engelman et al. 2006). Class III PI3K

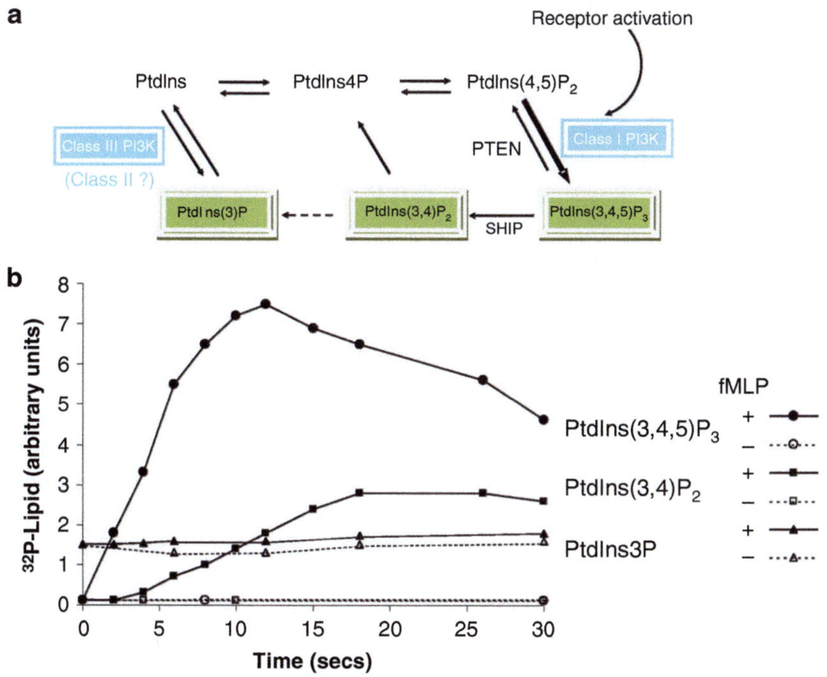

Fig. 2 fMLP-stimulates Class I PI3K. (**a**) Schematic diagram showing the major pathways of phosphoinositide metabolism in fMLP-stimulated neutrophils. (**b**) Changes in the concentrations of ^{32}P-labeled phosphoinositides in ^{32}Pi-prelabeled human neutrophils stimulated with fMLP (100 nM fMLP was added at time zero; data derived from Stephens et al. (1991))

phosphorylates PtdIns to PtdIns3P and is involved in the regulation of endosomal/lysosomal trafficking and in the control of autophagy (Fig. 2; Simonsen and Tooze 2009). Class II PI3Ks also probably phosphorylate PtdIns to PtdIns3P, but their cellular roles are far less clear and no functions have yet been ascribed to them in neutrophils. As in most tissues, the relative molecular concentrations and specific activities of these PI3Ks are uncertain, but p101/p110γ and p110δ are very highly expressed in neutrophils relative to other tissues, particularly non-hematopoietic cells, suggesting that they have a particularly important function in these cells.

Several G-protein coupled receptors (GPCRs) for soluble, inflammatory stimuli can activate the Class IB PI3K isoform to deliver rapid synthesis of PtdIns(3,4,5)P_3 and subsequent accumulation of its dephosphorylation product PtdIns(3,4)P_2 (Fig. 2) (Li et al. 2000; Hirsch et al. 2000; Sasaki et al. 2000; Stephens et al. 1991). Indeed, the activation of Class IB PI3K in neutrophils by Gi-coupled GPCRs provided the first evidence for PtdIns(3,4,5)P_3 formation in cells (Traynor-Kaplan et al. 1988) and the central paradigm for the reaction catalyzed by this enzyme (Stephens et al. 1991). Class IB PI3K in neutrophils is predominantly an obligate p101/p110γ heterodimer, with much smaller amounts of p84/p110γ (Suire et al. 2006). The lipid kinase activity of p101/p110γ can be directly and dramatically activated by Gβγ subunits in vitro (Stephens et al. 1997) and to a lesser extent by direct interaction with GTP-Ras (Suire et al. 2002). p101 is required for substantial activation by Gβγ but GTP-Ras requires only the p110γ subunit. A crystal structure is available for a complex between p110γ and GTP-Ras defining the points of contact (Pacold et al. 2000); no structure is yet available involving any part of p101. Recent use of a p101 knock-out mouse and a p110γ knock-in mutant specifically unable to bind GTP-Ras suggests that fMLP or C5a-stimulation of p101/p110γ in mouse bone marrow derived neutrophils requires simultaneous Gβγ-p101 and GTP-Ras regulatory inputs (Fig. 3) (Suire et al. 2006).

Class IA PI3Ks are generally considered to be activated by cell surface receptors which use tyrosine kinase coupled signal transduction pathways (Vanhaesebroeck et al. 2001), although in neutrophils these pathways are still relatively ill-defined. It is presumed by analogy to other systems that the SH2 domains of the regulatory subunits of these enzymes dock onto activating phosphotyrosines on "adaptor" proteins and this, together with an interaction between GTP-Ras and the p110 catalytic subunit, activates their lipid kinase activity (Fig. 3). Receptors known to activate Class IA PI3Ks in neutrophils include integrins, cytokines, antibody receptors, and GPCRs (Stephens et al. 1993; Corey et al. 1993; Utomo et al. 2006; Schymeinsky et al. 2007). In each of these cases, the molecular details of the adaptors to which these PI3Ks bind are unknown. Recent work has suggested that β2, β3 integrins and low affinity Fcγ receptors share common elements of signal transduction, namely the src family kinase (SFK) – driven phosphorylation of ITAM-containing adaptors (DAP12 and the Fcγ chain). These phosphorylated ITAM residues, in turn, recruit and activate the Syk tyrosine kinase and promote the phosphorylation of further adaptors and effectors (Fig. 3) (Jakus et al. 2007). Gi-coupled GPCRs activate Class IB PI3K (as detailed above), but they can also

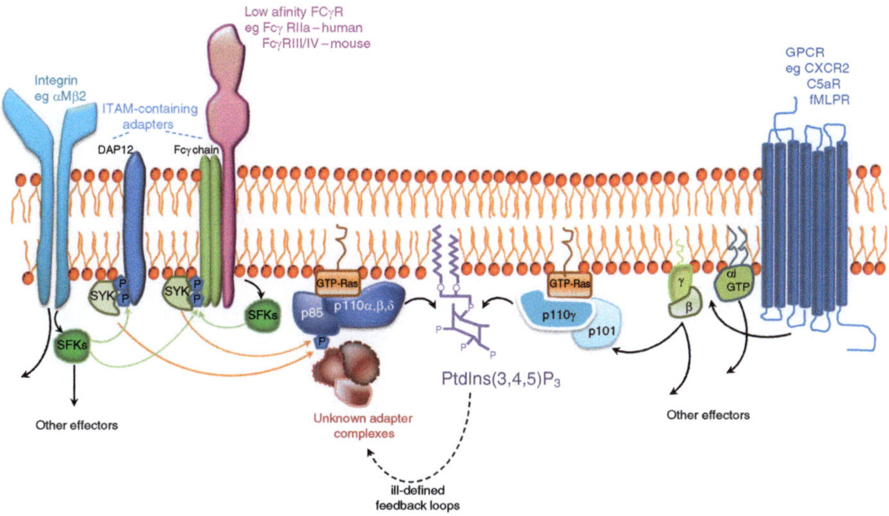

Fig. 3 Activation of Class I PI3Ks by GPCRs and ITAM-coupled receptors. A cartoon is shown depicting the major steps in the activation of Class I PI3Ks by GPCRs, FcRγs, and integrins in neutrophils. GPCRs activate Class IA PI3Ks by ill-defined mechanisms which may involve SFKs but may also involve further signals and receptors (e.g., in paracrine loops)

activate Class IA PI3Ks (Boulven et al. 2006), probably via direct or indirect activation of SFKs. In human neutrophils this GPCR-dependent activation of Class IA PI3Ks depends on prior activation of Class IB PI3K and may represent a mechanism for a "priming"-dependent augmentation of PtdIns(3,4,5)P_3 production at later times of stimulation, allowing the coupling of this pathway to further cellular responses, such as ROS production (Condliffe et al. 2005). In most cases the precise isoform of Class IA PI3K involved in a particular receptor-driven response is unknown, but there is now a body of evidence suggesting that p110δ is highly expressed in human and mouse neutrophils and probably plays an important role downstream of GPCRs for fMLP, MIP-2, IL-8, and LTB_4, possibly in conjunction with integrin activation (Condliffe et al. 2005; Puri et al. 2005; Liu et al. 2007; Randis et al. 2008).

Class III PI3K is activated during phagocytosis, delivering increased synthesis of PtdIns3P on the phagosomal membrane immediately after scission from the plasma membrane (Vieira et al. 2001; Ellson et al. 2001b). This synthesis of PtdIns3P appears to be a common feature of all forms of phagocytosis, irrespective of the precise cell surface receptors involved in the phagocytic event and whether they activate Class I PI3Ks or not. The molecular mechanisms which regulate the synthesis of PtdIns3P on the phagosome are unclear, but circumstantial evidence from other cells suggests that it may involve dynamin and the direct binding of GTP-Rab5 to the catalytic subunit of Class III PI3K (Shin et al. 2005; Kinchen et al. 2008).

3 Regulation of PtdIns(3,4,5)P_3

PtdIns(3,4,5)P_3 can be dephosphorylated in neutrophil lysates by both 3- and 5-phophatases (Stephens et al. 1991). These two routes of dephosphorylation lead to the formation of either PtdIns(3,4)P_2 or PtdIns(4,5)P_2, lipids with different signaling potential (see below), lending significance to the route of PtdIns(3,4,5) P_3 removal. Work involving other cells suggests that PTEN and SHIP1/2 are the major 3- and 5-phosphatases, respectively, although there remains the possibility that other enzymes play major roles. Recent studies using knock-out mice have confirmed a significant role for both PTEN (Sarraj et al. 2009; Heit et al. 2008a, b; Li et al. 2005) and SHIP1 (Vaillancourt et al. 2006; Nishio et al. 2007) in neutrophils, but there is still a lack of quantitative data defining the precise roles of each phosphatase in specific receptor-regulated synthesis of PtdIns(3,4,5)P_3. Furthermore, there is conflicting evidence as to their relative importance in neutrophil functional responses, such as GPCR-stimulated polarization and movement. SHIP1 is recruited through its SH2 domain to phosphotyrosines in the ITIM-motifs in the cytoplasmic tails of so-called inhibitory receptors, such as FcγRIIB (Nimmerjahn and Ravetch 2008). This suggests that SHIP1 plays a particularly important role in defining the threshold of neutrophil activation to protein tyrosine kinase-coupled receptors (Coggeshall et al. 2002; Helgason et al. 1998).

4 Effectors of PI3K Signaling Pathways

The membrane lipids synthesized by PI3Ks act as signals by binding to specific protein domains, characteristically PH domains for PtdIns(3,4,5)P_3/PtdIns(3,4)P_2 and PX or FYVE domains for PtdIns3P (Balla 2005). These interactions often cause a change in steady state localization of the effector protein, from predominantly soluble to substantially on the membrane. In this sense, the lipids can be viewed as "regulatable scaffolds," helping to define the presence and activity of several extrinsic membrane proteins. Usually, the binding of phosphoinositides is also accompanied by other regulatory interactions, for example, interaction of the protein targets with membrane localized GTPases, defining more precisely the location and activity of the effector. However, some of these domains appear to possess sufficient affinity to, and specificity for, their cognate lipid to be expressed as GFP-fusion probes in cells. This allows tracking of the levels and localization of the relevant lipid by fluorescent microscopy, for example, the GFP-PH domain of PKB for PtdIns (3,4,5)P_3/PtdIns(3,4)P_2 or the GFP-PX domain of p40^{phox} for PtdIns3P (Fig. 4).

A further general characteristic of PI3K signaling is that there are probably at least 10–20 individual effectors for PtdIns(3,4,5)P_3, PtdIns(3,4)P_2, and PtdIns3P in most mammalian cells. This creates a highly complex network of regulatory interactions downstream of these lipids (Hawkins et al. 2006). In addition, these effectors are often sensitive to more than one signaling pathway, making their activity highly context-dependent and hard to trace in the origins of complex cellular responses.

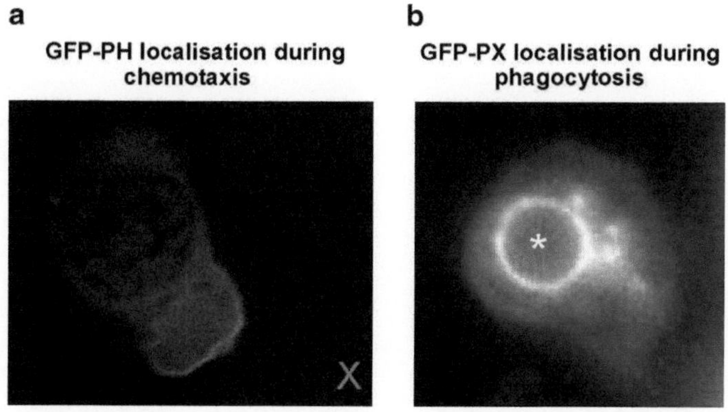

Fig. 4 Visualization of PI3K activity in neutrophils using GFP-domain probes. (**a**) PtdIns(3,4,5) P_3/PtdIns(3,4)P_2 is visualized in mouse bone marrow derived neutrophils through expression of a GFP-PH domain probe (these neutrophils were isolated from mice expressing the GFP-tagged PH domain of PKB, as described by Nishio et al. (2007) and Ferguson et al. (2007); confocal fluorescent image courtesy of Laura Milne, Babraham Institute, Cambridge). The neutrophil is migrating on glass towards a point source of fMLP (position of micropipette containing fMLP is marked by an "X"). There is enhanced fluorescence at the leading edge membrane indicating accumulation of Class I PI3K products at this location. (**b**) PtdIns3P is visualized in mouse bone marrow derived neutrophils through expression of a GFP-PX domain probe (mouse chimeras were created by injecting hematopoetic stem cells expressing the GFP-tagged PX domain of $p40^{phox}$ into irradiated mice; wide-field fluorescent image courtesy of Tamara Chessa, Babraham Institute, Cambridge). The neutrophil has phagocytosed an antibody-coated sheep red blood cell (position marked by an *asterisk*). There is enhanced fluorescence around the phagosome, indicating accumulation of the Class III PI3K product PtdIns3P

Further, the difficulties involved in generating mutants which specifically block lipid binding, combined with a lack of detailed knowledge of the surrounding signaling networks, mean it is still frustratingly difficult to explain PI3K-dependent cellular responses in terms of phosphoinositide levels, location, and the proteins they interact with. However, several convincing effectors of PI3K-generated lipids have now been identified in neutrophils through a combination of activity driven purification (Welch et al. 2002; Ellson et al. 2001a, b) and "fishing" with phosphoinositide-affinity matrices (Krugmann et al. 2002) (Fig. 5), and significant progress has been made in connecting a few of them to clear neutrophil functions.

5 Class I PI3K Regulation of Rho Family GTPases in the Context of Cell Spreading and Movement

A substantial number of effectors of PtdIns(3,4,5)P_3 and PtdIns(3,4)P_2 have been identified in neutrophils and a high proportion of these are GEFs and GAPs for small GTPases of the Rac, Rho, and Cdc42 family (Fig. 5). For several of these proteins, there is convincing evidence that their PH domains can bind PtdIns(3,4,5)P_3

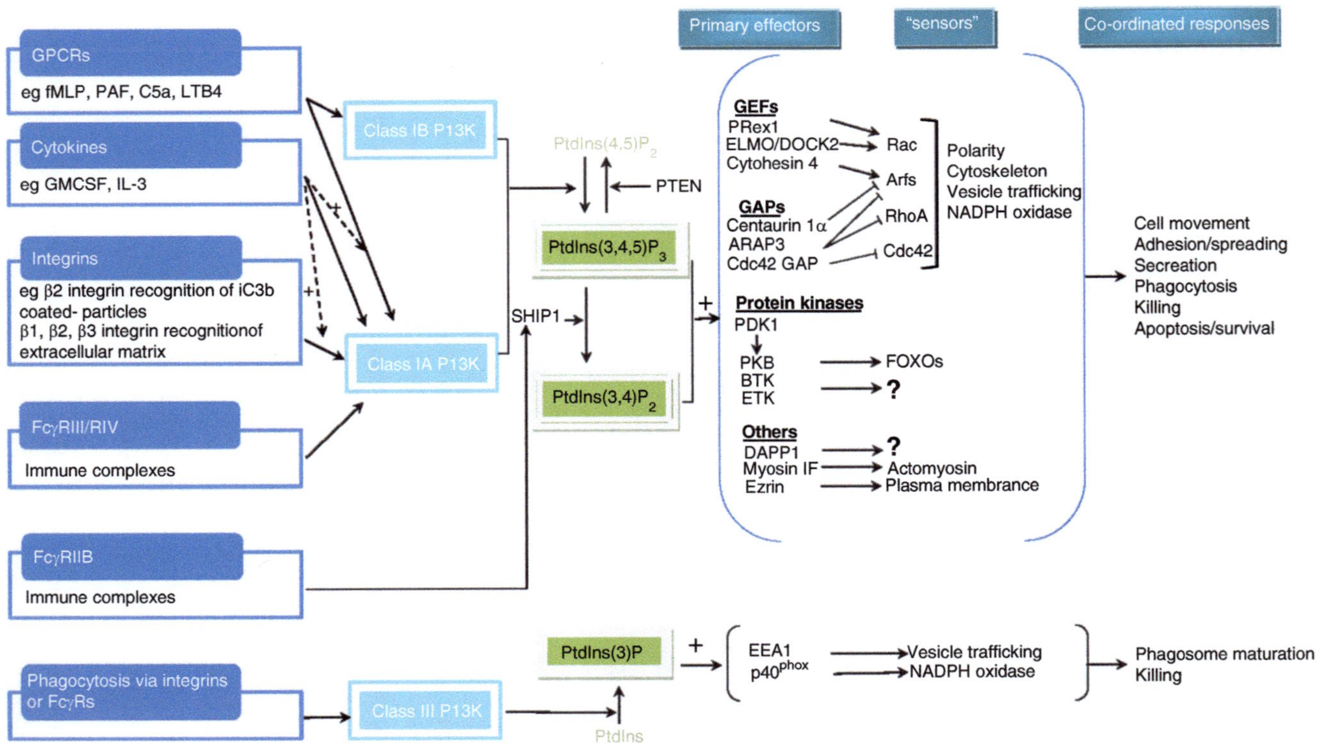

Fig. 5 Overview of PI3K effectors in neutrophils. Effectors are included where there is reasonable evidence of direct binding to the relevant phosphoinositide (Krugmann et al. 2002; Welch et al. 2002; Kunisaki et al. 2006; Ellson et al. 2001b; Gaullier et al. 2000; Stephens et al. 2008) plus our own unpublished observations indicating some of these proteins are expressed in neutrophils

(cytohesin 4, centaurin 1α) or PtdIns(3,4,5)P_3/PtdIns(3,4)P_2 (PRex1, ARAP3) with high affinity and selectivity, and in some cases lipid binding is demonstrated to cause a substantial change in localization (cytohesin 4, centaurin 1α, DOCK2) and activity (Rac GEF activity of PRex1 and Arf6 GAP activity of ARAP3) (Welch et al. 2002; Krugmann et al. 2002; Venkateswarlu et al. 2004; Kunisaki et al. 2006). The overall sense of these effects is that Class I PI3K activity promotes cycling of Arf family proteins, activation of Rac, and inhibition of Rho in areas of high Class I PI3K activity.

Evidence from a number of sources suggests that Arf6 is involved in the delivery of Rac proteins to sites of lamellipodia formation and the concerted action of Arf, Rac, and CDC42 is thought to co-ordinate membrane ruffling, protrusion, and spreading. Rho proteins, on the other hand, appear to be associated more often with the formation of actin stress fibers and focal adhesions. In neutrophils, these GTPases play major roles in the coordinated regulation of membrane delivery, cortical actin rearrangements, and actomyosin contraction that underlie polarity, spreading, cell movement, and the phagocytosis of large particles (Stephens et al. 2002; Insall and Machesky 2009). When neutrophils are stimulated by agonists for Gi-coupled GPCRs in vitro, such as fMLP and IL-8, they quickly polarize and, if an appropriate substratum is available, they will attach, spread, and move. We still do not understand the underlying mechanics of neutrophil movement, but it is likely to involve actin polymerization and protrusion at the front, coordinated with actomyosin contraction at the rear (Stephens et al. 2008). There is agreement that, under these circumstances, Class IB PI3K generates PtdIns(3,4,5)P_3/PtdIns(3,4)P_2 which is highly localized to the leading or "front" edge of the neutrophil (Fig. 4a) (Servant et al. 2000; Ferguson et al. 2007), making these lipids attractive candidates for spatial organization of small GTPase activity during movement. It has been suggested that "feed-forward" cycles of Class I PI3K/Rac1/CDC42 at the front and PTEN/RhoA at the back help define "frontness" and "backness" (Li et al. 2003; Nishikimi et al. 2009; Weiner et al. 2006; Wang et al. 2002; Van Keymeulen et al. 2006; Kunisaki et al. 2006) and are embedded in the mechanism of gradient sensing during chemotaxis (Bourne and Weiner 2002). However, the extent to which Class I PI3K signaling is really important in these processes is disputed, with recent work pointing to a context-dependent role for these enzymes in the efficiency of polarization and movement, rather than gradient sensing per se (Heit et al. 2008a, b; Ferguson et al. 2007; Stephens et al. 2008; Sai et al. 2008). Further work needs to done to resolve these issues and also to define more clearly which physiological process is being modeled by these in vitro experiments, for example, attachment and spreading on inflamed endothelium or chemotaxis through extracellular matrix.

6 Class I and Class III PI3K Regulation of the NADPH Oxidase

The neutrophil NADPH oxidase complex is a multisubunit enzyme capable of the vectoral transport of electrons from NADPH to molecular oxygen to generate superoxide anions, which are in turn converted by both enzyme and nonenzyme catalyzed reactions to form a collection of ROS (Quinn and Gauss 2004). The NADPH oxidase can be assembled at the plasma membrane in response to stimulation by soluble inflammatory stimuli (e.g., fMLP, C5a) or during spreading on activating surfaces (e.g., fungal hyphae, immobilized immune complexes, fibrin, and other extracellular matrices). The NADPH oxidase can also be assembled on the phagosomal membrane during phagocytosis of pathogens. The ROS produced by the oxidase are thought to play an essential role in the killing of certain bacteria and fungi through direct and indirect mechanisms (e.g., the generation of hypochlorous acid and the generation of electrochemical gradients that govern the movement of other ions into the phagosome) (Cross and Segal 2004; Hampton et al. 1998).

The catalytic core of the oxidase is thought to consist of two intrinsic membrane proteins ($gp91^{phox}$, $p22^{phox}$), collectively known as cytochrome b558, and four soluble proteins, Rac, $p47^{phox}$ and a heterodimer of $p40^{phox}$ and $p67^{phox}$. Activation of the complex involves assembly of the soluble subunits around the cytochrome at an appropriate membrane location (Fig. 6). Current evidence suggests that activity depends on the formation of a complex between cytochrome b558, $p67^{phox}$, and GTP-Rac, whereas the $p47^{phox}$ and $p40^{phox}$ subunits are generally seen as adaptors promoting efficient assembly at the correct time and place (Groemping and Rittinger 2005; Sumimoto 2008). The combined activity of multiple signal transduction pathways delivers guanine nucleotide exchange on Rac at the correct membrane location and multiple phosphorylation of the C-terminus of $p47^{phox}$ (which releases it from an autoinhibited conformation to a conformation which can bind $p22^{phox}$; Quinn and Gauss 2004; Cross and Segal 2004; Groemping and Rittinger 2005; Sumimoto 2008). Further mutual interactions between each of the components and the membrane promote formation of the active complex (Fig. 6). The general PI3K inhibitor wortmannin was initially discovered as a potent inhibitor of neutrophil ROS formation (Arcaro and Wymann 1993) and we now know that both Class I and Class III PI3Ks play different but important roles in oxidase activation (Perisic et al. 2004).

Class IB PI3K activity is required for the generation of significant extracellular ROS in response to Gi-coupled GPCR agonists (e.g., fMLP and C5a) (Li et al. 2000; Hirsch et al. 2000; Sasaki et al. 2000). In human neutrophils, there is also a priming-dependent development of this response involving Class IA PI3Kδ (Condliffe et al. 2005). As described above, Class I PI3Ks are implicated in the activation of Rac and there is good evidence that at least part of this role for Class I PI3Ks in oxidase activation occurs via the activation of guanine nucleotide exchange on Rac2 by the PtdIns(3,4,5)P_3 and Gβγ regulated GEF, PRex1 (Welch et al. 2005; Dong et al. 2005; Kim and Dinauer 2001). However, this is unlikely to represent a complete explanation. The effects of PRex1 deletion in the mouse vary according to priming conditions and a body of data now suggests that activation of Rac 1/2 and its effectors is highly compartmentalized, involving multiple GEFs, including Vavs and DOCKs (Kunisaki et al. 2006; Kim et al. 2003), and the relative PI3K input into each of these pathways has yet to be defined. There is also good evidence that PtdIns(3,4)P_2 and PtdOH can bind directly to different regions of the PX domain of

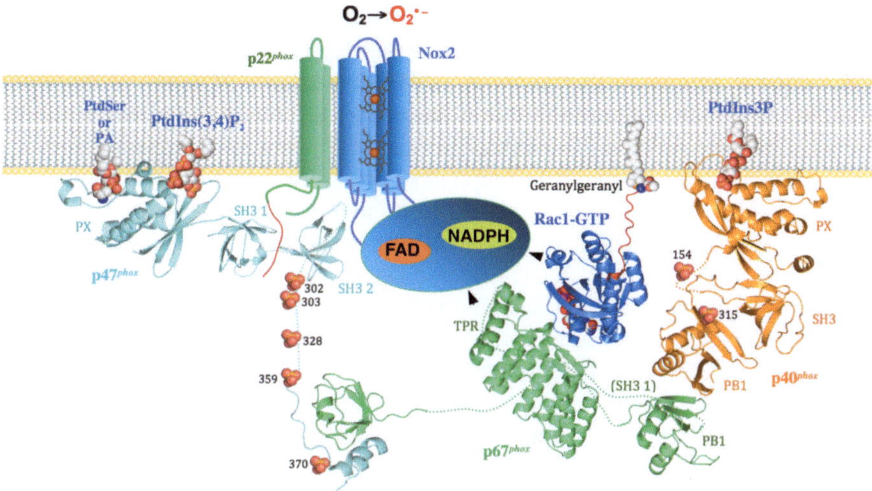

Fig. 6 The active NADPH oxidase complex. This figure is designed to illustrate key interactions in the assembly of an active oxidase complex, utilizing the available structural data and known points of interaction with membrane phospholipids. The scale and arrangement of structures was chosen for purposes of illustration based on biochemical requirements for NADPH oxidase activation. The phosphoryl symbol () indicates sites of phosphorylation that inhibit adjacent interactions. In p47^{phox}, several "activating" phosphorylations at the C-terminus (residues 302, 303, 328, 359) prevent binding of an adjacent polybasic domain to a groove between its own two SH3 domains, allowing this same groove to bind to the C-terminus of p22^{phox}; these phosphorylations are also thought to allow the PX domain to gain better access to phospholipids, but the mechanism of this effect is unknown. Phosphorylation at residue 370 is also shown but this is thought to be an "inhibitory" phosphorylation, preventing binding to p67^{phox}. The physiological function of phosphorylation at residues 154 and 315 in p40^{phox} is unknown. The structures used were p40^{phox} PX:PtdIns(3)P complex 1h6h.pdb (Bravo et al. 2001); p40^{phox} (Honbou et al. 2007); 2dyb.pdb, p47^{phox} PX 1o7k.pdb (Karathanassis et al. 2002); p67^{phox} TPR:Rac1-GTP complex, 1e96.pdb (Lapouge et al. 2002); p67^{phox} SH3: p47^{phox} polyproline tail, 1k4u.pdb (Kami et al. 2002; Massenet et al. 2005); p67^{phox} PB1: p40^{phox} PB1 complex, 1oey.pdb (Wilson et al. 2003); p47^{phox} tandem SH3 autoinhibitory complex 1ng2.pdb, and p47^{phox} tandem SH3: p22^{phox} complex, 1ov3.pdb (Groemping et al. 2003)

p47^{phox} (Fig. 6; Karathanassis et al. 2002), suggesting that these interactions may mediate both Class I PI3K and PLD regulation of the oxidase, although clear evidence for the importance of these interactions in intact cells is still lacking.

FcγR-activation of the oxidase requires Class I PI3K and may also involve PI3K-dependent regulation of Rac2 (Utomo et al. 2006; Kim and Dinauer 2001). This appears to be the case when these receptors are activated during phagocytosis of antibody-coated particles (oxidase is assembled at the cup and phagosome) or during spreading on immobilized antibody complexes ("frustrated phagocytosis"; oxidase assembled at the plasma membrane) (S Kulkarni, personal communication). However, the Rac GEFs most likely to be involved are members of the Vav family and the precise points of regulation by PtdIns(3,4,5)P_3/PtdIns(3,4)P_2 have yet to be fully defined. In contrast, phagocytosis of complement-opsonized bacteria and the consequent activation of the oxidase on the phagosome can proceed without Class I PI3K activity (Anderson et al. 2008); presumably in these circumstances guanine nucleotide exchange on Rac is organized differently (involving redundant use of both Rac1 and 2 isoforms; Karen Anderson, personal communication).

As described above, several groups have now shown that PtdIns3P appears to be universally synthesized on phagosomal membranes (Fig. 4b) and current evidence suggests that this is very likely to be generated by the activation of Class III PI3K (Vieira et al. 2001; Anderson et al. 2008). PtdIns3P binds with high affinity and specificity to the PX domain of $p40^{phox}$ and this seems to be important for phagosomal assembly of an active oxidase created via antibody-FcγR or iC3b-integrin recognition (Fig. 6; Ellson et al. 2006; Kanai et al. 2001; Suh et al. 2006). Neutrophils from mice carrying a point mutation in their $p40^{phox}$ PX domain which specifically prevents binding to PtdIns3P have significant defects in phagosomal oxidase activity (Ellson et al. 2006). Furthermore, a patient who carries an analogous mutation in the $p40^{phox}$ PX domain and whose neutrophils have severe defects in intracellular ROS reduction (Matute et al. 2009) has recently been identified. Taken together, these data provide the most definitive evidence for a PI3K-generated product playing a specific role in neutrophil physiology. However, the precise role played by PtdIns3P binding to $p40^{phox}$ is still unclear. PtdIns3P binding likely seems to increase the effective concentration of the soluble $p40^{phox}$/$p67^{phox}$ complex on the phagosomal membrane, but it may also promote further conformational effects (Tian et al. 2008).

7 PI3Ks in Neutrophils as Therapeutic Targets

Neutrophils clearly play an important role in the destruction of invading pathogens, but this carries with it the potential to cause significant damage to host tissue, for example, through the extraneous release of proinflammatory mediators, microbicidal proteases, and ROS. Thus, neutrophil activation must be very tightly controlled both in time and space. When this control breaks down and neutrophils are inappropriately activated they are thought to exacerbate several inflammatory pathologies, for example, ARDS, COPD, asthma, ischemia–reperfusion injury, rheumatoid arthritis, glomerulonephritis, and atherosclerosis (Moraes et al. 2006; Nathan 2006; Soehnlein and Weber 2009; Chen et al. 2009; Haynes 2007; Fougerat et al. 2009; Simpson et al. 2009). Therefore, limiting neutrophil activation in these circumstances is considered a plausible therapeutic approach. In this regard, inhibiting Class I PI3K signaling pathways has the potential to dampen neutrophil activation by a cocktail of different inflammatory stimuli in a manner that would be difficult to achieve by other means (e.g., antagonism of individual receptors). Support for this idea has come from studying the effects of genetic deletion of p110γ and p110δ in mouse models of inflammation, for example, in models of rheumatoid arthritis (Randis et al. 2008; Camps et al. 2005). Further, the ATP binding site of Class I PI3Ks is tractable to small molecule inhibition and early results with relatively selective p110γ and p110δ inhibitors appear promising (Rommel et al. 2007).

8 Conclusions

PI3K signaling pathways clearly play extensive and important roles in the regulation of neutrophil function. In particular, the elucidation of elements of GPCR signaling and oxidase regulation have revealed general principals of PI3K signaling that have proven applicable to many other cell types. However, the sheer complexity of these pathways, with their multiple effectors and regulators means it is hard to see past the use of multiple knock-in mutations and pharmacological inhibitors to deconvolute a satisfactory explanation of their functions. Further, the potential of these pathways to yield useful therapeutic targets remains to be tapped and there is clearly a need for substantial advances in our understanding of both in vitro and in vivo assays of neutrophil function before we can usefully connect perturbations in neutrophil signaling to whole animal physiology.

Acknowledgment The authors would like to acknowledge Su Kulkarni, Jatinder Juss and John Ferguson for critical reading of this manuscript.

References

Anderson KE, Boyle KB, Davidson K, Chessa TA, Kulkarni S, Jarvis GE, Sindrilaru A, Scharffetter-Kochanek K, Rausch O, Stephens LR, Hawkins PT (2008) CD18-dependent activation of the neutrophil NADPH oxidase during phagocytosis of *Escherichia coli* or *Staphylococcus aureus* is regulated by class III but not class I or II PI3Ks. Blood 112:5202–5211

Arcaro A, Wymann MP (1993) Wortmannin is a potent phosphatidylinositol 3-kinase inhibitor: the role of phosphatidylinositol 3,4,5-trisphosphate in neutrophil responses. Biochem J 296 (Pt 2):297–301

Balla T (2005) Inositol-lipid binding motifs: signal integrators through protein–lipid and protein–protein interactions. J Cell Sci 118:2093–2104

Boulven I, Levasseur S, Marois S, Pare G, Rollet-Labelle E, Naccache PH (2006) Class IA phosphatidylinositide 3-kinases, rather than p110 gamma, regulate formyl-methionyl-leucyl-phenylalanine-stimulated chemotaxis and superoxide production in differentiated neutrophil-like PLB-985 cells. J Immunol 176:7621–7627

Bourne HR, Weiner O (2002) A chemical compass. Nature 419:21

Bravo J, Karathanassis D, Pacold CM, Pacold ME, Ellson CD, Anderson KE, Butler PJ, Lavenir I, Perisic O, Hawkins PT, Stephens L, Williams RL (2001) The crystal structure of the PX domain from p40(phox) bound to phosphatidylinositol 3-phosphate. Mol Cell 8:829–839

Brinkmann V, Zychlinsky A (2007) Beneficial suicide: why neutrophils die to make NETs. Nat Rev Microbiol 5:577–582

Camps M, Ruckle T, Ji H, Ardissone V, Rintelen F, Shaw J, Ferrandi C, Chabert C, Gillieron C, Francon B, Martin T, Gretener D, Perrin D, Leroy D, Vitte PA, Hirsch E, Wymann MP, Cirillo R, Schwarz MK, Rommel C (2005) Blockade of PI3Kgamma suppresses joint inflammation and damage in mouse models of rheumatoid arthritis. Nat Med 11:936–943

Chen M, Kallenberg CG, Zhao MH (2009) ANCA-negative pauci-immune crescentic glomerulonephritis. Nat Rev Nephrol 5:313–318

Coggeshall KM, Nakamura K, Phee H (2002) How do inhibitory phosphatases work? Mol Immunol 39:521–529

Condliffe AM, Kitchen E, Chilvers ER (1998) Neutrophil priming: pathophysiological consequences and underlying mechanisms. Clin Sci (Lond) 94:461–471

Condliffe AM, Davidson K, Anderson KE, Ellson CD, Crabbe T, Okkenhaug K, Vanhaesebroeck B, Turner M, Webb L, Wymann MP, Hirsch E, Ruckle T, Camps M, Rommel C, Jackson SP, Chilvers ER, Stephens LR, Hawkins PT (2005) Sequential activation of class IB and class IA PI3K is important for the primed respiratory burst of human but not murine neutrophils. Blood 106:1432–1440

Corey S, Eguinoa A, Puyana-Theall K, Bolen JB, Cantley L, Mollinedo F, Jackson TR, Hawkins PT, Stephens LR (1993) Granulocyte macrophage-colony stimulating factor stimulates both association and activation of phosphoinositide 3OH-kinase and src-related tyrosine kinase(s) in human myeloid derived cells. EMBO J 12:2681–2690

Cross AR, Segal AW (2004) The NADPH oxidase of professional phagocytes–prototype of the NOX electron transport chain systems. Biochim Biophys Acta 1657:1–22

Dong X, Mo Z, Bokoch G, Guo C, Li Z, Wu D (2005) P-Rex1 is a primary Rac2 guanine nucleotide exchange factor in mouse neutrophils. Curr Biol 15:1874–1879

Ellson CD, Gobert-Gosse S, Anderson KE, Davidson K, Erdjument-Bromage H, Tempst P, Thuring JW, Cooper MA, Lim ZY, Holmes AB, Gaffney PR, Coadwell J, Chilvers ER, Hawkins PT, Stephens LR (2001a) PtdIns(3)P regulates the neutrophil oxidase complex by binding to the PX domain of p40(phox). Nat Cell Biol 3:679–682

Ellson CD, Anderson KE, Morgan G, Chilvers ER, Lipp P, Stephens LR, Hawkins PT (2001b) Phosphatidylinositol 3-phosphate is generated in phagosomal membranes. Curr Biol 11: 1631–1635

Ellson C, Davidson K, Anderson K, Stephens LR, Hawkins PT (2006) PtdIns3P binding to the PX domain of p40phox is a physiological signal in NADPH oxidase activation. EMBO J 25:4468–4478

Engelman JA, Luo J, Cantley LC (2006) The evolution of phosphatidylinositol 3-kinases as regulators of growth and metabolism. Nat Rev Genet 7:606–619

Ferguson GJ, Milne L, Kulkarni S, Sasaki T, Walker S, Andrews S, Crabbe T, Finan P, Jones G, Jackson S, Camps M, Rommel C, Wymann M, Hirsch E, Hawkins P, Stephens L (2007) PI(3) Kgamma has an important context-dependent role in neutrophil chemokinesis. Nat Cell Biol 9:86–91

Fougerat A, Gayral S, Malet N, Briand-Mesange F, Breton-Douillon M, Laffargue M (2009) Phosphoinositide 3-kinases and their role in inflammation: potential clinical targets in atherosclerosis? Clin Sci (Lond) 116:791–804

Gaullier JM, Ronning E, Gillooly DJ, Stenmark H (2000) Interaction of the EEA1 FYVE finger with phosphatidylinositol 3-phosphate and early endosomes. Role of conserved residues. J Biol Chem 275:24595–24600

Groemping Y, Rittinger K (2005) Activation and assembly of the NADPH oxidase: a structural perspective. Biochem J 386:401–416

Groemping Y, Lapouge K, Smerdon SJ, Rittinger K (2003) Molecular basis of phosphorylation-induced activation of the NADPH oxidase. Cell 113:343–355

Hampton MB, Kettle AJ, Winterbourn CC (1998) Inside the neutrophil phagosome: oxidants, myeloperoxidase, and bacterial killing. Blood 92:3007–3017

Hawkins PT, Anderson KE, Davidson K, Stephens LR (2006) Signalling through class I PI3Ks in mammalian cells. Biochem Soc Trans 34:647–662

Haynes DR (2007) Inflammatory cells and bone loss in rheumatoid arthritis. Arthritis Res Ther 9:104

Heit B, Tavener S, Raharjo E, Kubes P (2002) An intracellular signaling hierarchy determines direction of migration in opposing chemotactic gradients. J Cell Biol 159:91–102

Heit B, Liu L, Colarusso P, Puri KD, Kubes P (2008a) PI3K accelerates, but is not required for, neutrophil chemotaxis to fMLP. J Cell Sci 121:205–214

Heit B, Robbins SM, Downey CM, Guan Z, Colarusso P, Miller BJ, Jirik FR, Kubes P (2008b) PTEN functions to 'prioritize' chemotactic cues and prevent 'distraction' in migrating neutrophils. Nat Immunol 9:743–752

Helgason CD, Damen JE, Rosten P, Grewal R, Sorensen P, Chappel SM, Borowski A, Jirik F, Krystal G, Humphries RK (1998) Targeted disruption of SHIP leads to hemopoietic perturbations, lung pathology, and a shortened life span. Genes Dev 12:1610–1620

Hirsch E, Katanaev VL, Garlanda C, Azzolino O, Pirola L, Silengo L, Sozzani S, Mantovani A, Altruda F, Wymann MP (2000) Central role for G protein-coupled phosphoinositide 3-kinase gamma in inflammation. Science 287:1049–1053

Honbou K, Minakami R, Yuzawa S, Takeya R, Suzuki NN, Kamakura S, Sumimoto H, Inagaki F (2007) Full-length p40phox structure suggests a basis for regulation mechanism of its membrane binding. EMBO J 26:1176–1186

Insall RH, Machesky LM (2009) Actin dynamics at the leading edge: from simple machinery to complex networks. Dev Cell 17:310–322

Jakus Z, Fodor S, Abram CL, Lowell CA, Mocsai A (2007) Immunoreceptor-like signaling by beta 2 and beta 3 integrins. Trends Cell Biol 17:493–501

Kami K, Takeya R, Sumimoto H, Kohda D (2002) Diverse recognition of non-PxxP peptide ligands by the SH3 domains from p67(phox), Grb2 and Pex13p. EMBO J 21:4268–4276

Kanai F, Liu H, Field SJ, Akbary H, Matsuo T, Brown GE, Cantley LC, Yaffe MB (2001) The PX domains of p47phox and p40phox bind to lipid products of PI(3)K. Nat Cell Biol 3:675–678

Karathanassis D, Stahelin RV, Bravo J, Perisic O, Pacold CM, Cho W, Williams RL (2002) Binding of the PX domain of p47(phox) to phosphatidylinositol 3,4-bisphosphate and phosphatidic acid is masked by an intramolecular interaction. EMBO J 21:5057–5068

Kennedy AD, DeLeo FR (2009) Neutrophil apoptosis and the resolution of infection. Immunol Res 43:25–61

Kim C, Dinauer MC (2001) Rac2 is an essential regulator of neutrophil nicotinamide adenine dinucleotide phosphate oxidase activation in response to specific signaling pathways. J Immunol 166:1223–1232

Kim C, Marchal CC, Penninger J, Dinauer MC (2003) The hemopoietic Rho/Rac guanine nucleotide exchange factor Vav1 regulates *N*-formyl-methionyl-leucyl-phenylalanine-activated neutrophil functions. J Immunol 171:4425–4430

Kinchen JM, Doukoumetzidis K, Almendinger J, Stergiou L, Tosello-Trampont A, Sifri CD, Hengartner MO, Ravichandran KS (2008) A pathway for phagosome maturation during engulfment of apoptotic cells. Nat Cell Biol 10:556–566

Krugmann S, Anderson KE, Ridley SH, Risso N, McGregor A, Coadwell J, Davidson K, Eguinoa A, Ellson CD, Lipp P, Manifava M, Ktistakis N, Painter G, Thuring JW, Cooper MA, Lim ZY, Holmes AB, Dove SK, Michell RH, Grewal A, Nazarian A, Erdjument-Bromage H, Tempst P, Stephens LR, Hawkins PT (2002) Identification of ARAP3, a novel PI3K effector regulating both Arf and Rho GTPases, by selective capture on phosphoinositide affinity matrices. Mol Cell 9:95–108

Kunisaki Y, Nishikimi A, Tanaka Y, Takii R, Noda M, Inayoshi A, Watanabe K, Sanematsu F, Sasazuki T, Sasaki T, Fukui Y (2006) DOCK2 is a Rac activator that regulates motility and polarity during neutrophil chemotaxis. J Cell Biol 174:647–652

Lapouge K, Smith SJ, Groemping Y, Rittinger K (2002) Architecture of the p40–p47–p67phox complex in the resting state of the NADPH oxidase. A central role for p67phox. J Biol Chem 277:10121–10128

Ley K, Laudanna C, Cybulsky MI, Nourshargh S (2007) Getting to the site of inflammation: the leukocyte adhesion cascade updated. Nat Rev Immunol 7:678–689
Li Z, Jiang H, Xie W, Zhang Z, Smrcka AV, Wu D (2000) Roles of PLC-beta2 and -beta3 and PI3Kgamma in chemoattractant-mediated signal transduction. Science 287:1046–1049
Li Z, Hannigan M, Mo Z, Liu B, Lu W, Wu Y, Smrcka AV, Wu G, Li L, Liu M, Huang CK, Wu D (2003) Directional sensing requires G beta gamma-mediated PAK1 and PIX alpha-dependent activation of Cdc42. Cell 114:215–227
Li Z, Dong X, Wang Z, Liu W, Deng N, Ding Y, Tang L, Hla T, Zeng R, Li L, Wu D (2005) Regulation of PTEN by Rho small GTPases. Nat Cell Biol 7:399–404
Liu L, Puri KD, Penninger JM, Kubes P (2007) Leukocyte PI3Kgamma and PI3Kdelta have temporally distinct roles for leukocyte recruitment in vivo. Blood 110:1191–1198
Massenet C, Chenavas S, Cohen-Addad C, Dagher MC, Brandolin G, Pebay-Peyroula E, Fieschi F (2005) Effects of p47phox C terminus phosphorylations on binding interactions with p40phox and p67phox. Structural and functional comparison of p40phox and p67phox SH3 domains. J Biol Chem 280:13752–13761
Matute JD, Arias AA, Wright NA, Wrobel I, Waterhouse CC, Li XJ, Marchal CC, Stull ND, Lewis DB, Steele M, Kellner JD, Yu W, Meroueh SO, Nauseef WM, Dinauer MC (2009) A new genetic subgroup of chronic granulomatous disease with autosomal recessive mutations in p40 phox and selective defects in neutrophil NADPH oxidase activity. Blood 114:3309–3315
Moraes TJ, Zurawska JH, Downey GP (2006) Neutrophil granule contents in the pathogenesis of lung injury. Curr Opin Hematol 13:21–27
Nathan C (2006) Neutrophils and immunity: challenges and opportunities. Nat Rev Immunol 6:173–182
Nauseef WM (2007) How human neutrophils kill and degrade microbes: an integrated view. Immunol Rev 219:88–102
Nimmerjahn F, Ravetch JV (2008) Fcgamma receptors as regulators of immune responses. Nat Rev Immunol 8:34–47
Nishikimi A, Fukuhara H, Su W, Hongu T, Takasuga S, Mihara H, Cao Q, Sanematsu F, Kanai M, Hasegawa H, Tanaka Y, Shibasaki M, Kanaho Y, Sasaki T, Frohman MA, Fukui Y (2009) Sequential regulation of DOCK2 dynamics by two phospholipids during neutrophil chemotaxis. Science 324:384–387
Nishio M, Watanabe K, Sasaki J, Taya C, Takasuga S, Iizuka R, Balla T, Yamazaki M, Watanabe H, Itoh R, Kuroda S, Horie Y, Forster I, Mak TW, Yonekawa H, Penninger JM, Kanaho Y, Suzuki A, Sasaki T (2007) Control of cell polarity and motility by the PtdIns(3,4,5)P3 phosphatase SHIP1. Nat Cell Biol 9:36–44
Pacold ME, Suire S, Perisic O, Lara-Gonzalez S, Davis CT, Walker EH, Hawkins PT, Stephens L, Eccleston JF, Williams RL (2000) Crystal structure and functional analysis of Ras binding to its effector phosphoinositide 3-kinase gamma. Cell 103:931–943
Perisic O, Wilson MI, Karathanassis D, Bravo J, Pacold ME, Ellson CD, Hawkins PT, Stephens L, Williams RL (2004) The role of phosphoinositides and phosphorylation in regulation of NADPH oxidase. Adv Enzyme Regul 44:279–298
Pinho V, de Castro RR, Amaral FA, de Sousa LP, Barsante MM, de Souza DG, Alves-Filho JC, Cara DC, Hayflick JS, Rommel C, Ruckle T, Rossi AG, Teixeira MM (2007) Tissue- and stimulus-dependent role of phosphatidylinositol 3-kinase isoforms for neutrophil recruitment induced by chemoattractants in vivo. J Immunol 179:7891–7898
Puri KD, Doggett TA, Huang CY, Douangpanya J, Hayflick JS, Turner M, Penninger J, Diacovo TG (2005) The role of endothelial PI3Kgamma activity in neutrophil trafficking. Blood 106:150–157
Quinn MT, Gauss KA (2004) Structure and regulation of the neutrophil respiratory burst oxidase: comparison with nonphagocyte oxidases. J Leukoc Biol 76:760–781
Randis TM, Puri KD, Zhou H, Diacovo TG (2008) Role of PI3Kdelta and PI3Kgamma in inflammatory arthritis and tissue localization of neutrophils. Eur J Immunol 38:1215–1224
Rommel C, Camps M, Ji H (2007) PI3K delta and PI3K gamma: partners in crime in inflammation in rheumatoid arthritis and beyond? Nat Rev Immunol 7:191–201

Sai J, Raman D, Liu Y, Wikswo J, Richmond A (2008) Parallel phosphatidylinositol 3-kinase (PI3K)-dependent and Src-dependent pathways lead to CXCL8-mediated Rac2 activation and chemotaxis. J Biol Chem 283:26538–26547

Sarraj B, Massberg S, Li Y, Kasorn A, Subramanian K, Loison F, Silberstein LE, von Andrian U, Luo HR (2009) Myeloid-specific deletion of tumor suppressor PTEN augments neutrophil transendothelial migration during inflammation. J Immunol 182:7190–7200

Sasaki T, Irie-Sasaki J, Jones RG, Oliveira-dos-Santos AJ, Stanford WL, Bolon B, Wakeham A, Itie A, Bouchard D, Kozieradzki I, Joza N, Mak TW, Ohashi PS, Suzuki A, Penninger JM (2000) Function of PI3Kgamma in thymocyte development, T cell activation, and neutrophil migration. Science 287:1040–1046

Schymeinsky J, Then C, Sindrilaru A, Gerstl R, Jakus Z, Tybulewicz VL, Scharffetter-Kochanek K, Walzog B (2007) Syk-mediated translocation of PI3Kdelta to the leading edge controls lamellipodium formation and migration of leukocytes. PLoS One 2:e1132

Servant G, Weiner OD, Herzmark P, Balla T, Sedat JW, Bourne HR (2000) Polarization of chemoattractant receptor signaling during neutrophil chemotaxis. Science 287:1037–1040

Shin HW, Hayashi M, Christoforidis S, Lacas-Gervais S, Hoepfner S, Wenk MR, Modregger J, Uttenweiler-Joseph S, Wilm M, Nystuen A, Frankel WN, Solimena M, De Camilli P, Zerial M (2005) An enzymatic cascade of Rab5 effectors regulates phosphoinositide turnover in the endocytic pathway. J Cell Biol 170:607–618

Simonsen A, Tooze SA (2009) Coordination of membrane events during autophagy by multiple class III PI3-kinase complexes. J Cell Biol 186:773–782

Simpson JL, Phipps S, Gibson PG (2009) Inflammatory mechanisms and treatment of obstructive airway diseases with neutrophilic bronchitis. Pharmacol Ther 124:86–95

Smith DF, Deem TL, Bruce AC, Reutershan J, Wu D, Ley K (2006) Leukocyte phosphoinositide-3 kinase gamma is required for chemokine-induced, sustained adhesion under flow in vivo. J Leukoc Biol 80:1491–1499

Soehnlein O, Weber C (2009) Myeloid cells in atherosclerosis: initiators and decision shapers. Semin Immunopathol 31:35–47

Steinberg BE, Grinstein S (2008) Pathogen destruction versus intracellular survival: the role of lipids as phagosomal fate determinants. J Clin Invest 118:2002–2011

Stephens LR, Hughes KT, Irvine RF (1991) Pathway of phosphatidylinositol (3,4,5)-trisphosphate synthesis in activated neutrophils. Nature 351:33–39

Stephens L, Eguinoa A, Corey S, Jackson T, Hawkins PT (1993) Receptor stimulated accumulation of phosphatidylinositol (3,4,5)-trisphosphate by G-protein mediated pathways in human myeloid derived cells. EMBO J 12:2265–2273

Stephens LR, Eguinoa A, Erdjument-Bromage H, Lui M, Cooke F, Coadwell J, Smrcka AS, Thelen M, Cadwallader K, Tempst P, Hawkins PT (1997) The G beta gamma sensitivity of a PI3K is dependent upon a tightly associated adaptor, p101. Cell 89:105–114

Stephens L, Ellson C, Hawkins P (2002) Roles of PI3Ks in leukocyte chemotaxis and phagocytosis. Curr Opin Cell Biol 14:203–213

Stephens L, Milne L, Hawkins P (2008) Moving towards a better understanding of chemotaxis. Curr Biol 18:R485–R494

Suh CI, Stull ND, Li XJ, Tian W, Price MO, Grinstein S, Yaffe MB, Atkinson S, Dinauer MC (2006) The phosphoinositide-binding protein p40phox activates the NADPH oxidase during FcgammaIIA receptor-induced phagocytosis. J Exp Med 203:1915–1925

Suire S, Hawkins P, Stephens L (2002) Activation of phosphoinositide 3-kinase gamma by Ras. Curr Biol 12:1068–1075

Suire S, Condliffe AM, Ferguson GJ, Ellson CD, Guillou H, Davidson K, Welch H, Coadwell J, Turner M, Chilvers ER, Hawkins PT, Stephens L (2006) Gbetagammas and the Ras binding domain of p110gamma are both important regulators of PI(3)Kgamma signalling in neutrophils. Nat Cell Biol 8:1303–1309

Sumimoto H (2008) Structure, regulation and evolution of Nox-family NADPH oxidases that produce reactive oxygen species. FEBS J 275:3249–3277

Tian W, Li XJ, Stull ND, Ming W, Suh CI, Bissonnette SA, Yaffe MB, Grinstein S, Atkinson SJ, Dinauer MC (2008) Fc{gamma}R-stimulated activation of the NADPH oxidase: phosphoinositide-binding protein p40phox regulates NADPH oxidase activity after enzyme assembly on the phagosome. Blood 112:3867–3877

Traynor-Kaplan AE, Harris AL, Thompson BL, Taylor P, Sklar LA (1988) An inositol tetrakisphosphate-containing phospholipid in activated neutrophils. Nature 334:353–356

Utomo A, Cullere X, Glogauer M, Swat W, Mayadas TN (2006) Vav proteins in neutrophils are required for FcgammaR-mediated signaling to Rac GTPases and nicotinamide adenine dinucleotide phosphate oxidase component p40(phox). J Immunol 177:6388–6397

Vaillancourt M, Levasseur S, Tremblay ML, Marois L, Rollet-Labelle E, Naccache PH (2006) The Src homology 2-containing inositol 5-phosphatase 1 (SHIP1) is involved in CD32a signaling in human neutrophils. Cell Signal 18:2022–2032

Van Keymeulen A, Wong K, Knight ZA, Govaerts C, Hahn KM, Shokat KM, Bourne HR (2006) To stabilize neutrophil polarity, PIP3 and Cdc42 augment RhoA activity at the back as well as signals at the front. J Cell Biol 174:437–445

Vanhaesebroeck B, Leevers SJ, Ahmadi K, Timms J, Katso R, Driscoll PC, Woscholski R, Parker PJ, Waterfield MD (2001) Synthesis and function of 3-phosphorylated inositol lipids. Annu Rev Biochem 70:535–602

Venkateswarlu K, Brandom KG, Lawrence JL (2004) Centaurin-alpha1 is an in vivo phosphatidylinositol 3,4,5-trisphosphate-dependent GTPase-activating protein for ARF6 that is involved in actin cytoskeleton organization. J Biol Chem 279:6205–6208

Vieira OV, Botelho RJ, Rameh L, Brachmann SM, Matsuo T, Davidson HW, Schreiber A, Backer JM, Cantley LC, Grinstein S (2001) Distinct roles of class I and class III phosphatidylinositol 3-kinases in phagosome formation and maturation. J Cell Biol 155:19–25

Wang F, Herzmark P, Weiner OD, Srinivasan S, Servant G, Bourne HR (2002) Lipid products of PI(3)Ks maintain persistent cell polarity and directed motility in neutrophils. Nat Cell Biol 4:513–518

Weiner OD, Rentel MC, Ott A, Brown GE, Jedrychowski M, Yaffe MB, Gygi SP, Cantley LC, Bourne HR, Kirschner MW (2006) Hem-1 complexes are essential for Rac activation, actin polymerization, and myosin regulation during neutrophil chemotaxis. PLoS Biol 4:e38

Welch HC, Coadwell WJ, Ellson CD, Ferguson GJ, Andrews SR, Erdjument-Bromage H, Tempst P, Hawkins PT, Stephens LR (2002) P-Rex1, a PtdIns(3, 4, 5)P3- and Gbetagamma-regulated guanine-nucleotide exchange factor for Rac. Cell 108:809–821

Welch HC, Condliffe AM, Milne LJ, Ferguson GJ, Hill K, Webb LM, Okkenhaug K, Coadwell WJ, Andrews SR, Thelen M, Jones GE, Hawkins PT, Stephens LR (2005) P-Rex1 regulates neutrophil function. Curr Biol 15:1867–1873

Wilson MI, Gill DJ, Perisic O, Quinn MT, Williams RL (2003) PB1 domain-mediated heterodimerization in NADPH oxidase and signaling complexes of atypical protein kinase C with Par6 and p62. Mol Cell 12:39–50

PI 3-Kinase p110β Regulation of Platelet Integrin $\alpha_{IIb}\beta_3$

Shaun P. Jackson and Simone M. Schoenwaelder

Contents

Abstract Hemopoietic cells express relatively high levels of the type I phosphoinositide (PI) 3-kinase isoforms, with p110δ and γ exhibiting specialized signaling functions in neutrophils, monocytes, mast cells, and lymphocytes. In platelets, p110β appears to be the dominant PI 3-kinase isoform regulating platelet activation, irrespective of the nature of the primary platelet activating stimulus. Based on findings

S.P. Jackson (✉) and S.M. Schoenwaelder
Australian Centre for Blood Diseases, Monash University, 6th Level Burnet Building, Alfred Medical Research and Education Precinct (AMREP), 89 Commercial Road, Melbourne, VIC 3004, Australia
e-mail: Shaun.Jackson@med.monash.edu.au

C. Rommel et al. (eds.). *Phosphoinositide 3-kinase in Health and Disease*, Volume 1
Current Topics in Microbiology and Immunology 346, DOI 10.1007/82_2010_61

with isoform-selective p110β pharmacological inhibitors and more recently with p110β–deficient platelets, p110β appears to primarily signal downstream of G_i- and tyrosine kinase-coupled receptors. Functionally, inhibition of p110β kinase function leads to a marked defect in integrin $\alpha_{IIb}\beta_3$ adhesion and reduced platelet thrombus formation in vivo. This defect in platelet adhesive function is not associated with increased bleeding, suggesting that therapeutic targeting of p110β may represent a safe approach to reduce thrombotic complications in patients with cardiovascular disease.

Abbreviations

3-PPIs	3-Phosphorylated phosphoinositides
ADP	Adenosine diphosphate
CalDAG-GEF1	Calcium and DAG-dependent guanine nucleotide exchange factor
DAG	Diacylglycerol
fMLP	formyl-Met-Leu-Phe
GP	Glycoprotein
GPCR	G-protein coupled receptor
IP3	Inositol (1,4,5)-triphosphate
ITAM	Immunoreceptor tyrosine-based activation motif
PAR	Protease-activated receptor
PH	Pleckstrin homology
PI	Phosphoinositide
$PI(3,4)P_2$	Phosphoinositide (3,4)-biphosphate
$PI(3,4,5)P_3$	Phosphoinositide (3,4,5)-triphosphate
PKC	Protein kinase C
PLC	Phospholipase C
RIAM	Rap1-GTP interacting adapter molecule
TXA2	Thromboxane A_2
VWF	von Willebrand factor

1 Introduction

The platelet is one of the most extensively investigated and well-characterized blood cells, particularly in the context of stimulus-dependent regulation of cell adhesion. This focus reflects the specialized adhesive function of platelets as well as their central role in promoting arterial thrombus formation and cardiovascular disease. For example, platelet-rich arterial thrombi are the principal pathological process underlying the development of the acute coronary syndromes and ischemic stroke (Bhatt and Topol 2003), the two leading causes of death and disability in the developed world. Platelets play a central role in thrombus development through their ability to adhere to injured blood vessels (primary platelet adhesion) and to other platelets (platelet cohesion or aggregation) at sites of vascular injury (Ruggeri 2009). In this brief report, we focus on the role of PI 3-kinase signaling processes in regulating the adhesive function of platelets, specifically the major platelet integrin, $\alpha_{IIb}\beta_3$ (also termed glycoprotein (GP) IIb-IIIa), which is the key receptor mediating stable platelet aggregation and thrombus growth. We begin with a brief description of normal platelet adhesion and its regulation by intracellular signaling pathways, discuss the molecular basis by which PI 3-kinases regulate integrin $\alpha_{IIb}\beta_3$ adhesive function, and conclude by discussing the role of PI 3-kinases in regulating thrombus development and how this may be exploited therapeutically.

2 Basic Principles of Platelet Adhesion and Activation

The fundamental function of platelets is to "plug" leakages in blood vessels, thereby preventing excessive blood loss from the circulatory system. Platelets perform this vital function through their ability to rapidly adhere to subendothelial matrix proteins and to other activated platelets at sites of vascular injury, forming the primary hemostatic plug (Jackson 2007; Ruggeri 2009; Varga-Szabo et al. 2008). The formation of these plugs must occur rapidly and in all areas of the vasculature to effectively stem the flow of blood into traumatized tissues. This unique capacity of platelets requires the synergistic contribution of multiple adhesion receptor–ligand interactions that have sufficient tensile strength to resist the detaching effects of blood flow (Frojmovic 1998; Goto et al. 1998; Kulkarni et al. 2000; Ruggeri et al. 1999).

The adhesion of platelets to sites of vascular injury involves a multi-step adhesion process that is conceptually similar to leukocyte adhesion to endothelial cells at sites of inflammation (Lawrence and Springer 1991; Wagner and Frenette 2008). Platelet adhesion requires an initial tethering step that is mediated by the binding of the platelet GPIb receptor to the vascular adhesive protein von Willebrand factor (VWF) (Ruggeri 2009; Tschopp et al. 1974, Weiss et al. 1978). Individuals deficient in either VWF or GPIb have a severe bleeding problem, underlying the importance of this interaction to normal platelet function (Nurden and Caen 1975; Nurden and Nurden 2008). Once captured to the blood vessel wall, a second adhesion event is required to promote firm platelet adhesion and thrombus development. The β_1-integrins and GPVI are primarily responsible for promoting platelet arrest to the injured vessel wall (Baumgartner 1977; Baumgartner and Haudenschild 1972; Nieswandt et al. 2001; Savage et al. 1998), whereas the β_3-integrin ($\alpha_{IIb}\beta_3$) promotes stable platelet aggregation (Jackson 2007). Each of the major platelet adhesion receptors, including GPIb, GPVI, integrin $\alpha_{IIb}\beta_3$, and the β_1-integrins have been demonstrated to transduce activating signals following engagement of their respective ligands (Canobbio et al. 2004; Nieswandt and Watson 2003; Ozaki et al. 2005; Varga-Szabo et al. 2008; Watson et al. 2005), as a prerequisite step for global platelet activation induced by soluble agonists. As will be discussed below, despite their structural and functional diversity, platelet adhesion receptors appear to utilize a similar repertoire of signaling molecules to activate platelets, consisting of proximal receptor-associated Src family kinases, Syk, and one or more signaling adaptor proteins that induce activation of PI 3-kinase and phospholipase C (PLC)γ (predominantly PLCγ2) (Varga-Szabo et al. 2008). The subsequent generation of 3-phosphoinositides (PPIs) (Jackson et al. 2004; Kasirer-Friede et al. 2004; Yap et al. 2002; Zhang et al. 1996), calcium generation (Goto et al. 2006; Kasirer-Friede et al. 2004; Mazzucato et al. 2002; Nesbitt et al. 2002, 2003) and activation of Protein Kinase C (PKC) (Giuliano et al. 2003, Harper and Poole 2007, Kasirer-Friede et al. 2004) plays an important role in inducing cytoskeletal remodeling and the upregulation of integrin affinity, particularly integrin $\alpha_{IIb}\beta_3$.

3 Integrin $\alpha_{IIb}\beta_3$: The Key Receptor Mediating Stable Platelet Aggregation and Thrombus Development

Integrin $\alpha_{IIb}\beta_3$ is typically maintained in a resting (low affinity) conformation on the surface of resting platelets. As with all integrins, the adhesive function of the receptor is tightly regulated by signals generated from within the cell (inside-out signaling), leading to conformational changes in the receptors' extracellular domains, increasing ligand binding affinity for adhesive proteins such as fibrinogen and VWF (Shattil and Newman 2004). The importance of integrin $\alpha_{IIb}\beta_3$ in supporting the hemostatic function of platelets is underscored by the severe bleeding disorder suffered by individuals deficient in this receptor (termed Glanzmann's Thrombasthenia) (Nurden 2005, 2006). The conversion of integrin $\alpha_{IIb}\beta_3$ to a fully activated state is critically dependent on the generation of one or more soluble agonists at the sites of vascular injury, including adenosine diphosphate (ADP), thromboxane A2 (TXA2), and thrombin (Gachet and Hechler 2005; Hirano 2007; Ruggeri 2002; Shankar et al. 2006). Each of these agonists bind to specific G-protein coupled receptors (GPCRs) and stimulate platelet activation through well-defined signaling pathways (Fig. 1).

3.1 G_q

Central to platelet activation are the G_q-coupled receptors, including the human thrombin receptors, PAR1 and PAR4, the ADP purinergic receptor, $P2Y_1$, and the TXA2 receptors, TPα and TPβ (Martorell et al. 2008; Murugappan et al. 2004). Deficiency of G_q is associated with a major defect in platelet activation by a broad range of physiological agonists, effectively eliminating platelet aggregation and

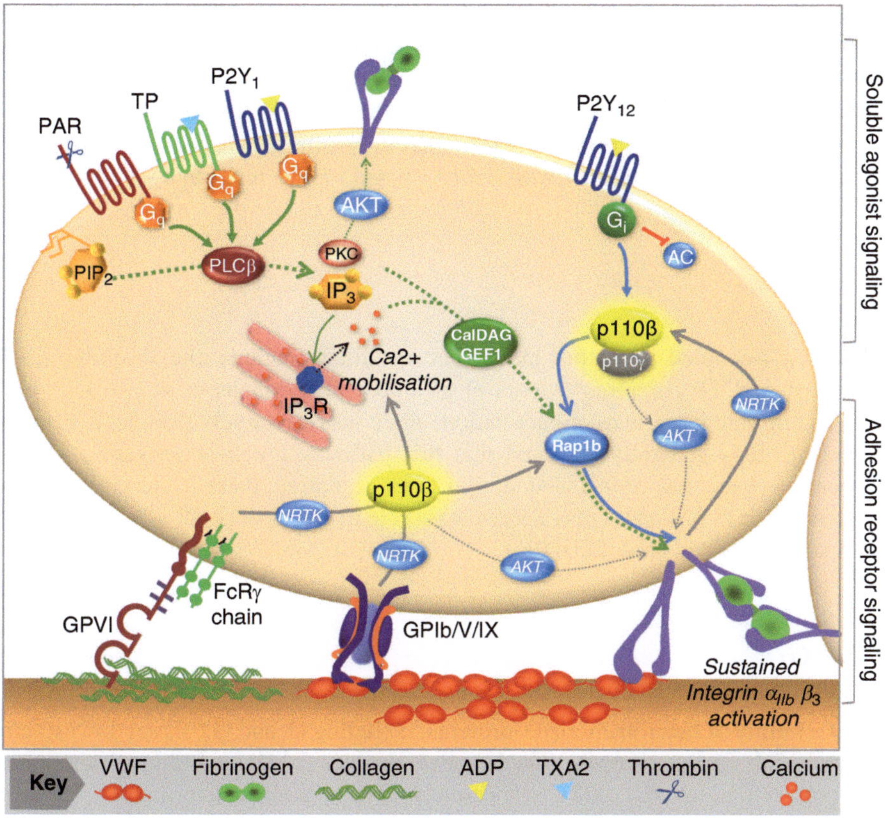

Fig. 1 A major role for the Type I PI 3-kinase p110β in regulating integrin $\alpha_{IIb}\beta_3$ adhesive function downstream from GPCRs and adhesion receptors. PI 3-kinase has been identified as a key signaling molecule required to sustain the major platelet integrin $\alpha_{IIb}\beta_3$ in a high-affinity state, and is therefore considered an important regulator of platelet adhesive function. *Adhesion receptor signaling leading to PI 3-kinase activation:* Ligation of GPIb/V/IX and GPVI with their ligands, VWF and collagen, respectively, initiates intracellular signaling via non-receptor tyrosine kinases (NRTK) leading to the activation of PI 3-kinase. Similarly, ligation of integrin $\alpha_{IIb}\beta_3$ with VWF or fibrinogen utilizes a similar signaling repertoire to stimulate PI 3-kinase. *GPCR signaling and activation of PI 3-kinase:* The local release and/or generation of soluble agonists at the sites of thrombus growth act to further promote platelet activation and adhesive function. Receptors for thrombin (PAR), TXA2 (TP) and ADP ($P2Y_1$) are G-protein coupled receptors linked to G_q. While stimulation of G_q-coupled receptors in platelets is considered a weak activator of PI 3-kinase, they are nevertheless pivotal for the initial upregulation of integrin $\alpha_{IIb}\beta_3$ affinity. Activation of G_q stimulates PLCβ activation, the generation of inositol (1,4,5)-triphosphate (IP_3) and diacylglycerol (DAG), leading to mobilization of intracellular calcium (Ca^{2+}) and the activation of PKC, respectively. Both PKC and calcium mobilization stimulate the guanine nucleotide exchange factor (GEF), CalDAG-GEF1, leading to Rap1b activation. PKC has also been proposed to directly activate AKT. ADP also signals through a second purinergic receptor $P2Y_{12}$, linked to G_i. Signaling via G_i-coupled receptors such as $P2Y_{12}$ primarily serves to amplify platelet activation, through the inhibition of adenylyl cyclase (AC), and the activation of PI 3-kinase. While both Type Ia p110β and Type Ib p110γ are activated downstream from GPCRs, p110β has been demonstrated to play a major role in sustaining integrin $\alpha_{IIb}\beta_3$ activation downstream from $P2Y_{12}$. Once activated, PI 3-kinase leads to sustained integrin $\alpha_{IIb}\beta_3$ activation, primarily through the activation of Rap1b. Other effectors of PI 3-kinase signaling include AKT, which has been suggested to directly phosphorylate the cytoplasmic tail of the integrin, as well as regulators of intraplatelet calcium concentration (for specific details, see Fig. 3)

leading to a severe bleeding tendency (Offermanns et al. 1997a). G_q-dependent PLCβ activation and the subsequent generation of inositol (1,4,5)-triphosphate (IP_3) and diacylglycerol (DAG), is a pivotal pathway regulating integrin $\alpha_{IIb}\beta_3$ affinity, through well-defined signaling intermediates (Fig. 1). As will be discussed below, there is currently considerable controversy whether G_q can stimulate significant levels of PI 3-kinase activation directly and induce integrin $\alpha_{IIb}\beta_3$ activation.

3.2 $G_{12/13}$

The $G_{12/13}$-coupled receptors (Moers et al. 2003) include the thrombin PAR receptors and TP receptors (Moers et al. 2004; Offermanns et al. 1994), which can also initiate integrin $\alpha_{IIb}\beta_3$ activation, albeit less effectively than G_q-dependent signaling. It is generally assumed that this pathway involves activation of Rho kinases (Klages et al. 1999), and similar to G_q signaling, there is currently limited evidence that PI 3-kinases play a significant role in $G_{12/13}$-dependent integrin $\alpha_{IIb}\beta_3$ activation (Jackson et al. 2004).

3.3 G_i

The G_i family of heterotrimeric G-proteins, including G_i and G_z, signal downstream of the ADP $P2Y_{12}$ receptor (Gachet and Hechler 2005) and the α_2-adrenergic receptor (Brass et al. 2007), respectively. These pathways primarily serve to amplify platelet activation, regardless of the nature of the primary activating stimulus. G_i-dependent PI 3-kinase activation is one of the major pathways regulating the production of 3-PPIs and this signaling pathway plays a major role in sustaining integrin $\alpha_{IIb}\beta_3$ activation (Gratacap et al. 2000; Kauffenstein et al. 2001; Selheim et al. 2000a; Trumel et al. 1999), primarily through the activation of Rap1b (Fig. 2) (Lova et al. 2002, 2003; Schoenwaelder et al. 2007).

3.4 PI 3-Kinase Signaling Downstream of G_q-Coupled Receptors?

It is often cited that PI 3-kinases play an important role in inducing integrin $\alpha_{IIb}\beta_3$ activation in response to platelet stimulation by soluble agonists such as thrombin and TXA2. In support of this, both agonists induce PI 3-kinase activation, the generation of 3-PPI and the phosphorylation of AKT (Goel et al. 2004; Kucera and Rittenhouse 1990). Furthermore, inhibition of either PI 3-kinases or AKT decreases integrin $\alpha_{IIb}\beta_3$ activation in response to low concentrations of thrombin and TXA2, leading to defective platelet aggregation (Jackson et al. 2005; Kovacsovics et al. 1995; Lauener et al. 1999; Schoenwaelder et al. 2007). Notably, AKT phosphorylation is abolished in G_q-deficient platelets, but not $G_i^{-/-}$ platelets,

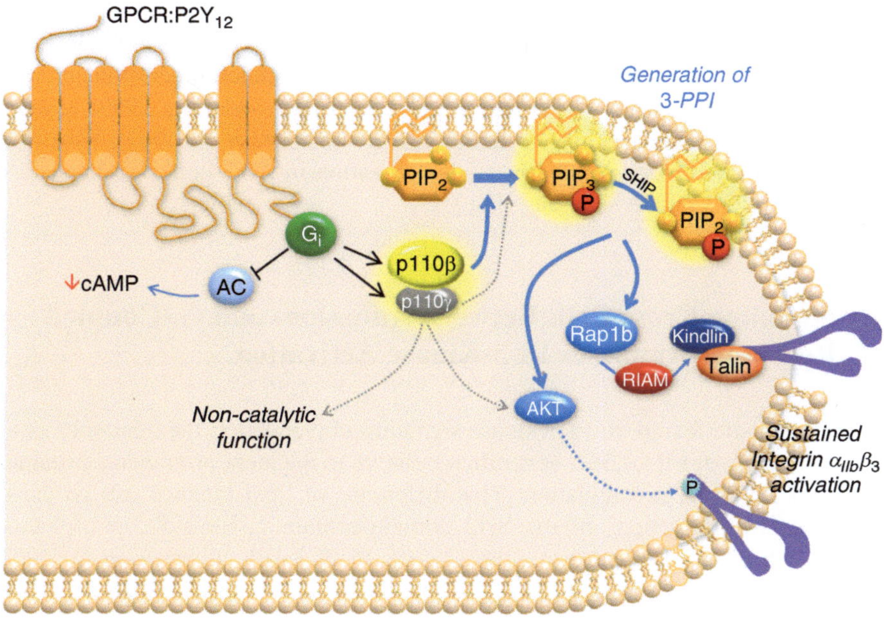

Fig. 2 PI 3-kinase signaling downstream from G_i-coupled receptors. Signaling downstream from G_i-linked receptors (including $P2Y_{12}$) represents the major signaling pathway leading to PI 3-kinase activation downstream from GPCRs, and is critical for sustained integrin $\alpha_{IIb}\beta_3$ activation. Ligation of $P2Y_{12}$ leads to the inhibition of adenylyl cyclase (AC) lowering cAMP, and the activation of PI 3-kinase, leading to the generation of 3-phosphorylated phosphoinositides (3-PPIs – PIP_3). An initial small transient phase of PI(3,4,5)P_3 (PIP_3) is followed by a more prominent, sustained phase of PI(3,4)P_2 (PIP_2) accumulation, partly because of the actions of the SH2 domain-containing inositol-5-phosphatase SHIP1. The generation of 3-PPIs leads to the activation of Rap1b, and its association with Rap1-GTP interacting adapter molecule (RIAM). This interaction in turn unmasks a talin-binding site on the integrin β_3 subunit, allowing talin to bind and activate integrin β_3, an event enhanced by the co-activator kindlin molecules. A potential non-catalytic function for p110γ has also been proposed downstream of G_i; however, the signaling intermediates involved remain unclear

suggesting a potentially important role for PI 3-kinase signaling processes downstream of G_q (Woulfe et al. 2004). However, these findings are not easily reconciled with the observation that PI 3-kinase activation in response to thrombin and TXA2 is largely eliminated by blocking feedback amplification by released ADP (signaling through G_i) and by inhibiting fibrinogen binding to integrin $\alpha_{IIb}\beta_3$ (Kim et al. 2002, 2004, 2006). Furthermore, most of the reported defects in integrin $\alpha_{IIb}\beta_3$ function are associated with defects in sustained receptor activation, rather than primary defects in the initiation of integrin $\alpha_{IIb}\beta_3$ activation.

A potential explanation for these apparent discrepancies may be due to the presence of PI 3-kinase-dependent and -independent pathways regulating AKT phosphorylation. For example, it has recently been demonstrated that initial rapid AKT activation downstream of G_q-coupled PAR receptors occurs independent of released ADP and PI 3-kinase, whereas sustained AKT phosphorylation is dependent on both ADP and PI 3-kinase (Resendiz et al. 2007). PI 3-kinase-independent AKT phosphorylation has been reported in mammary epithelial cells via calcium/calmodulin-dependent protein kinase kinase (CaMKK) (Soderling 1999) or alternatively, through direct phosphorylation by PKC (Kroner et al. 2000). Thus, it is possible that the defect in integrin $\alpha_{IIb}\beta_3$ activation in AKT-deficient platelets is unrelated to PI 3-kinase signaling (Fig. 1).

4 Signaling Cross Talk Between Adhesion- and G_i-Coupled Receptors Amplifies PI 3-Kinase Activation

A characteristic feature of PI 3-kinase signaling in platelets is the relatively small transient phase of $PI(3,4,5)P_3$ generation, relative to the more prominent, sustained phase of $PI(3,4)P_2$ accumulation. This difference in lipid kinetics can be partly explained by the actions of the SH2 domain-containing inositol-5-phosphatase SHIP1, which rapidly hydrolyses $PI(3,4,5)P_3$ to $PI(3,4)P_2$ (Giuriato et al. 2003; Maxwell et al. 2004; Pasquet et al. 2000). It has also been proposed that these temporal differences may be explained by the action of a Type II PI 3-kinase, converting PI(4)P to $PI(3,4)P_2$ (Zhang et al. 1998) or, alternatively, through inhibition of $PI(3,4)P_2$ 4-phosphatase, leading to the cellular accumulation of $PI(3,4)P_2$ (Munday et al. 1999), although there is currently limited experimental evidence supporting these latter hypotheses. Regardless of the mechanisms controlling 3-PPI metabolism, it is clear that robust PI 3-kinase activation in platelets requires the cooperative input of multiple activating stimuli, typically involving G_i- and adhesion receptor signaling. For example, following platelet stimulation with thrombin, TXA2 or ADP, the generation of PI 3-kinase lipid products is markedly reduced by inhibiting the G_i-coupled $P2Y_{12}$ receptor and by blocking ligand binding to integrin $\alpha_{IIb}\beta_3$ (Guinebault et al. 1993, 1995; Selheim et al. 1999; Sultan et al. 1991) Similarly, adhesion-dependent activation of PI 3-kinase is markedly reduced by the elimination of released ADP or the blockade of its $P2Y_{12}$ receptor (Dorsam et al. 2002; Dorsam and Kunapuli 2004; Gironcel et al. 1996; Gratacap et al. 2000; Jackson et al. 2003; Larson et al. 2003; Moake et al. 1988; Moritz et al. 1983; Oda et al. 1995). Thus, on the basis of currently available evidence, it is reasonable to conclude that PI 3-kinases are primarily regulated downstream of G_i- and tyrosine kinase-coupled receptors in platelets (Fig. 1).

5 Functional Role of Type I PI 3-Kinases in Platelets

As with most cell types, the initial evidence supporting an important role for PI 3-kinases in platelet function was derived from the studies using the pan-PI 3-kinase inhibitors, LY294002 and wortmannin. These inhibitors modulate a broad range of functional platelets responses, most notably sustained integrin $\alpha_{IIb}\beta_3$ activation and stable platelet aggregation (Jackson et al. 2004; Selheim et al. 2000b). The kinetics of $PI(3,4)P_2$ accumulation (Kucera and Rittenhouse 1990; Nolan and Lapetina 1990; Sultan et al. 1990; Trumel et al. 1999), correlating with Raplb activation (Lova et al. 2003; Schoenwaelder et al. 2007; Woulfe et al. 2002), suggest a potentially important role for this lipid in maintaining high affinity integrin $\alpha_{IIb}\beta_3$ bonds (Kovacsovics et al. 1995). PI 3-kinase activation and $PI(3,4)P_2$ production also occurs downstream of integrin $\alpha_{IIb}\beta_3$ where it participates in a number of integrin $\alpha_{IIb}\beta_3$ outside-in signaling events linked to cytoskeletal remodeling (Hartwig et al. 1996), platelet spreading (Heraud et al. 1998; Yap et al. 2002) and the cellular transmission of contractile forces (Martin et al. 2010; Schoenwaelder et al. 2010).

Whilst studies with wortmannin and LY294002 have been informative in delineating the roles of PI 3-kinases in regulating specific platelet functional responses, they have provided limited insight into the role of individual PI 3-kinase isoforms in these processes. The first definitive evidence for an important role for Type Ia PI 3-kinase isoforms in platelet function was derived from the study of p85-deficient mice (Watanabe et al. 2003). Deletion of the p85 regulatory subunit has no significant effect on the expression of the other type Ia regulatory subunits; however, it leads to a major reduction in the levels of the p110α catalytic subunit, along with greatly reduced p110β and δ levels (Suzuki et al. 1999; Terauchi et al. 1999; Watanabe et al. 2003). Significantly, these platelets retain a normal platelet aggregation response to soluble agonists, including ADP, TxA2, and thrombin; however, platelet activation downstream of the Immunoreceptor Tyrosine-based Activation Motif (ITAM)-based adhesion receptor, GPVI, was significantly reduced (Watanabe et al. 2003). Functional studies on $p110\delta^{-/-}$ platelets have revealed a normal platelet aggregation response to soluble agonists (ADP and thrombin) with a mild defect in GPVI signaling, a finding consistent with the low-level expression of p110δ in platelets (Schoenwaelder et al. 2007; Senis et al. 2005). These studies had demonstrated a potentially important role for Type Ia PI 3-kinases in regulating ITAM signaling in platelets (Senis et al. 2005).

The role of Type Ib PI 3-kinase isoform p110γ in platelets has been investigated using $p110\gamma^{-/-}$ and $p110\gamma^{KD}$ mice (Canobbio et al. 2009; Hirsch et al. 2001; Jackson et al. 2005; Lian et al. 2005; Schoenwaelder et al. 2007). Platelets from these mice demonstrate normal ATP secretion, F-actin assembly, and normal aggregation profiles in response to thrombin and high-dose collagen (Hirsch et al. 2001; Lian et al. 2005); however; they have an impaired aggregation response to ADP and threshold concentrations of collagen or convulxin (Canobbio et al. 2009; Hirsch et al. 2001; Lian et al. 2005; Schoenwaelder et al. 2007) and a reduced spreading response on immobilized fibrinogen (Lian et al. 2005). These defects may in part be explained by impaired ADP-induced integrin $\alpha_{IIb}\beta_3$ activation. Detailed biochemical analysis of $p110\gamma^{-/-}$ platelets have revealed a defect in signaling through the ADP purinergic receptor, $P2Y_{12}$, leading to impaired Akt activation (Hirsch et al. 2001).

5.1 Roles of p110β and p110γ in G_i-Signaling

The $Gi_{\beta\gamma}$ dimers are highly effective at inducing the activation of p110β and p110γ; however, the relative contribution of these isoforms to G_i signal transduction appears to be cell type and stimulus-dependent (Murga et al. 1998, 2000; Condliffe et al. 2005; Van Kolen and Slegers 2004). Recent direct comparisons of the roles of p110β and p110γ in platelet G_i-signaling has demonstrated a major role for p110β in mediating $PI(3,4)_2$ and PI(3,4,5)P3 generation (Martin et al. 2010; Schoenwaelder et al. 2007), AKT phosphorylation and activation of Rap1b (Schoenwaelder et al. 2007; Canobbio et al. 2009), It has previously been proposed that given the limited tissue distribution of p110γ, p110β is likely to be the major PI3-kinase isoform transducing G_i-dependent signals in most cell types (Guillermet-Guibert et al. 2008; Murga et al. 2000). However, the situation in hemopoietic cells is potentially different as these cells contain abundant levels of both p110β and p110γ. Furthermore, on the basis of studies in formyl-Met-Leu-Phe (fMLP) stimulated $p110\gamma^{-/-}$ neutrophils (Condliffe et al. 2005), p110γ appears to play the major role in G_i-dependent superoxide formation and $PI(3,4,5)P_3$ production (Condliffe et al. 2005). The situation in platelets appears to be different in that p110γ appears to play a relatively minor role in $PI(3,4)P_2$ production, an unexpected finding given the well-defined ability of $Gi_{\beta\gamma}$ dimers to promote p110γ catalytic function (Maier et al. 1999). Notably, the level of 3-PPIs induced by G_i-coupled agonists in platelets is considerably lower than in neutrophils, perhaps reflecting a reduced level of G_i stimulation in the former cell types. It is possible that the number of free $Gi_{\beta\gamma}$ dimers is the critical determinant controlling the catalytic function of p110γ and p110β; however, it is also possible that other cell-type specific molecules may influence the activation state of individual PI 3-kinase isoforms (Schoenwaelder et al. 2007) .

5.2 Contribution of PI 3-Kinase Isoforms to Adhesion Receptor Signaling

The contribution of individual PI 3-kinase isoforms to adhesion receptor signaling is currently under investigation, although based on preliminary findings it appears that p110β is the major isoform operating downstream of the major platelet adhesion receptors, including GPIb, integrin $\alpha_{IIb}\beta_3$ and GPVI (Fig. 1) (Canobbio et al. 2009; Jackson et al. 2005; Martin et al. 2010). Furthermore, although accurate analysis of the relative abundance of individual PI 3-kinase isoforms has not been definitively established in platelets, most of the p85-associated PI 3-kinase activity in platelets appears to be due to p110β (unpublished observations, SMS and SPJ). Thus, the relatively high level expression of p110β may partly explain its relative signaling dominance over other PI 3-kinase isoforms in platelets.

The mechanisms by which adhesion receptors activate PI 3-kinases in platelets have been most clearly defined from the investigation of GPVI signal transduction

(Varga-Szabo et al. 2008; Watson et al. 2005) (Fig. 3). GPVI binding to collagen results in the clustering of ITAM-bearing FcR γ-chains. Phosphorylation of the cytoplasmic ITAM domains by the Src kinases, $p59^{fyn}$ and $p53/56^{lyn}$, provides binding sites for Syk (Briddon and Watson 1999; Ezumi et al. 1998; Quek et al. 2000; Suzuki-Inoue et al. 2002, 2004). Subsequent recruitment of the adaptor proteins, SLP-76 and LAT (Linker for Activation of T cells), promotes accumulation of numerous signaling proteins (Clements et al. 1999; Pasquet et al. 1999b), including PI 3-kinases, Grb2, Vav, WASP, and PKC (Watson et al. 2005). Ultimately, this signaling cascade leads to the phosphorylation and activation of PLCγ2 generating a robust calcium signal that promotes efficient platelet activation

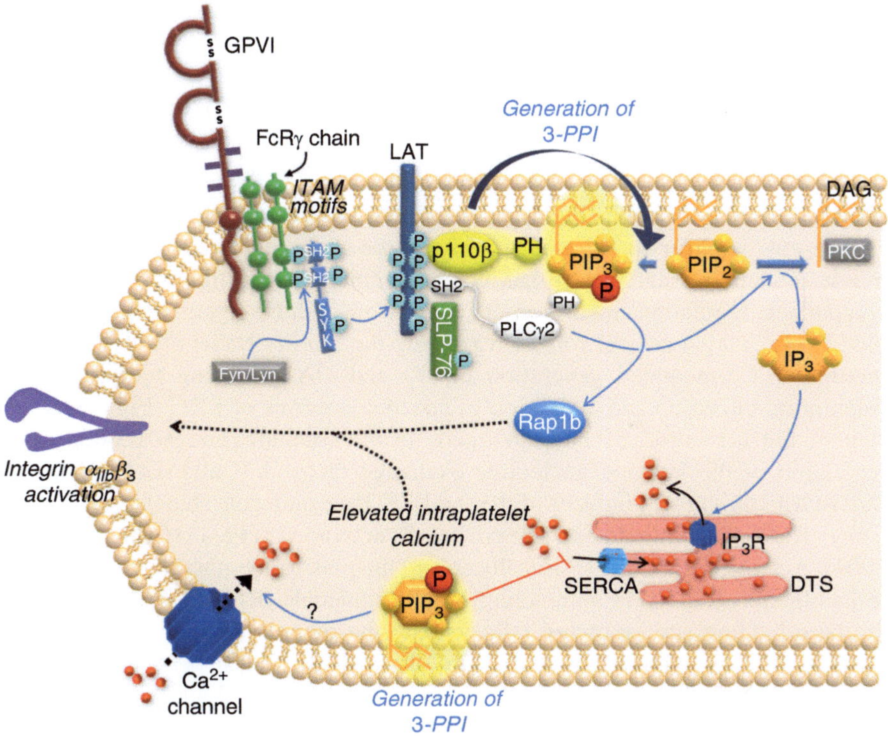

Fig. 3 PI 3-kinase signaling downstream from GPVI. GPVI binding to collagen results in the clustering of ITAM-bearing FcR γ-chains. Phosphorylation of ITAM domains by the Src kinases, Fyn and Lyn (Fyn/Lyn), induces the binding of Syk, accumulation of the adaptor proteins, SLP-76 and LAT (Linker for Activation of T cells), and recruitment and activation of PLCγ2 and PI 3-kinase. PI 3-kinases and the generation of 3-PPIs contribute to integrin $\alpha_{IIb}\beta_3$ regulation at several key steps in this pathway. For example, PI3K enhances the signaling function of PLCγ2, potentiating cytosolic calcium flux downstream of GPVI. 3-PPIs bind the PH domain of PLCγ2, bringing it into closer proximity with its lipid substrates at the plasma membrane. PI 3-kinases have also been proposed to enhance transmembrane calcium flux (Ca^{2+} channel) and sustain intracellular calcium levels by inhibiting re-uptake through the sarco(endo)plasmic reticulum Ca^{2+} ATPase (SERCA)

(Gratacap et al. 1998; Pasquet et al. 1999a). 3-PPI bind the pleckstrin homology (PH) domains of Tec kinases which increase the phosphorylation and activation of PLCγ2 (Quek et al. 1998). Similarly, 3-PPI bind the PH domain of PLCγ2, thereby enhancing its localization to the plasma membrane in close proximity to its lipid substrates (Falasca et al. 1998). Studies employing pharmacological inhibition of p110β or p110β^{KD} mice have confirmed an important role for this isoform in GPVI-mediated aggregation and AKT phosphorylation (Canobbio et al. 2009; Gilio et al. 2009; Kim et al. 2009; Martin et al. 2010). PI 3-kinase signaling pathways have also been demonstrated to enhance transmembrane calcium flux (Pasquet et al. 2000) and promote activation of Rap1b downstream of GPVI (Canobbio et al. 2009; Gilio et al. 2009), two additional signaling mechanisms that can enhance sustained integrin $\alpha_{IIb}\beta_3$ activation.

5.3 PI 3-Kinase Modulation of Integrin $\alpha_{IIb}\beta_3$ Affinity

Before describing the potential mechanisms by which PI 3-kinases regulate integrin $\alpha_{IIb}\beta_3$ affinity, we will highlight the recent progress in the understanding of "inside-out" integrin $\alpha_{IIb}\beta_3$ regulation. From the study of genetically manipulated mouse models, a clear picture has begun to emerge as to the critical signaling components regulating integrin $\alpha_{IIb}\beta_3$ activation. Soluble agonist receptors coupled to Gα_q play a critical role in regulating the affinity status of integrin $\alpha_{IIb}\beta_3$ through PLCβ-mediated generation of IP_3 and DAG, leading to intracellular calcium mobilization and activation of specific isoforms of PKC (Fig. 1). A key step in the activation of integrin $\alpha_{IIb}\beta_3$ is the calcium- and DAG-dependent regulation of the guanine nucleotide exchange factor 1 (CalDAG-GEF1). CalDAG-GEF1 is a potent inducer of the small GTPase molecule Rap1b, which plays a key effector role in inducing integrin $\alpha_{IIb}\beta_3$ activation (Fig. 1) (Crittenden et al. 2004). Activated Rap1b-GTP translocates to the plasma membrane where it binds to its effector, Rap1-GTP interacting adapter molecule (RIAM). RIAM unmasks the talin-binding site, allowing talin to bind the cytoplasmic tail of integrin β_3 subunits, thereby activating $\alpha_{IIb}\beta_3$ (Tadokoro et al. 2003). The activation of integrins by talin is enhanced by the co-activators, kindlin-2 and kindlin-3. Kindlins bind to the cytoplasmic tail of β-integrin at sites distinct from those bound by talin (Fig. 2). The C-terminal region of kindlin-3 aids conformational activation of $\alpha_{IIb}\beta_3$ by connecting it to the cytoskeletal protein actin (Fig. 2) (Siegel et al. 2003; Ussar et al. 2006). The importance of this signaling pathway in regulating $\alpha_{IIb}\beta_3$ adhesive function is underscored by the severe bleeding defects experienced by mice lacking either Gα_q (Offermanns et al. 1997b), CalDAG-GEF1 (Bergmeier et al. 2007), Rap1b (Chrzanowska-Wodnicka et al. 2005), talin-1 (Nieswandt et al. 2007) or kindlin-3 (Moser et al. 2008).

PI 3-kinases can contribute to integrin $\alpha_{IIb}\beta_3$ regulation at several key steps in this pathway. As mentioned earlier, it can enhance the signaling function of PLCγ2 and potentiate cytosolic calcium flux downstream of platelet adhesion receptors (Fig. 1). PI 3-kinases can also enhance transmembrane calcium flux (Lu et al. 1998; Pasquet et al. 2000), necessary for sustained integrin $\alpha_{IIb}\beta_3$ activation (Fig. 3). PI 3-kinase activation of Rap1b is likely to represent a major mechanism sustaining integrin $\alpha_{IIb}\beta_3$ activation, particularly downstream of the $P2Y_{12}$ G_i-coupled receptor (Fig. 2). Analysis of isoform selectivity has indicated a major role for p110β in this role (Canobbio et al. 2009; Schoenwaelder et al. 2007). Finally, PI 3-kinase phosphorylation of AKT has also been proposed to stimulate integrin $\alpha_{IIb}\beta_3$ activation, although whether this is through direct or indirect phosphorylation of β-integrin cytoplasmic tails remains to be established.

5.4 *Potential Role for PI 3-Kinases in Regulating Integrin $\alpha_{IIb}\beta_3$ Avidity*

Whilst the vast majority of studies on PI 3-kinase signaling in platelets have focused on integrin $\alpha_{IIb}\beta_3$ affinity regulation, the formation of stable integrin adhesion bonds is also dependent on receptor avidity (adhesion strength). Integrin avidity is influenced by the intrinsic binding kinetics of individual ligand–receptor bonds (affinity) as well as the total number of bonds (valency) (Carman and Springer 2003). Recent studies have demonstrated that PI 3-kinase signaling processes play a potentially important role in stabilizing high-affinity integrin $\alpha_{IIb}\beta_3$ bonds on a fibrinogen matrix (Schoenwaelder et al. 2010), with isoform selective inhibitors and genetic mouse models again identifying p110β as the major isoform responsible for this effect (Schoenwaelder et al. 2010). A defect in receptor avidity in the presence of pan PI 3-kinase or p110β inhibitors is most clearly manifest under conditions of sparse ligand density, suggesting a potentially important role for p110β in promoting adhesion strengthening. Several factors are known to influence the valency of integrin bonds, including the density of both receptors and ligands at the adhesive surface, the geometric arrangement of these surfaces, as well as the mobility of the receptor and/or ligand (Carman and Springer 2003). Of these potential mechanisms, one possibility is that PI 3-kinase p110β is regulating integrin receptor distribution as a result of changes in cytoskeletal reorganization. PI 3-kinase inhibitors reduce the cytoskeletal anchorage of integrin receptors in platelets (Schoenwaelder et al. 2010), a process that is essential for the ability of actin cables to cluster integrins into focal adhesion-like complexes and transmit contractile forces to extracellular matrices (Burridge and Chrzanowska-Wodnicka 1996; Martin et al. 2010; Schoenwaelder et al. 2010). Similar findings have been reported in nucleated cells, including PI 3-kinase-dependent clustering of integrin $\alpha_V\beta_1$ (in human umbilical vein endothelial cells), $\alpha_L\beta_2$ (in leukocytes) and $\alpha_6\beta_4$ (in breast cell carcinoma), leading to enhanced receptor avidity (Dormond et al. 2004; Gilcrease et al. 2004; Jones et al. 1998; Krauss and Altevogt 1999; Lynch et al. 2005).

5.5 PI 3-Kinase and Shear-Resistant Platelet Adhesion

Given the importance of PI 3-kinase signaling processes in sustaining high affinity integrin $\alpha_{IIb}\beta_3$ bonds, in combination with its potential role in regulating receptor clustering and avidity, it is not surprising that PI 3-kinase inhibitors are highly effective at reducing the stability of integrin $\alpha_{IIb}\beta_3$ adhesion contacts, leading to unstable platelet aggregation. This is particularly prominent under conditions of rapid blood flow, in which hemodynamic shear stresses impose considerable tensile forces on adhesive bonds, leading to a marked defect in shear-resistant platelet adhesion. In both in vitro and in vivo thrombosis models, inhibition of PI 3-kinase undermines the stability of platelet aggregates, leading to unstable thrombus development (Cosemans et al. 2006; Hirsch et al. 2001; Jackson et al. 2005; Lian et al. 2005; Schoenwaelder et al. 2007; Yap et al. 2002). Given the important role of p110β in integrin $\alpha_{IIb}\beta_3$ regulation, isoform-selective inhibitors against p110β are particularly effective at undermining the stability of platelet aggregates in vivo and preventing thrombotic occlusion of injured arteries in experimental animal models (Jackson et al. 2005; Martin et al. 2010). Perhaps more surprising is the relatively weak effect these inhibitors have on haemostatic platelet plug formation (Canobbio et al. 2009; Jackson and Schoenwaelder 2006; Jackson et al. 2005; Martin et al. 2010), a finding that may be partly explained by the limited effect of p110β inhibitors on initial integrin $\alpha_{IIb}\beta_3$ activation in response to soluble agonist stimulation (Jackson et al. 2005; Schoenwaelder et al. 2007).

6 PI 3-Kinase Inhibitors as Potential Antithrombotic Agents

The interest in targeting PI 3-kinase as a novel antithrombotic strategy is based on two important findings; first, genetic deficiency of individual PI 3-kinase isoforms or inhibition of their kinase function does not lead to a substantial increased risk of bleeding (Canobbio et al. 2009; Jackson et al. 2005; Martin et al. 2010); and second, p110β inhibitors can be combined with anticoagulant agents (drugs that limit blood coagulation and fibrin generation), without substantially increasing bleeding time (Jackson et al. 2005). This is a particularly attractive feature since commonly used antiplatelet therapies, such as aspirin and clopidogrel markedly increase bleeding time when combined with anticoagulants. Combination antithrombotic therapies are increasingly used in the clinic, partly because of the growing awareness of antithrombotic drug resistance (Jackson and Schoenwaelder 2003); however, more intensive antithrombotic approaches inevitably lead to an increased risk of bleeding, sometimes fatal in high-risk patients. This is particularly relevant to combined antiplatelet and anticoagulant therapies; therefore, there is a pressing need for safer combinations of antithrombotic drugs.

While a strong case can be made for the clinical evaluation of p110β inhibitors, the long-term tolerance of p110β inhibition remains uncertain, with concern over the ubiquitous expression of this isoform and the metabolic disturbances evident in p110β null mice (Jia et al. 2008). Whether partial inhibition of the kinase by pharmacological approaches will be more effectively tolerated than complete deficiency of the kinase remains to be established. From an antithrombotic perspective, rapidly acting short-term therapies for the management of acute coronary syndromes and ischemic stroke are important and the possibility of using PI3K inhibitors in combination with other antithrombotic agents without increasing bleeding risk, particularly intracerebral hemorrhage, is an attractive option. If p110β inhibitors perform well in such situations, strategies for the safe long-term use of these compounds will most likely be developed. Foremost among these is the development of irreversible p110β inhibitors. Platelets are anucleate and therefore have limited capacity to synthesize new protein, thus irreversible inhibitors typically block their therapeutic target for the life of the platelet ($t_{1/2}$ of ~4 days for human platelets), whereas all other nucleated cells have the capacity to synthesize new protein and overcome the effects of the inhibitor. This has been successfully exploited with aspirin, which blocks the function of its ubiquitous target, cyclo-oxygenase I (COX-1), however given aspirin's short half-life (<30 mins) and the ability of all other tissues to synthesize COX-1, the deleterious effects of aspirin on these tissues is minimalized. Such features partly explain the excellent long term tolerance of aspirin.

7 Conclusions

Considerable progress has been made in defining the mechanisms by which PI 3-kinases regulate integrin $\alpha_{IIb}\beta_3$ adhesive function and platelet thrombus development. Given that the mechanism by which p110β inhibitors dampen the platelet thrombus formation is distinct from that employed by conventional antithrombotic approaches, it is likely that p110β inhibitors may provide additional clinical benefit when combined with existing therapies. The new insight on integrin $\alpha_{IIb}\beta_3$ regulation provided by p110β inhibitors should be helpful in designing future therapeutic strategies to reduce pathological thrombus development whilst minimizing the impact on physiological hemostasis.

References

Baumgartner HR (1977) Platelet interaction with collagen fibrils in flowing blood. I. Reaction of human platelets with alpha chymotrypsin-digested subendothelium. Thromb Haemost 37:1–16

Baumgartner HR, Haudenschild C (1972) Adhesion of platelets to subendothelium. Ann N Y Acad Sci 201:22–36

Bergmeier W, Goerge T, Wang H, Crittenden JR, Baldwin ACW, Cifuni SM, Housman DE, Graybiel AM, Wagner DD (2007) Mice lacking the signaling molecule CalDAG-GEFI represent a model for leukocyte adhesion deficiency type III. J Clin Invest 117:1699

Bhatt DL, Topol EJ (2003) Scientific and therapeutic advances in antiplatelet therapy. Nat Rev Drug Discov 2:15–28

Brass L, Stalker T, Zhu L, Woulfe D (2007) Signal transduction during platelet plug formation. In: Michelson A (ed) Platelets. Academic/Elsevier, Burlington, MA, pp 319–346

Briddon SJ, Watson SP (1999) Evidence for the involvement of p59fyn and p53/56lyn in collagen receptor signalling in human platelets. Biochem J 338:203–209
Burridge K, Chrzanowska-Wodnicka M (1996) Focal adhesions, contractility, and signaling. Annu Rev Cell Dev Biol 12:463–518
Canobbio I, Balduini C, Torti M (2004) Signalling through the platelet glycoprotein Ib-V-IX complex. Cell Signal 16:1329–1344
Canobbio I, Stefanini L, Cipolla L, Ciraolo E, Gruppi C, Balduini C, Hirsch E, Torti M (2009) Genetic evidence for a predominant role of PI3K{beta} catalytic activity in ITAM- and integrin-mediated signaling in platelets. Blood 114:2193–2196
Carman CV, Springer TA (2003) Integrin avidity regulation: are changes in affinity and conformation underemphasized? Curr Opin Cell Biol 15:547–556
Chrzanowska-Wodnicka M, Smyth SS, Schoenwaelder SM, Fischer TH, White GC (2005) Rap1b is required for normal platelet function and hemostasis in mice. J Clin Invest 115:680
Clements JL, Lee JR, Gross B, Yang B, Olson JD, Sandra A, Watson SP, Lentz SR, Koretzky GA (1999) Fetal hemorrhage and platelet dysfunction in SLP-76-deficient mice. J Clin Invest 103:19–25
Condliffe AM, Davidson K, Anderson KE, Ellson CD, Crabbe T, Okkenhaug K, Vanhaesebroeck B, Turner M, Webb L, Wymann MP, Hirsch E, Ruckle T, Camps M, Rommel C, Jackson SP, Chilvers ER, Stephens LR, Hawkins PT (2005) Sequential activation of class IB and class IA PI3K is important for the primed respiratory burst of human but not murine neutrophils. Blood 106:1432–1440
Cosemans J, Munnix I, Wetzker R, Heller R, Jackson SP, Heemskerk JW (2006) Continuous signaling via PI3K isoforms and is required for platelet ADP receptor function in dynamic thrombus stabilization. Blood 108:3045–3052
Crittenden J, Bergmeier W, Zhang Y, Piffath C, Liang Y, Wagner D, Housman D, Graybiel A (2004) CalDAG-GEFI integrates signaling for platelet aggregation and thrombus formation. Nat Med 10:982–986
Dormond O, Ponsonnet L, Hasmim M, Foletti A, Ruegg C (2004) Manganese-induced integrin affinity maturation promotes recruitment of alpha V beta 3 integrin to focal adhesions in endothelial cells: evidence for a role of phosphatidylinositol 3-kinase and Src. Thromb Haemost 92:151–161
Dorsam RT, Kunapuli SP (2004) Central role of the P2Y12 receptor in platelet activation. J Clin Invest 113:340–345
Dorsam RT, Kim S, Jin J, Kunapuli SP (2002) Coordinated signaling through both G12/13 and G(i) pathways is sufficient to activate GPIIb/IIIa in human platelets. J Biol Chem 277:47588–47595
Ezumi Y, Shindoh K, Tsuji M, Takayama H (1998) Physical and functional association of the Src family kinases Fyn and Lyn with the collagen receptor glycoprotein VI-Fc receptor gamma chain complex on human platelets. J Exp Med 188:267–276
Falasca M, Logan SK, Lehto VP, Baccante G, Lemmon MA, Schlessinger J (1998) Activation of phospholipase C gamma by PI 3-kinase-induced PH domain-mediated membrane targeting. Embo J 17:414–422
Frojmovic MM (1998) Platelet aggregation in flow: differential roles for adhesive receptors and ligands. Am Heart J 135:S119–S131
Gachet C, Hechler B (2005) The platelet P2 receptors in thrombosis. Semin Thromb Hemost 31:162–167
Gilcrease MZ, Zhou X, Welch K (2004) Adhesion-independent alpha6beta4 integrin clustering is mediated by phosphatidylinositol 3-kinase. Cancer Res 64:7395–7398
Gilio K, Munnix IC, Mangin P, Cosemans JM, Feijge MA, van der Meijden PE, Olieslagers S, Chrzanowska-Wodnicka MB, Lillian R, Schoenwaelder S et al (2009) Non-redundant roles of phosphoinositide 3-kinase isoforms alpha and beta in glycoprotein VI-induced platelet signaling and thrombus formation. J Biol Chem 284:33750–33762
Gironcel D, Racaud-Sultan C, Payrastre B, Haricot M, Borchert G, Kieffer N, Breton M, Chap H (1996) alphaIIb beta 3-integrin mediated adhesion of human platelets to a fibrinogen matrix

triggers phospholipase C activation and phosphatidylinositol 3′,4′-biphosphate accumulation. FEBS Lett 389:253–256

Giuliano S, Nesbitt WS, Rooney M, Jackson SP (2003) Bidirectional integrin alphaIIbbeta3 signalling regulating platelet adhesion under flow: contribution of protein kinase C. Biochem J 372:163–172

Giuriato S, Pesesse X, Bodin S, Sasaki T, Viala C, Marion E, Penninger J, Schurmans S, Erneux C, Payrastre B (2003) SH2-containing inositol 5-phosphatases 1 and 2 in blood platelets: their interactions and roles in the control of phosphatidylinositol 3,4,5-trisphosphate levels. Biochem J 376:199–207

Goel R, Phillips-Mason PJ, Gardner A, Raben DM, Baldassare JJ (2004) Alpha-thrombin-mediated phosphatidylinositol 3-kinase activation through release of Gbetagamma dimers from Galphaq and Galphai2. J Biol Chem 279:6701–6710

Goto S, Ikeda Y, Saldivar E, Ruggeri ZM (1998) Distinct mechanisms of platelet aggregation as a consequence of different shearing flow conditions. J Clin Invest 101:479–486

Goto S, Tamura N, Ishida H, Ruggeri ZM (2006) Dependence of platelet thrombus stability on sustained glycoprotein IIb/IIIa activation through adenosine 5’-diphosphate receptor stimulation and cyclic calcium signaling. J Am Coll Cardiol 47:155–162

Gratacap MP, Payrastre B, Viala C, Mauco G, Plantavid M, Chap H (1998) Phosphatidylinositol 3,4,5-trisphosphate-dependent stimulation of phospholipase C-gamma2 is an early key event in FcgammaRIIA-mediated activation of human platelets. J Biol Chem 273:24314–24321

Gratacap MP, Herault JP, Viala C, Ragab A, Savi P, Herbert JM, Chap H, Plantavid M, Payrastre B (2000) FcgammaRIIA requires a Gi-dependent pathway for an efficient stimulation of phosphoinositide 3-kinase, calcium mobilization, and platelet aggregation. Blood 96:3439–3446

Guillermet-Guibert J, Bjorklof K, Salpekar A, Gonella C, Ramadani F, Bilancio A, Meek S, Smith AJ, Okkenhaug K, Vanhaesebroeck B (2008) The p110beta isoform of phosphoinositide 3-kinase signals downstream of G protein-coupled receptors and is functionally redundant with p110gamma. Proc Natl Acad Sci USA 105:8292–8297

Guinebault C, Payrastre B, Sultan C, Mauco G, Breton M, Levy-Toledano S, Plantavid M, Chap H (1993) Tyrosine kinases and phosphoinositide metabolism in thrombin-stimulated human platelets. Biochem J 292(Pt 3):851–856

Guinebault C, Payrastre B, Racaud-Sultan C, Mazarguil H, Breton M, Mauco G, Plantavid M, Chap H (1995) Integrin-dependent translocation of phosphoinositide 3-kinase to the cytoskeleton of thrombin-activated platelets involves specific interactions of p85 alpha with actin filaments and focal adhesion kinase. J Cell Biol 129:831–842

Harper MT, Poole AW (2007) Isoform-specific functions of protein kinase C: the platelet paradigm. Biochem Soc Trans 35:1005–1008

Hartwig JH, Kung S, Kovacsovics T, Janmey PA, Cantley LC, Stossel TP, Toker A (1996) D3 phosphoinositides and outside-in integrin signaling by glycoprotein IIb-IIIa mediate platelet actin assembly and filopodial extension induced by phorbol 12-myristate 13-acetate. J Biol Chem 271:32986–32993

Heraud JM, Racaud-Sultan C, Gironcel D, Albiges-Rizo C, Giacomini T, Roques S, Martel V, Breton-Douillon M, Perret B, Chap H (1998) Lipid products of phosphoinositide 3-kinase and phosphatidylinositol 4′,5′-bisphosphate are both required for ADP-dependent platelet spreading. J Biol Chem 273:17817–17823

Hirano K (2007) The roles of proteinase-activated receptors in the vascular physiology and pathophysiology. Arterioscler Thromb Vasc Biol 27:27–36

Hirsch E, Bosco O, Tropel P, Laffargue M, Calvez R, Altruda F, Wymann M, Montrucchio G (2001) Resistance to thromboembolism in PI3Kgamma-deficient mice. Faseb J 15:2019–2021

Jackson SP (2007) The growing complexity of platelet aggregation. Blood 109:5087–5095

Jackson SP, Schoenwaelder SM (2003) Antiplatelet therapy: in search of the ‘magic bullet’. Nat Rev Drug Discov 2:775–789

Jackson SF, Schoenwaelder SM (2006) Type I phosphoinositide 3-kinases: potential antithrombotic targets? Cell Mol Life Sci 63:1085–1090

Jackson SP, Nesbitt WS, Kulkarni S (2003) Signaling events underlying thrombus formation. J Thromb Haemost 1:1602–1612

Jackson SP, Yap CL, Anderson KE (2004) Phosphoinositide 3-kinases and the regulation of platelet function. Biochem Soc Trans 32:387–392

Jackson SP, Schoenwaelder SM, Goncalves I, Nesbitt WS, Yap CL, Wright CE, Kenche V, Anderson KE, Dopheide SM, Yuan Y, Sturgeon SA, Prabaharan H, Thompson PE, Smith GD, Shepherd PR, Daniele N, Kulkarni S, Abbott B, Saylik D, Jones C, Lu L, Giuliano S, Hughan SC, Angus JA, Robertson AD, Salem HH (2005) PI 3-kinase p110beta: a new target for antithrombotic therapy. Nat Med 11:507–514

Jia S, Liu Z, Zhang S, Liu P, Zhang L, Lee SH, Zhang J, Signoretti S, Loda M, Roberts TM, Zhao JJ (2008) Essential roles of PI(3)K-p110beta in cell growth, metabolism and tumorigenesis. Nature 454:776–779

Jones SL, Wang J, Turck CW, Brown EJ (1998) A role for the actin-bundling protein L-plastin in the regulation of leukocyte integrin function. Proc Natl Acad Sci USA 95:9331–9336

Kasirer-Friede A, Cozzi MR, Mazzucato M, De Marco L, Ruggeri ZM, Shattil SJ (2004) Signaling through GP Ib-IX-V activates alpha IIb beta 3 independently of other receptors. Blood 103:3403–3411

Kauffenstein G, Bergmeier W, Eckly A, Ohlmann P, Leon C, Cazenave JP, Nieswandt B, Gachet C (2001) The P2Y(12) receptor induces platelet aggregation through weak activation of the alpha(IIb)beta(3) integrin–a phosphoinositide 3-kinase-dependent mechanism. FEBS Lett 505:281–290

Kim S, Foster C, Lecchi A, Quinton TM, Prosser DM, Jin J, Cattaneo M, Kunapuli SP (2002) Protease-activated receptors 1 and 4 do not stimulate G(i) signaling pathways in the absence of secreted ADP and cause human platelet aggregation independently of G(i) signaling. Blood 99:3629–3636

Kim S, Jin J, Kunapuli SP (2004) Akt activation in platelets depends on Gi signaling pathways. J Biol Chem 279:4186–4195

Kim S, Jin J, Kunapuli SP (2006) Relative contribution of G-protein-coupled pathways to protease-activated receptor-mediated Akt phosphorylation in platelets. Blood 107:947–954

Kim S, Mangin P, Dangelmaier C, Lillian R, Jackson SP, Daniel JL, Kunapuli SP (2009) Role of phosphoinositide 3-kinase beta in glycoprotein VI-mediated Akt activation in platelets. J Biol Chem 284:33763–33772

Klages B, Brandt U, Simon MI, Schultz G, Offermanns S (1999) Activation of G12/G13 results in shape change and Rho/Rho-kinase-mediated myosin light chain phosphorylation in mouse platelets. J Cell Biol 144:745–754

Kovacsovics TJ, Bachelot C, Toker A, Vlahos CJ, Duckworth B, Cantley LC, Hartwig JH (1995) Phosphoinositide 3-kinase inhibition spares actin assembly in activating platelets but reverses platelet aggregation. J Biol Chem 270:11358–11366

Krauss K, Altevogt P (1999) Integrin leukocyte function-associated antigen-1-mediated cell binding can be activated by clustering of membrane rafts. J Biol Chem 274:36921–36927

Kroner C, Eybrechts K, Akkerman JW (2000) Dual regulation of platelet protein kinase B. J Biol Chem 275:27790–27798

Kucera GL, Rittenhouse SE (1990) Human platelets form 3-phosphorylated phosphoinositides in response to alpha-thrombin, U46619, or GTP gamma S. J Biol Chem 265:5345–5348

Kulkarni S, Dopheide SM, Yap CL, Ravanat C, Freund M, Mangin P, Heel KA, Street A, Harper IS, Lanza F, Jackson SP (2000) A revised model of platelet aggregation. J Clin Invest 105:783–791

Larson MK, Chen H, Kahn ML, Taylor AM, Fabre JE, Mortensen RM, Conley PB, Parise LV (2003) Identification of P2Y12-dependent and -independent mechanisms of glycoprotein VI-mediated Rap1 activation in platelets. Blood 101:1409–1415

Lauener RW, Stevens CM, Sayed MR, Salari H, Duronio V (1999) A role for phosphatidylinositol 3-kinase in platelet aggregation in response to low, but not high, concentrations of PAF or thrombin. Biochim Biophys Acta 1452:197–208

Lawrence MB, Springer TA (1991) Leukocytes roll on a selectin at physiologic flow rates: distinction from and prerequisite for adhesion through integrins. Cell 65:859–873

Lian L, Wang Y, Draznin J, Eslin D, Bennett JS, Poncz M, Wu D, Abrams CS (2005) The relative role of PLCbeta and PI3Kgamma in platelet activation. Blood 106:110–117

Lova P, Paganini S, Sinigaglia F, Balduini C, Torti M (2002) A Gi-dependent pathway is required for activation of the small GTPase Rap1B in human platelets. J Biol Chem 277:12009–12015

Lova P, Paganini S, Hirsch E, Barberis L, Wymann M, Sinigaglia F, Balduini C, Torti M (2003) A selective role for phosphatidylinositol 3,4,5-trisphosphate in the Gi-dependent activation of platelet Rap1B. J Biol Chem 278:131–138

Lu PJ, Hsu AL, Wang DS, Chen CS (1998) Phosphatidylinositol 3,4,5-trisphosphate triggers platelet aggregation by activating Ca2+ influx. Biochemistry 37:9776–9783

Lynch L, Vodyanik PI, Boettiger D, Guvakova MA (2005) Insulin-like growth factor I controls adhesion strength mediated by alpha5beta1 integrins in motile carcinoma cells. Mol Biol Cell 16:51–63

Maier U, Babich A, Nurnberg B (1999) Roles of non-catalytic subunits in gbetagamma-induced activation of class I phosphoinositide 3-kinase isoforms beta and gamma. J Biol Chem 274:29311–29317

Martin V, Guillermet-Guibert J, Chicanne G, Cabou C, Jandrot-Perrus M, Plantavid M, Vanhaesebroeck B, Payrastre B, Gratacap MP (2010) Deletion of the p110beta isoform of phosphoinositide 3-kinase in platelets reveals its central role in Akt activation and thrombus formation in vitro and in vivo. Blood 115:2008–2013

Martorell L, Martinez-Gonzalez J, Rodriguez C, Gentile M, Calvayrac O, Badimon L (2008) Thrombin and protease-activated receptors (PARs) in atherothrombosis. Thromb Haemost 99:305–315

Maxwell MJ, Yuan Y, Anderson KE, Hibbs ML, Salem HH, Jackson SP (2004) SHIP1 and Lyn Kinase Negatively Regulate Integrin alpha IIb beta 3 signaling in platelets. J Biol Chem 279:32196–32204

Mazzucato M, Pradella P, Cozzi MR, De Marco L, Ruggeri ZM (2002) Sequential cytoplasmic calcium signals in a 2-stage platelet activation process induced by the glycoprotein Ibalpha mechanoreceptor. Blood 100:2793–2800

Moake JL, Turner NA, Stathopoulos NA, Nolasco L, Hellums JD (1988) Shear-induced platelet aggregation can be mediated by vWF released from platelets, as well as by exogenous large or unusually large vWF multimers, requires adenosine diphosphate, and is resistant to aspirin. Blood 71:1366–1374

Moers A, Nieswandt B, Massberg S, Wettschureck N, Gruner S, Konrad I, Schulte V, Aktas B, Gratacap MP, Simon MI, Gawaz M, Offermanns S (2003) G13 is an essential mediator of platelet activation in hemostasis and thrombosis. Nat Med 9:1418–1422

Moers A, Wettschureck N, Offermanns S (2004) G13-mediated signaling as a potential target for antiplatelet drugs. Drug News Perspect 17:493–498

Moritz MW, Reimers RC, Baker RK, Sutera SP, Joist JH (1983) Role of cytoplasmic and releasable ADP in platelet aggregation induced by laminar shear stress. J Lab Clin Med 101:537–544

Moser M, Nieswandt B, Ussar S, Pozgajova M, Fässler R (2008) Kindlin-3 is essential for integrin activation and platelet aggregation. Nat Med 14:325–330

Munday AD, Norris FA, Caldwell KK, Brown S, Majerus PW, Mitchell CA (1999) The inositol polyphosphate 4-phosphatase forms a complex with phosphatidylinositol 3-kinase in human platelet cytosol. Proc Natl Acad Sci USA 96:3640–3645

Murga C, Laguinge L, Wetzker R, Cuadrado A, Gutkind JS (1998) Activation of Akt/protein kinase B by G protein-coupled receptors. A role for alpha and beta gamma subunits of heterotrimeric G proteins acting through phosphatidylinositol-3-OH kinasegamma. J Biol Chem 273:19080–19085

Murga C, Fukuhara S, Gutkind JS (2000) A novel role for phosphatidylinositol 3-kinase beta in signaling from G protein-coupled receptors to Akt. J Biol Chem 275:12069–12073

Murugappan S, Shankar H, Kunapuli SP (2004) Platelet receptors for adenine nucleotides and thromboxane A2. Semin Thromb Hemost 30:411–418

Nesbitt WS, Kulkarni S, Giuliano S, Goncalves I, Dopheide SM, Yap CL, Harper IS, Salem HH, Jackson SP (2002) Distinct glycoprotein Ib/V/IX and integrin alpha IIbbeta 3-dependent calcium signals cooperatively regulate platelet adhesion under flow. J Biol Chem 277: 2965–2972

Nesbitt WS, Giuliano S, Kulkarni S, Dopheide SM, Harper IS, Jackson SP (2003) Intercellular calcium communication regulates platelet aggregation and thrombus growth. J Cell Biol 160:1151–1161

Nieswandt B, Watson SP (2003) Platelet-collagen interaction: is GPVI the central receptor? Blood 102:449–461

Nieswandt B, Brakebusch C, Bergmeier W, Schulte V, Bouvard D, Mokhtari-Nejad R, Lindhout T, Heemskerk JW, Zirngibl H, Fassler R (2001) Glycoprotein VI but not alpha2beta1 integrin is essential for platelet interaction with collagen. Embo J 20:2120–2130

Nieswandt B, Moser M, Pleines I, Varga-Szabo D, Monkley S, Critchley D, Fassler R (2007) Loss of talin1 in platelets abrogates integrin activation, platelet aggregation, and thrombus formation in vitro and in vivo. J Exp Med 204:3113

Nolan RD, Lapetina EG (1990) Thrombin stimulates the production of a novel polyphosphoinositide in human platelets. J Biol Chem 265:2441–2445

Nurden AT (2005) Qualitative disorders of platelets and megakaryocytes. J Thromb Haemost 3:1773–1782

Nurden AT (2006) Glanzmann thrombasthenia. Orphanet J Rare Dis 1:10

Nurden AT, Caen JP (1975) Specific roles for platelet surface glycoproteins in platelet function. Nature 255:720–722

Nurden P, Nurden AT (2008) Congenital disorders associated with platelet dysfunctions. Thromb Haemost 99:253–263

Oda A, Yokoyama K, Murata M, Tokuhira M, Nakamura K, Handa M, Watanabe K, Ikeda Y (1995) Protein tyrosine phosphorylation in human platelets during shear stress-induced platelet aggregation (SIPA) is regulated by glycoprotein (GP) Ib/IX as well as GP IIb/IIIa and requires intact cytoskeleton and endogenous ADP. Thromb Haemost 74:736–742

Offermanns S, Laugwitz KL, Spicher K, Schultz G (1994) G proteins of the G12 family are activated via thromboxane A2 and thrombin receptors in human platelets. Proc Natl Acad Sci USA 91:504–508

Offermanns S, Toombs CF, Hu YH, Simon MI (1997a) Defective platelet activation in G alpha(q)-deficient mice. Nature 389:183–186

Offermanns S, Toombs CF, Hu YH, Simon MI (1997b) Defective platelet activation in Ga?q?-deficient mice. Nature(London) 389:183–186

Ozaki Y, Asazuma N, Suzuki-Inoue K, Berndt MC (2005) Platelet GPIb-IX-V-dependent signaling. J Thromb Haemost 3:1745–1751

Pasquet JM, Bobe R, Gross B, Gratacap MP, Tomlinson MG, Payrastre B, Watson SP (1999a) A collagen-related peptide regulates phospholipase Cgamma2 via phosphatidylinositol 3-kinase in human platelets. Biochem J 342(Pt 1):171–177

Pasquet JM, Gross B, Quek L, Asazuma N, Zhang W, Sommers CL, Schweighoffer E, Tybulewicz V, Judd B, Lee JR, Koretzky G, Love PE, Samelson LE, Watson SP (1999b) LAT is required for tyrosine phosphorylation of phospholipase cgamma2 and platelet activation by the collagen receptor GPVI. Mol Cell Biol 19:8326–8334

Pasquet JM, Quek L, Stevens C, Bobe R, Huber M, Duronio V, Krystal G, Watson SP (2000) Phosphatidylinositol 3,4,5-trisphosphate regulates Ca(2+) entry via btk in platelets and megakaryocytes without increasing phospholipase C activity. Embo J 19:2793–2802

Quek LS, Bolen J, Watson SP (1998) A role for Bruton's tyrosine kinase (Btk) in platelet activation by collagen. Curr Biol 8:1137–1140

Quek LS, Pasquet JM, Hers I, Cornall R, Knight G, Barnes M, Hibbs ML, Dunn AR, Lowell CA, Watson SP (2000) Fyn and Lyn phosphorylate the Fc receptor gamma chain downstream of glycoprotein VI in murine platelets, and Lyn regulates a novel feedback pathway. Blood 96:4246–4253

Resendiz JC, Kroll MH, Lassila R (2007) Protease-activated receptor-induced Akt activation–regulation and possible function. J Thromb Haemost 5:2484–2493

Ruggeri ZM (2002) Platelets in atherothrombosis. Nat Med 8:1227–1234

Ruggeri ZM (2009) Platelet adhesion under flow. Microcirculation 16:58–83

Ruggeri ZM, Dent JA, Saldivar E (1999) Contribution of distinct adhesive interactions to platelet aggregation in flowing blood. Blood 94:172–178

Savage B, Almus-Jacobs F, Ruggeri ZM (1998) Specific synergy of multiple substrate-receptor interactions in platelet thrombus formation under flow. Cell 94:657–666

Schoenwaelder SM, Ono A, Sturgeon S, Chan SM, Mangin P, Maxwell MJ, Turnbull S, Mulchandani M, Anderson K, Kauffenstein G, Rewcastle GW, Kendall J, Gachet C, Salem HH, Jackson SP (2007) Identification of a unique co-operative phosphoinositide 3-kinase signaling mechanism regulating integrin alpha IIb beta 3 adhesive function in platelets. J Biol Chem 282:28648–28658

Schoenwaelder SM, Ono A, Warwick Nesbitt, Joanna Lim, Kate Jarman, Shaun P Jackson (2010) PI 3-kinase p110β regulates integrin $\alpha_{IIb}\beta_3$ avidity and the cellular transmission of contractile forces. J Biol Chem 285:2886–2896

Selheim F, Idsoe R, Fukami MH, Holmsen H, Vassbotn FS (1999) Formation of PI 3-kinase products in platelets by thrombin, but not collagen, is dependent on synergistic autocrine stimulation, particularly through secreted ADP. Biochem Biophys Res Commun 263: 780–785

Selheim F, Froyset AK, Strand I, Vassbotn FS, Holmsen H (2000a) Adrenaline potentiates PI 3-kinase in platelets stimulated with thrombin and SFRLLN: role of secreted ADP. FEBS Lett 485:62–66

Selheim F, Holmsen H, Vassbotn FS (2000b) PI 3-kinase signalling in platelets: the significance of synergistic, autocrine stimulation. Platelets 11:69–82

Senis YA, Atkinson BT, Pearce AC, Wonerow P, Auger JM, Okkenhaug K, Pearce W, Vigorito E, Vanhaesebroeck B, Turner M, Watson SP (2005) Role of the p110delta PI 3-kinase in integrin and ITAM receptor signalling in platelets. Platelets 16:191–202

Shankar H, Kahner B, Kunapuli SP (2006) G-protein dependent platelet signaling–perspectives for therapy. Curr Drug Targets 7:1253–1263

Shattil SJ, Newman PJ (2004) Integrins: dynamic scaffolds for adhesion and signaling in platelets. Blood 104:1606–1615

Siegel D, Ashton G, Penagos H, Lee J, Feiler H, Wilhelmsen K, South A, Smith F, Prescott A, Wessagowit V (2003) Loss of kindlin-1, a human homolog of the Caenorhabditis elegans actin–extracellular-matrix linker protein UNC-112, causes Kindler syndrome. Am J Hum Genet 73:174–187

Soderling TR (1999) The Ca2+–calmodulin-dependent protein kinase cascade. Trends Biochem Sci 24:232–236

Sultan C, Breton M, Mauco G, Grondin P, Plantavid M, Chap H (1990) The novel inositol lipid phosphatidylinositol 3,4-bisphosphate is produced by human blood platelets upon thrombin stimulation. Biochem J 269:831–834

Sultan C, Plantavid M, Bachelot C, Grondin P, Breton M, Mauco G, Levy-Toledano S, Caen JP, Chap H (1991) Involvement of platelet glycoprotein IIb-IIIa (alpha IIb-beta 3 integrin) in thrombin-induced synthesis of phosphatidylinositol 3′,4′-bisphosphate. J Biol Chem 266:23554–23557

Suzuki H, Terauchi Y, Fujiwara M, Aizawa S, Yazaki Y, Kadowaki T, Koyasu S (1999) Xid-like immunodeficiency in mice with disruption of the p85alpha subunit of phosphoinositide 3-kinase. Science 283:390–392

Suzuki-Inoue K, Tulasne D, Shen Y, Bori-Sanz T, Inoue O, Jung SM, Moroi M, Andrews RK, Berndt MC, Watson SP (2002) Association of Fyn and Lyn with the proline-rich domain of glycoprotein VI regulates intracellular signaling. J Biol Chem 277:21561–21566

Suzuki-Inoue K, Wilde JI, Andrews RK, Auger JM, Siraganian RP, Sekiya F, Rhee SG, Watson SP (2004) Glycoproteins VI and Ib-IX-V stimulate tyrosine phosphorylation of tyrosine kinase Syk and phospholipase Cgamma2 at distinct sites. Biochem J 378:1023–1029
Tadokoro S, Shattil SJ, Eto K, Tai V, Liddington RC, de Pereda JM, Ginsberg MH, Calderwood DA (2003) Talin binding to integrin ß tails: a final common step in integrin activation. Am Assoc Adv Sci 302:103–106
Terauchi Y, Tsuji Y, Satoh S, Minoura H, Murakami K, Okuno A, Inukai K, Asano T, Kaburagi Y, Ueki K, Nakajima H, Hanafusa T, Matsuzawa Y, Sekihara H, Yin Y, Barrett JC, Oda H, Ishikawa T, Akanuma Y, Komuro I, Suzuki M, Yamamura K, Kodama T, Suzuki H, Koyasu S, Aizawa S, Tobe K, Fukui Y, Yazaki Y, Kadowaki T (1999) Increased insulin sensitivity and hypoglycaemia in mice lacking the p85 alpha subunit of phosphoinositide 3-kinase. Nat Genet 21:230–235
Trumel C, Payrastre B, Plantavid M, Hechler B, Viala C, Presek P, Martinson EA, Cazenave JP, Chap H, Gachet C (1999) A key role of adenosine diphosphate in the irreversible platelet aggregation induced by the PAR1-activating peptide through the late activation of phosphoinositide 3-kinase. Blood 94:4156–4165
Tschopp TB, Weiss HJ, Baumgartner HR (1974) Decreased adhesion of platelets to subendothelium in von Willebrand's disease. J Lab Clin Med 83:296–300
Ussar S, Wang H, Linder S, Fässler R, Moser M (2006) The Kindlins: subcellular localization and expression during murine development. Exp Cell Res 312:3142–3151
Van Kolen K, Slegers H (2004) P2Y12 receptor stimulation inhibits beta-adrenergic receptor-induced differentiation by reversing the cyclic AMP-dependent inhibition of protein kinase B. J Neurochem 89:442–453
Varga-Szabo D, Pleines I, Nieswandt B (2008) Cell adhesion mechanisms in platelets. Arterioscler Thromb Vasc Biol 28:403–412
Wagner DD, Frenette PS (2008) The vessel wall and its interactions. Blood 111:5271–5281
Watanabe N, Nakajima H, Suzuki H, Oda A, Matsubara Y, Moroi M, Terauchi Y, Kadowaki T, Koyasu S, Ikeda Y, Handa M (2003) Functional phenotype of phosphoinositide 3-kinase p85alpha-null platelets characterized by an impaired response to GP VI stimulation. Blood 102:541–548
Watson SP, Auger JM, McCarty OJ, Pearce AC (2005) GPVI and integrin alphaIIb beta3 signaling in platelets. J Thromb Haemost 3:1752–1762
Weiss HJ, Turitto VT, Baumgartner HR (1978) Effect of shear rate on platelet interaction with subendothelium in citrated and native blood. I. Shear rate–dependent decrease of adhesion in von Willebrand's disease and the Bernard-Soulier syndrome. J Lab Clin Med 92:750–764
Woulfe D, Jiang H, Mortensen R, Yang J, Brass LF (2002) Activation of Rap1B by G(i) family members in platelets. J Biol Chem 277:23382–23390
Woulfe D, Jiang H, Morgans A, Monks R, Birnbaum M, Brass LF (2004) Defects in secretion, aggregation, and thrombus formation in platelets from mice lacking Akt2. J Clin Invest 113:441–450
Yap CL, Anderson KE, Hughan SC, Dopheide SM, Salem HH, Jackson SP (2002) Essential role for phosphoinositide 3-kinase in shear-dependent signaling between platelet glycoprotein Ib/V/IX and integrin alpha(IIb)beta(3). Blood 99:151–158
Zhang J, Shattil SJ, Cunningham MC, Rittenhouse SE (1996) Phosphoinositide 3-kinase gamma and p85/phosphoinositide 3-kinase in platelets. Relative activation by thrombin receptor or beta-phorbol myristate acetate and roles in promoting the ligand-binding function of alphaIIb-beta3 integrin. J Biol Chem 271:6265–6272
Zhang J, Banfic H, Straforini F, Tosi L, Volinia S, Rittenhouse SE (1998) A type II phosphoinositide 3-kinase is stimulated via activated integrin in platelets. A source of phosphatidylinositol 3-phosphate. J Biol Chem 273:14081–14084

Regulatory Subunits of Class IA PI3K

David A. Fruman

Contents

Abstract All class I PI3K enzymes are obligate heterodimers, consisting of a catalytic subunit tightly bound to a regulatory subunit. The regulatory subunit influences the subcellular location, binding partners, and activity of the catalytic subunit. Regulatory subunits also possess adaptor functions in cellular signaling, which are largely independent of their role in regulating PI3K activity. This chapter reviews the structure and function of PI3K regulatory subunits, focusing on the class IA subgroup.

1 Introduction

The mammalian genome encodes four class I enzymes: p110α, p110β, p110γ, and p110δ. These proteins catalyze the same chemical reaction, converting phosphatidylinositol-4,5-bisphosphate (PIP_2) into phosphatidylinositol-3,4,5-trisphosphate

D.A. Fruman
Department of Molecular Biology & Biochemistry, Institute for Immunology, University of California, Irvine, CA 92697-3900, USA
e-mail: dfruman@uci.edu

C. Rommel et al. (eds.). *Phosphoinositide 3-kinase in Health and Disease*, Volume 1
Current Topics in Microbiology and Immunology 346, DOI 10.1007/82_2010_39

(PIP_3). How is this crucial cellular activity controlled in time and space? There is some evidence that different p110s preferentially interact with different signaling components, for example, distinct isoforms of the Ras GTPase (Rodriguez-Viciana et al. 2004). However, it is the regulatory subunits that seem to be primarily responsible for spatiotemporal control of PI3K activation. While the catalytic subunits have received widespread attention as attractive "druggable" targets, studies focusing on the regulatory subunits have revealed fascinating biology that could potentially be exploited for therapeutic applications.

The class I PI3Ks are generally subdivided into class IA and class IB, depending on which subsets of regulatory subunits can be associated. The class IA enzymes (p110α, p110β, and p110δ) bind to one of five regulatory subunits (p85α, p55α, p50α, p85β, p55γ). These are encoded by three genes: *PIK3R1* (p85α, p55α, p50α, produced from mRNAs transcribed using alternative promoters), *PIK3R2* (p85β), and *PIK3R3* (p55γ). The class IB enzyme, p110γ, binds to either p101 (encoded by *PIK3R5*) or p84/87 (encoded by *PIK3R6*). This chapter will focus on the class IA regulatory subunits, with the class IB regulatory subunits covered by (Hawkins et al. 2010).

2 Structure and PI3K Enzyme Regulation

2.1 Modular Domains

The class IA regulatory subunits possess a modular structure, with a series of protein domains whose structure and general function have been studied extensively. Each isoform possesses two Src-homology-2 (SH2) domains in the C-terminal portion, flanking a domain termed the inter-SH2 or iSH2 region (Fig. 1). The N-terminal SH2 domain is referred to as N-SH2, and the other as C-SH2. The N-SH2 sequences are highly conserved among all isoforms, as are the C-SH2 domains between isoforms. The homology between N-SH2 and C-SH2 is lower but both have selective affinity for phosphorylated tyrosines with methionine at position +3 (pYXXM). There are many cellular signaling proteins that possess YXXM motifs, and a number of these have been shown to specifically recruit class IA PI3K dimers. YXXM motifs often are present in tandem with a short spacer, which could allow engagement of both SH2 domains. The 3D structures of p85α N-SH2 and C-SH2 alone and in complex with phosphotyrosyl peptides have been determined (PDB ID numbers 2IUG, 1QAD, 1H9O, 1BFJ). The N-SH2 domain of p85α also contacts the helical domain of p110α through opposing charge interactions (Miled et al. 2007) (Fig. 1). Oncogenic mutations in p110α are common in human cancers, and the residues within the helical region that contact N-SH2 are a "hot-spot" for frequent point mutations (Gymnopoulos et al. 2007; Kang et al. 2005). This supports the idea that the N-SH2/helical domain contact serves to help maintain the dimer in a low activity state (discussed further in Sect. 2.2).

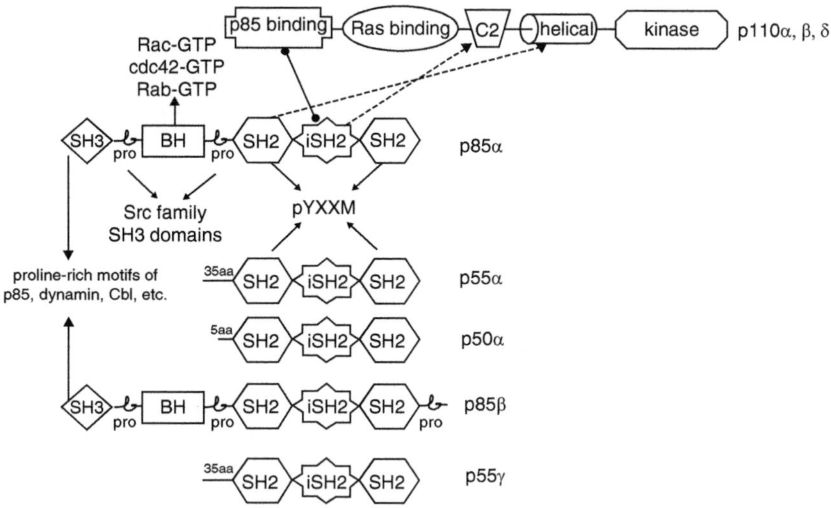

Fig. 1 Schematic diagram of domain structure of class IA regulatory (p85α, p55α, p50α, p85β, p55γ) and catalytic (p110α, p110β, p110δ) isoforms. The stable, constitutive dimer interface of p85 and p110 is indicated by the *thick line with circular ends* ("barbell"). Other regulated intersubunit interactions (N-SH2/helical, iSH2/C2) are indicated by *dashed arrows*. The primary protein–protein interactions mediated by different modular domains within the regulatory subunits are shown by *solid arrows*. Most of these have been defined using domains isolated from p85α. The interactions involving C-terminal domains, namely iSH2-p110 and SH2-pYXXM, occur similarly for each of the five isoforms shown. p85β has an additional proline-rich motif at the C-terminus as indicated

The iSH2 region has at least two critical functions. First, it mediates stable dimer formation with p110 subunits through an interaction with the p110 adaptor-binding domain (ABD) (Fig. 1). The three-dimensional structure of the p110/iSH2 interface has been reported in two X-ray crystallography studies (PDB ID numbers 2V1Y and 2RDO) (Miled et al. 2007; Huang et al. 2007). These showed that the iSH2 domain has a coiled-coil topology, as predicted by earlier approaches using spin labeling and electron paramagnetic resonance along with homology modeling (Fu et al. 2003). In-frame deletion of a short segment of the iSH2 region results in a protein (Δp85) unable to bind to catalytic subunits (Dhand et al. 1994), demonstrating that this portion of the interaction interface is essential for stable binding. A second function that has been identified by crystallography is a contact between the iSH2 and the C2 domain of p110α (Huang et al. 2007) (Fig. 1). A rare subset of p110α oncogenic mutations alters Asn345 in the C2 domain, which normally contacts p85α residues Asp560 and Asn564 within the iSH2. More frequently, cancer cells carry mutations or deletions in *PIK3R1* that delete the iSH2 residues important for contacting the p110-C2 domain (The Cancer Genome Atlas Research Network 2008; Parsons et al. 2008). Recent mutational analysis further supports the model that the iSH2/C2 contact is one of the interactions that contribute to stabilizing the inactive conformation (Wu et al. 2009). A third possible function of the iSH2 region is to form an intramolecular interaction with the N-SH2 domain (Wu et al. 2009). However, this has not yet been proved by X-ray structures, in which the N-SH2 domain was absent (Miled et al. 2007; Huang et al. 2007). In addition, mutation analysis has shown that disrupting the putative nSH2/iSH2 contact residues does not relieve p110 inhibition (Wu et al. 2009).

The N-terminus of p55α consists of 35 amino acids that are homologous to the N-terminus of p55γ, whereas p50α possesses a unique sequence of five amino acids (Fig. 1). The structure and function of these short extensions is poorly understood. The N-terminal portions of p85α and p85β have a similar domain arrangement with a Src-homology-3 (SH3) domain and two proline-rich regions flanking a domain with homology to GTPase-activating proteins (GAPs) for Rho family G proteins (Fig. 1). In vitro, the RhoGAP homology domain of p85α lacks GTPase activity towards Rho/Rac/cdc42 but can bind to Rac-GTP and cdc42-GTP (Fruman et al. 1998; Vanhaesebroeck et al. 2001). This domain has been given different names including Breakpoint Cluster Region homology (BH domain, used in Fig. 1), and Rac-binding domain. The crystal structure of the p85α BH domain was determined (PDB ID number 1PBW). Several functions have been proposed for the BH domain, by virtue of its ability to bind GTP-bound forms of Rac and cdc42 without accelerating GTP hydrolysis: (1) Prolongation of Rac/cdc42 activation by competing with active GAPs; (2) Positioning small G proteins near membrane signaling complexes; (3) An additional mechanism for positioning class IA PI3Ks near the membrane and possibly activating the associated catalytic subunit. There is experimental evidence that class IA PI3K can act either upstream or downstream of Rac and cdc42 during cellular responses, so all of the mechanisms might apply in certain situations. The BH domain of p85α was also reported to possess GAP activity towards Rab small G proteins (Chamberlain et al. 2004).

The three-dimensional structure of the SH3 domain of p85α, alone or bound to high affinity proline-rich peptide, has been determined by X-ray crystallography and NMR (PDB ID numbers 1PHT, 1PNJ). Many potential binding partners for the p85α-SH3 domain have been reported (Fig. 1), but the physiological significance of these interactions is unclear. The p85α-SH3 domain can form intramolecular interactions with the proline-rich motifs within the same protein, but are unlikely to mediate intermolecular multimerization since p85/p110 complexes purify as simple dimers. The proline-rich motifs of p85 proteins can bind to SH3 domains of Src family kinases (Fig. 1), and the molecular basis for this interaction has been studied by NMR (PDB ID number 1AZG). This interaction has been proposed to regulate PI3K activation downstream of lymphocyte antigen receptors (Pleiman et al. 1994).

There is growing evidence that the N-terminal domains of p85α and p85β carry out adaptor functions of the regulatory subunits, which are independent of PI3K enzyme activation. Examples are discussed in Sect. 4.2.

2.2 Regulation of the Catalytic Subunit

Each of the five regulatory isoforms forms dimers with one of the class IA catalytic subunits p110α, p110β, or p110δ (Fig. 1). There is no apparent preference for pair-wise dimer formation. These dimers are of such high affinity that they cannot be dissociated even by high concentrations of urea. As discussed in Sect. 2.1, the structural basis of this dimer formation involves the iSH2 region of regulatory subunits and the N-terminal ABD of catalytic subunits (indicated by "barbell" in Fig. 1). In the absence of regulatory subunits, the catalytic isoform p110α is thermally unstable and is rapidly degraded (Yu et al. 1998). It is likely that the same is true for the other class IA catalytic proteins, though this has not been directly demonstrated. Importantly, free p110α has higher intrinsic kinase activity (Yu et al. 1998). Hence, regulatory subunits stabilize the catalytic subunits while maintaining them in a low activity state. Suppression of kinase function is mediated by secondary, regulated interactions between other domains in the catalytic and regulatory subunits. As discussed in Sect. 2.1, one interaction involves the N-SH2 domain of regulatory subunits and the helical region of catalytic subunits (Fig. 2) (Miled et al. 2007). In addition, the C2 domain of p110α contacts a surface of the p85α iSH2 domain that is distinct from the portion that binds tightly to the p110α N-terminus.

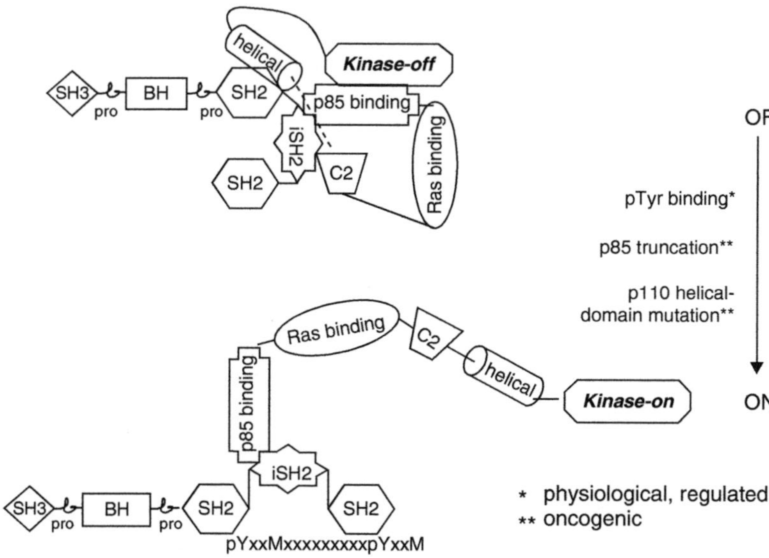

Fig. 2 Model for regulation of class IA PI3K activation, from X-ray structures (Miled et al. 2007; Huang et al. 2007). In the low activity state, the N-SH2 domain contacts the helical region of p110 catalytic subunits. A contact between the p85-iSH2 and p110-C2 domains, and between the p85-binding (ABD) and kinase domains of p110, also exists in the inactive conformation. Release of these interactions switches the enzyme to the high activity state. There are least three mechanisms to accomplish this "OFF–ON" switch, as indicated. Binding of SH2 domains to pTyr peptides is a physiological mechanism during tyrosine kinase signaling. Oncogenic mutations in p85 (truncation, iSH2 deletions) and p110 (point mutation in helical domain or C2 domain) can also promote the active conformation. Reprinted with permission from *The Handbook of Cellular Signaling*, Elsevier, 2009

There are several mechanisms by which the inactive conformation can be disrupted, leading to increased kinase activity. The normal physiological mechanism is the binding of the SH2 domains and/or other domains of regulatory subunits to partner proteins (Fig. 2). This relieves the inhibitory intersubunit interactions, without disrupting the constitutive dimer interface of iSH2/ABD, allowing increased activity of the enzymatic subunit. This might be due in part to exposure of basic surfaces of the iSH2 domain that mediate binding to membranes (Miled et al. 2007; End et al. 1993). In addition, most of the partner proteins for regulatory subunits are located near cellular membranes, where the catalytic subunits can gain access to lipid substrates and additional regulatory proteins (e.g., Ras). Importantly, the ability of Ras-GTP to augment p110 catalytic activity depends on concomitant binding of the p85 subunit to phosphorylated tyrosine (pTyr) residues (Jimenez et al. 2002). A second mechanism that disrupts the inactive dimer is through mutations in p85 that occur frequently in cancer cells. Deletion of the C-terminus of the regulatory subunit at residue 572, including the region of iSH2 that contacts the p110-C2 domain, produces a protein that mediates constitutive kinase activity (Borlado et al. 2000; Jimenez et al. 1998). Analogous mutations occur in human tumors (The Cancer Genome Atlas Research Network 2008; Parsons et al. 2008) and relieve p110 inhibition while maintaining p110 stabilization (Jaiswal et al. 2009). A third mechanism involves mutation of the residues in the p110 helical region that contact the N-SH2 domain, as mentioned above.

3 Expression and Localization

The three protein products of the *PIK3R1* gene are encoded by mRNAs that are transcribed from three different promoters with distinct activity in different tissues. The p85α protein is expressed ubiquitously, whereas p55α and p50α display a restricted tissue distribution as determined by Northern and immunoblot analyses in mice and rats (Fruman et al. 1996; Inukai et al. 1996). In some cells, such as T lymphocytes, p50α expression is equivalent to p85α. Confocal imaging of cells expressing fluorescent p85α fusion proteins has shown that in unstimulated cells, p85α has a diffuse cytoplasmic distribution but becomes partly associated with membranes following cell activation. In human T cells stimulated through the antigen receptor, p85α accumulates at the contact site with antigen-presenting cells (known as the immunological synapse) (Fabre et al. 2005). The p50α subunit preferentially accumulates at the contact site of previously activated T cells (Fos et al. 2008). In CHO cells stimulated with insulin-like growth factor-1 (IGF-1), p85α is localized to intracellular foci along with insulin receptor substrate (IRS) proteins (Luo et al. 2005a). Stimulation of CHO cells with platelet-derived growth factor (PDGF) induces a different localization pattern, with transient recruitment of p85α to plasma membrane patches and ruffles (Luo et al. 2005a).

The p85β isoform is also expressed in most tissues. Based on immunoblotting with "pan-p85" antibodies that cross-react with p85α and p85β, most cells and tissues appear to express relatively lower amounts of p85β. Further supporting this idea, loss of p85α causes a marked reduction in p110 isoform expression in most tissues (presumably due to degradation of p110 monomers), whereas loss of p85β has lesser effects. In fibroblasts, p85β appears to be more highly expressed and loss of either p85α or p85β causes a partial loss of p110 expression (Fruman et al. 1999). A quantitative mass spectrometry study concluded that p85α is expressed more highly in all tissues examined except the brain (Geering et al. 2007). The p55γ protein is expressed at highest levels in brain and testes of adult mice, and in embryonic tissues (Pons et al. 1995).

For further information about PI3K gene expression patterns, the reader is referred to a detailed review article (Kok et al. 2009).

4 Function in Health and Disease

Class IA regulatory isoforms have two general functions. First, they control the location and activation status of class IA catalytic isoforms. Second, they contribute to signaling complex assembly through the adaptor functions of their modular domains.

4.1 Biochemical Mechanisms of Class IA PI3K Activation

The best-known mechanism of class IA PI3K activation occurs downstream of tyrosine kinases, which trigger tyrosine phosphorylation of proteins that subsequently bind to the SH2 domains of class IA regulatory isoforms (Fruman et al. 1998; Vanhaesebroeck et al. 2001) (Fig. 2). Growth factor receptors, oncoproteins, antigen receptors, and adhesion receptors can all initiate tyrosine kinase cascades leading to class IA PI3K activation. Ras-GTP is also elevated under many of these conditions, contributing to class IA catalytic activity.

Given the high conservation of the SH2 domains between the five proteins encoded by *PIK3R1*, *PIK3R2*, and *PIK3R3*, it seems likely that signaling specificity arises from domains in the N-terminal portions of the proteins, which are less conserved. It is possible that the BH, SH3, and proline-rich motifs do not act alone, yet play a supporting role in stabilizing interactions initiated by SH2-pTyr binding events. There is indirect evidence that the BH domains of p85α and p85β have roles in PI3K enzyme activation, particularly in lymphocyte antigen receptor signaling. In both T cells and B cells, deletion of Rac GTPases or upstream activators of the Vav family leads to impaired PI3K activation (Reynolds et al. 2002; Vigorito et al. 2004; Walmsley et al. 2003). This suggests that binding of p85 BH domains to Rac-GTP in antigen receptor signaling complexes contributes to PI3K recruitment and activation. Another potentially important interaction involves the SH3 domain of p85 proteins and ubiquitin ligases of the Cbl family (Hartley et al. 1995; Soltoff and Cantley 1996). In addition, the proline-rich motifs of p85 proteins interact with SH3 domains of Src family tyrosine kinases, which could stabilize the active conformation of both the tyrosine and lipid kinases in the complex (Fruman et al. 1998). To test further the functions of BH, SH3, and proline-rich motifs, it would be helpful to generate mice with knock-in mutations that inactivate specific domains without altering protein expression.

4.2 Adaptor Functions

There is now solid evidence that class IA regulatory isoforms can have adaptor functions independent of their roles in coordinating the activation and localization of their catalytic partner (reviewed in Okkenhaug and Vanhaesebroeck 2001; Vanhaesebroeck et al. 2005). The potential adaptor functions of domains from p85α have been studied much more extensively than the analogous domains within p85β. The BH domain has been implicated in the regulation of the actin cytoskeleton, likely due to interactions with the small GTPases Rac and/or cdc42. In some cellular systems, overexpression of N-terminal domains of p85α has similar functional effects on actin dynamics as expression of the full-length protein (Hill et al. 2001; Jimenez et al. 2000). It has also been shown that expression of Δp85, which cannot associate with catalytic isoforms but retains other domains, can restore Jnk activation in fibroblasts lacking p85α (Ueki et al. 2002a). p85α also possesses an adaptor function during cytokinesis, via an interaction with cdc42 that regulates septin localization in the cytokinetic cleavage furrow (Garcia et al. 2006). Two recent studies showed that p85 proteins interact with the protein XBP-1 and regulate the unfolded protein response. The interaction is independent of PI3K catalytic subunits and requires the BH domain of p85 (Park et al. 2010; Winnay et al. 2010).

4.3 Gene-Targeting Studies

Much of our current knowledge of regulatory subunit function comes from studies of mouse strains with targeted mutations in *Pik3r1* or *Pik3r2* (Table 1). Null mutations have shown that neither *Pik3r1* nor *Pik3r2* is required for early embryonic development or fibroblast proliferation (Ueki et al. 2002a; Brachmann et al. 2005; Fruman et al. 2000a; Suzuki et al. 1999). However, combined deletion of *Pik3r1* and *Pik3r2* causes embryonic lethality in mid-gestation, with a phenotype similar to mice lacking the PDGF receptor-α (Brachmann et al. 2005). Fibroblasts from these embryos have reduced responsiveness to several growth factors. These results illustrate that class IA regulatory subunits as a group have essential functions downstream of receptor tyrosine kinases, but that these functions are partially redundant. Mice lacking *Pik3r3* have not been reported, so the redundant and unique functions of p55γ remain unknown. However, there is evidence that this protein can substitute for other class IA regulatory isoforms in some contexts.

Table 1 Summary of mouse knockout alleles and publications

PI3K isoform (s) targeted	Name and details of allele	Original reference	Additional references
p85α	*Pik3r1*^{tm1} (exon 1A deleted in germline)	Terauchi et al. (1999)	Suzuki et al. (1999), Donahue et al. (2004), Hess et al. (2004), Suzuki et al. (2003), Fukao et al. (2002a), Fukao et al. (2002b), Watanabe et al. (2003)
p85α/p55α/p50α	*Pik3r1*^{tm2} (exon 7 deleted in germline)	Fruman et al. (1999)	Garcia et al. (2006), Brachmann et al. (2005), Fruman et al. (2000a), Mouta-Bellum et al. (2009), Mauvais-Jarvis et al. (2002), Lu-Kuo et al. (2000), Huddleston et al. (2003), Munugalavadla et al. (2007), Kharas et al. (2004), Luo et al. (2005b)
p55α/p50α	*Pik3r1*^{tm3} (exon 1B and C deleted in germline)	Chen et al. (2004)	
p85α/p55α/p50α	*Pik3r1*^{tm4} (exon 7 flanked by loxP sites)	Luo et al. (2006)	Kharas et al. (2008), Taniguchi et al. (2007), Taniguchi et al. (2006b), Deane et al. (2007), Oak et al. (2006)
p85β	*Pik3r2*^{tm1} (exon 1 deleted in germline)	Ueki et al. (2002a)	Deane et al. (2004), Alcazar et al. (2009), Oak et al. (2009)

In fibroblasts and leukemia cells lacking *Pik3r1* and *Pik3r2*, p55γ is often upregulated and this correlates with maintenance of PI3K signaling function (Brachmann et al. 2005; Kharas et al. 2008).

4.3.1 Defects in Mouse Development

Pik3r1 gene products do play some unique role in mammalian development as *Pik3r1*^{tm2} mice die soon after birth, displaying several abnormalities including liver defects and chylous ascites (Fruman et al. 2000a). Embryos and/or newborns of this knockout strain exhibit massive hepatocellular necrosis and elevated liver enzymes in the plasma, which are not observed in mice lacking only p85α or p55α/p50α. Since both p85α and p50α are highly expressed in mouse liver (Fruman et al. 1996), it seems likely that these isoforms have a redundant function in cellular homeostasis of developing hepatocytes. Brown fat necrosis in mice lacking p85α/p55α/p50α (Fruman et al. 2000a) might also occur through loss of homeostatic mechanisms in brown adipocytes. Newborn mice lacking p85α/p55α/p50α also exhibit calcinosis in cardiac tissue (Fruman et al. 2000a), which might result from similar necrotic processes occurring early in embryogenesis. Chylous ascites in these mice appears to result from impaired development of lymphatic vessels (Mouta-Bellum et al. 2009). A similar phenotype was reported in mice lacking VEGF-C or with a point mutation in the Ras-binding domain of p110α (Gupta et al. 2007; Karkkainen et al. 2004). This suggests that signaling through receptors for VEGF-C in developing lymphatic endothelial cells activates PI3K through a mechanism that requires both Ras binding to p110α, and p85α or a smaller *Pik3r1* variant as the regulatory subunit.

4.3.2 Defects in Glucose Homeostasis

Pik3r1^{tm2} heterozygotes are viable but show reduced plasma glucose levels and increased insulin sensitivity (Fruman et al. 2000a; Mauvais-Jarvis et al. 2002). A similar metabolic phenotype is seen in *Pik3r1*^{tm1} mutant mice lacking only p85α (Terauchi et al. 1999), in *Pik3r1*^{tm3} mutant mice lacking only p55α and p50α (Chen et al. 2004), and in *Pik3r2*^{tm1} mice lacking p85β (Ueki et al. 2002a). The increased insulin sensitivity in *Pik3r1*^{tm2} heterozygous mice prompted an investigation of the effects of this mutation in mouse models of insulin resistance. Remarkably, the approximately 50% reduction in p85α/p55α/p50α protein expression in *Pik3r1*^{tm2} heterozygotes protected IR$^{+/-}$/IRS1$^{+/-}$ mice from diabetes (Mauvais-Jarvis et al. 2002).

The increased insulin sensitivity in mice with targeted deletion of class IA regulatory subunits has been an unexpected finding. Based on in vitro experiments with cell lines, it was predicted that inactivation of class IA PI3K genes would reduce insulin sensitivity. Indeed, mice expressing a single allele of kinase-inactive p110α show decreased insulin signaling and associated changes in metabolism (Foukas et al. 2006). The physiological basis of increased insulin sensitivity and glucose tolerance in mice lacking regulatory subunits is still not clear, and many models have been proposed (Luo et al. 2005a; Vanhaesebroeck et al. 2005; Luo and Cantley 2005; Taniguchi et al. 2006a, 2007). One possibility (Fig. 3, *top*) is that there is an excess of free regulatory subunits in wild-type cells but not in knockout cells (Ueki et al. 2002b). In this scenario, signaling is more efficient in knockout cells because p85/p110 dimers do not compete with free p85 monomers. This model has been challenged by a mass spectrometry study showing that p85 and p110 subunits are obligate heterodimers (Geering et al. 2007). Another model is that regulatory subunits participate in feedback inhibition of PI3K signaling (Fig. 3, *bottom*). One mechanism is through activation of the stress kinase JNK, and another through activation of the lipid phosphatase PTEN, which opposes PI3K signaling by dephosphorylating PtdIns-3,4,5-P_3 (Taniguchi et al. 2006a, 2007; Ueki et al. 2003). It was also reported that monomeric p85α subunits sequester insulin-regulated signaling complexes in an intracellular location, away from the plasma membrane (Luo et al. 2005a). The binding of p85 proteins to XBP-1 is also regulated by insulin (Park et al. 2010; Winnay et al. 2010). Each of these models appears to involve adaptor functions of the regulatory subunit(s). It is possible that different regulatory isoforms have both unique and shared adaptor functions in limiting insulin signaling output. Further complicating matters, studies using conditional gene deletion have confirmed the initial hypothesis that class IA regulatory subunits have a positive role in insulin signaling. Deletion of *Pik3r1* in muscle cells

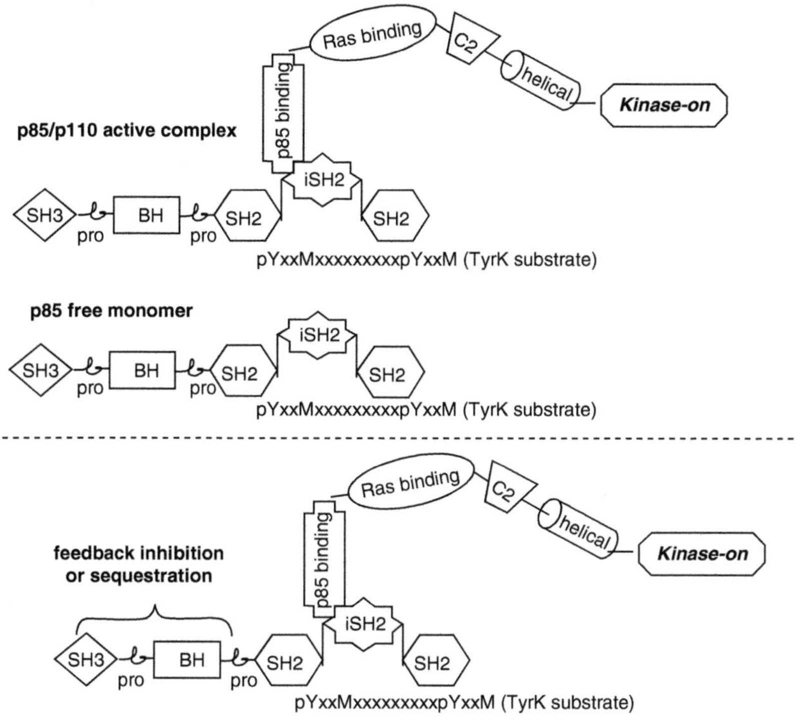

Fig. 3 Models for negative regulation of insulin signaling by class IA regulatory subunits. (*top*) Free monomers of p85 (or p55/p50) compete with active heterodimeric PI3K. (*bottom*) N-terminal domains of regulatory subunits engage in feedback inhibitory pathways, or sequester signaling complexes. Reprinted with permission from *The Handbook of Cellular Signaling*, Elsevier, 2009

of mice lacking *Pik3r2* causes muscle-specific insulin resistance and whole-body glucose intolerance (Luo et al. 2006). Similarly, combined deletion of *Pik3r1* and *Pik3r2* in liver cells, or liver-specific expression of a dominant-negative form of p85α, causes hyperinsulinemia and glucose intolerance (Miyake et al. 2002; Taniguchi et al. 2006b). Therefore, one can make the general conclusion that class IA regulatory subunits have both positive and negative roles in insulin signaling, such that partial reductions in regulatory subunit expression can actually increase insulin sensitivity.

The insulin-sensitive phenotype of mice lacking regulatory subunits raises the interesting question of whether modulating p85 expression could be used as a therapeutic approach in insulin resistance syndromes such as type II diabetes. Insulin resistance in humans with type II diabetes (Bandyopadhyay et al. 2005) or gestational diabetes (Kirwan et al. 2004) is associated with overexpression of regulatory subunits. Polymorphisms in the human *Pik3r1* gene have been reported, with apparently increased incidence in a population of Pima native American women prone to type II diabetes (Baier et al. 1998). However, the functional effects of these mutations are not yet clear. It is likely that the balance of expression of p85α with other PI3K isoforms and other proteins in the insulin receptor signaling pathway all contribute to determining the functional outcome of altered p85α expression.

4.3.3 Defects in Immune Function

The functions of *Pik3r1* gene products in hematopoietic cells have been studied extensively, and reviewed in detail (Deane and Fruman 2004; Deane et al. 2007; Koyasu 2003; Okkenhaug and Vanhaesebroeck 2003). *Pik3r1*^{tm1} mice, lacking only p85α but retaining p55α and p50α, have significant immune defects. B cell development and function is severely impaired, in part because of defective Ca^{2+} mobilization downstream of tyrosine kinases activated by the B cell receptor (Fruman et al. 1999; Suzuki et al. 1999, 2003; Donahue et al. 2004; Hess et al. 2004). Together with other observations this has led to the model that class IA PI3K and its lipid products play an integral role in the assembly of a large "signalosome" that promotes the activation of phospholipase-Cγ in B cells (Fruman et al. 2000b). Mice lacking p85α also have impaired mast cell development and function, and aberrant cytokine production by innate immune cells (dendritic cells, macrophages) responding to pathogen-derived products (Fukao et al. 2002a, b; Lu-Kuo et al. 2000). Erythropoiesis and platelet function is also impaired in mice lacking *Pik3r1* gene products (Huddleston et al. 2003; Watanabe et al. 2003). Surprisingly, T cell function is largely normal in the absence of *Pik3r1* (Fruman et al. 1999; Suzuki et al. 1999; Deane et al. 2004). However, combined inactivation of *Pik3r1* and *Pik3r2* in T cells by Cre-lox-mediated deletion causes partial defects in T cell function that are similar in some ways to T cells from mice lacking functional p110δ (Deane et al. 2007; Okkenhaug et al. 2006). Despite impaired T cell activation, mice with T cells lacking *Pik3r1* and *Pik3r2* develop an autoimmune syndrome (Deane et al. 2007; Oak et al. 2006). These mice exhibit destruction of the lacrimal glands and accumulate antinuclear antibodies with age. Overall, the phenotype resembles the human disease Sjögren's Syndrome. The mechanism of autoimmunity is not yet clear but might involve a defect in regulatory T cells (Oak et al. 2006).

Mice lacking *Pik3r2* (p85β) have somewhat subtle yet interesting immune phenotypes. T and B cells develop normally in these mice, and the mice are not overtly immunocompromised (Deane et al. 2004). However, T cells lacking p85β show an elevated primary response to antigen along with a severely impaired secondary response (Deane et al. 2004; Alcazar et al. 2009). A clue to the mechanism underlying these phenotypes is that p85β in T cells selectively associates with the CD28 costimulatory receptor and with Cbl ubiquitin ligases, and is required for CD28-mediated Cbl downregulation (Alcazar et al. 2009). B cells lacking p85β also display modestly greater proliferation following B cell receptor crosslinking, and this correlates with elevated phosphorylation of Erk (Oak et al. 2009). In addition to this negative regulatory signaling function in B cells, p85β also has a redundant protective role in B cell survival that is unmasked when p85α/p55α/p50α are absent (Oak et al. 2009). However, the function of p85α in the B cell Ca^{2+} signalosome is unique and cannot be fulfilled by p85β (Hess et al. 2004; Oak et al. 2009).

4.3.4 Role in Cancer

PIK3R1 and *PIK3R2* both have oncogenic potential. Mutations that lead to truncation of the C-terminus, or deletion of sequences in the iSH2 region, enable the associated catalytic subunit to be constitutively active (Jimenez et al. 1998; Jaiswal et al. 2009; Philp et al. 2001). A mutated *Pik3r1* encoding a truncated form of p85α, termed p65, was cloned from a thymoma isolated from an irradiated mouse (Jimenez et al. 1998). p65 cooperates with v-raf to transform fibroblasts and causes lymphoproliferation when expressed as a transgene in mouse T cells (Borlado et al. 2000; Jimenez et al. 1998). Lymphoproliferation in p65 transgenic mice is associated with a lupus-like autoimmune disease (Borlado et al. 2000; Barber et al. 2006). Deletions in the iSH2 region have also been found in human colon and ovarian cancers (Philp et al. 2001), and are present in approximately 10% of human glioblastomas (The Cancer Genome Atlas Research Network 2008; Parsons et al. 2008). *PIK3R2* is not frequently mutated in cancer, but a fusion product of p85β with part of HUMORF8 (product of *USP8* gene) was isolated from blood of a patient with myeloproliferative disease (Janssen et al. 1998). Similar to *PIK3R1* oncogenic mutants, this p85β fusion protein lacks a C-terminal portion that includes the iSH2 residues important for contacting the p110-C2 domain.

Reducing expression of regulatory subunits inhibits proliferation and survival of cells expressing diverse oncogenes that activate class IA PI3K. For example, mastocytosis driven by oncogenic KIT mutations is reduced in the absence of p85α (Munugalavadla et al. 2007). Pre-B cell leukemia driven by BCR-ABL is partially reduced by loss of p85α (Kharas et al. 2004) and nearly abolished by combined deletion of p85α/p55α/p50α/p85β (Kharas et al. 2008). In osteoclasts with elevated Ras activity due to haploinsufficiency of the NF1 RasGAP, hyperresponsiveness to growth factors is suppressed by loss of p85α (Yang et al. 2006). RNAi-based knockdown of p85α sensitizes colon cancer cell lines to TRAIL-induced apoptosis (Rychahou et al. 2005). However, reduced expression of wild-type p85α in some cell contexts might have the opposite effect of contributing to tumorigenesis. In mice heterozygous for PTEN, which develop lymphoproliferation and hyperplasia of epithelial tissues, there is an increased incidence of intestinal polyp phenotypes when one allele of the *Pik3r1*^{tm2} mutation is also present (Luo et al. 2005b). Overexpression of p85α does not seem to oppose growth signaling as long as the catalytic subunits are also overexpressed. Thus, human lung cancers often display overexpression of p85α in concert with p110 overexpression relative to normal lung tissue (Lin et al. 2001). Elevated expression of p85α and p110α also correlated with invasiveness of prostatic tumors (Shukla et al. 2007) and head/neck squamous cell carcinoma (Stahl et al. 2004).

5 Conclusions

Class IA PI3K acts in diverse signaling pathways, downstream of numerous cell surface receptors, and dictates physiological responses at both the cellular and organismal levels. To understand how specificity is achieved in this complex system, it is essential to define the unique and shared functions of class IA regulatory isoforms. In the 1990s, the genes encoding these proteins were cloned and we began to learn the general functions of modular domains and to catalog interacting partners. In the last 10 years, gene-targeting studies have shown that p85α and p85β have unique functions despite the similar overall structure of these proteins. The unique functions of the smaller isoforms, p55α/p50α/p55γ, are less well defined, but several studies have used biochemical and cellular approaches to address this question.

As of early 2010, we are not aware of any drugs in development that act on class IA regulatory subunits. The focus of pharmaceutical research in this field has been to target the catalytic subunits (Rommel et al. 2007; Bilancio et al. 2006; Wetzker and Rommel 2004; Wymann et al. 2003). Kinases are enzymes that are more classical pharmaceutical targets. However, as more is learned about the structure of class IA heterodimers and their interacting partners, it might be possible to design molecules that interfere with specific interactions for therapeutic benefit. If it were possible to disrupt the interface of iSH2/ABD, this should decrease the stability and disrupt the localization of catalytic isoforms. Knowledge gained from recent crystallography studies (Miled et al. 2007; Huang et al. 2007) should facilitate this goal. However, it is not clear whether this approach would add any benefit over pharmacological inhibition of the catalytic active site. Perhaps a more attractive goal would be blocking or enhancing interactions that regulate PI3K localization or adaptor functions in specific cell contexts. The ability of different regulatory isoforms to transmit distinct signals is likely due to their N-terminal domains rather than the more highly conserved SH2-iSH2-SH2 module in the C-terminus. Thus, agents that target the SH3 or BH domain, or proline-rich regions, might have interesting physiological effects. It is essential to obtain a greater understanding of these domains, through structural studies in the context of the intact full-length protein (preferably in complex with p110), as well as through genetic disruption of domain function.

Acknowledgment PI3K studies in my lab have been funded by grants from the National Institutes of Health.

References

Alcazar I, Cortes I, Zaballos A, Hernandez C, Fruman DA, Barber DF, Carrera AC (2009) p85Beta phosphoinositide 3-kinase regulates CD28 coreceptor function. Blood 113:3198–3208

Baier LJ, Wiedrich C, Hanson RL, Bogardus C (1998) Variant in the regulatory subunit of phosphatidylinositol 3-kinase (p85alpha): preliminary evidence indicates a potential role of this variant in the acute insulin response and type 2 diabetes in Pima women. Diabetes 47:973–975

Bandyopadhyay GK, Yu JG, Ofrecio J, Olefsky JM (2005) Increased p85/55/50 expression and decreased phosphotidylinositol 3-kinase activity in insulin-resistant human skeletal muscle. Diabetes 54:2351–2359

Barber DF, Bartolome A, Hernandez C, Flores JM, Fernandez-Arias C, Rodriguez-Borlado L, Hirsch E, Wymann M, Balomenos D, Carrera AC (2006) Class IB-phosphatidylinositol 3-kinase (PI3K) deficiency ameliorates IA-PI3K-induced systemic lupus but not T cell invasion. J Immunol 176:589–593

Bilancio A, Okkenhaug K, Camps M, Emery JL, Ruckle T, Rommel C, Vanhaesebroeck B (2006) Key role of the p110delta isoform of PI3K in B-cell antigen and IL-4 receptor signaling: comparative analysis of genetic and pharmacologic interference with p110delta function in B cells. Blood 107:642–650

Borlado LR, Redondo C, Alvarez B, Jimenez C, Criado LM, Flores J, Marcos MA, Martinez AC, Balomenos D, Carrera AC (2000) Increased phosphoinositide 3-kinase activity induces a lymphoproliferative disorder and contributes to tumor generation in vivo. FASEB J 14:895–903

Brachmann SM, Yballe CM, Innocenti M, Deane JA, Fruman DA, Thomas SM, Cantley LC (2005) Role of phosphoinositide 3-kinase regulatory isoforms in development and actin rearrangement. Mol Cell Biol 25:2593–2606

Chamberlain MD, Berry TR, Pastor MC, Anderson DH (2004) The p85alpha subunit of phosphatidylinositol 3′-kinase binds to and stimulates the GTPase activity of Rab proteins. J Biol Chem 279:48607–48614

Chen D, Mauvais-Jarvis F, Bluher M, Fisher SJ, Jozsi A, Goodyear LJ, Ueki K, Kahn CR (2004) p50alpha/p55alpha phosphoinositide 3-kinase knockout mice exhibit enhanced insulin sensitivity. Mol Cell Biol 24:320–329

Deane JA, Fruman DA (2004) Phosphoinositide 3-kinase: diverse roles in immune cell activation. Annu Rev Immunol 22:563–598

Deane JA, Trifilo MJ, Yballe CM, Choi S, Lane TE, Fruman DA (2004) Enhanced T cell proliferation in mice lacking the p85beta subunit of phosphoinositide 3-kinase. J Immunol 172:6615–6625

Deane JA, Kharas MG, Oak JS, Stiles LN, Luo J, Moore TI, Ji H, Rommel C, Cantley LC, Lane TE, Fruman DA (2007) T cell function is partially maintained in the absence of class IA phosphoinositide 3-kinase signaling. Blood 109:2894–2902

Dhand R, Hara K, Hiles I, Bax B, Gout I, Panayotou G, Fry MJ, Yonezawa K, Kasuga M, Waterfield MD (1994) PI 3-kinase: structural and functional analysis of intersubunit interactions. EMBO J 13:511–521

Donahue AC, Hess KL, Ng KL, Fruman DA (2004) Altered splenic B cell subset development in mice lacking phosphoinositide 3-kinase p85alpha. Int Immunol 16:1789–1798

End P, Gout I, Fry MJ, Panayotou G, Dhand R, Yonezawa K, Kasuga M, Waterfield MD (1993) A biosensor approach to probe the structure and function of the p85 alpha subunit of the phosphatidylinositol 3-kinase complex. J Biol Chem 268:10066–10075

Fabre S, Lang V, Harriague J, Jobart A, Unterman TG, Trautmann A, Bismuth G (2005) Stable activation of phosphatidylinositol 3-kinase in the T cell immunological synapse stimulates Akt signaling to FoxO1 nuclear exclusion and cell growth control. J Immunol 174:4161–4171

Fos C, Salles A, Lang V, Carrette F, Audebert S, Pastor S, Ghiotto M, Olive D, Bismuth G, Nunes JA (2008) ICOS ligation recruits the p50alpha PI3K regulatory subunit to the immunological synapse. J Immunol 181:1969–1977

Foukas LC, Claret M, Pearce W, Okkenhaug K, Meek S, Peskett E, Sancho S, Smith AJ, Withers DJ, Vanhaesebroeck B (2006) Critical role for the p110alpha phosphoinositide-3-OH kinase in growth and metabolic regulation. Nature 441:366–370

Fruman DA, Cantley LC, Carpenter CL (1996) Structural organization and alternative splicing of the murine phosphoinositide 3-kinase p85 alpha gene. Genomics 37:113–121

Fruman DA, Meyers RE, Cantley LC (1998) Phosphoinositide kinases. Annu Rev Biochem 67:481–507

Fruman DA, Snapper SB, Yballe CM, Davidson L, Yu JY, Alt FW, Cantley LC (1999) Impaired B cell development and proliferation in absence of phosphoinositide 3-kinase p85alpha. Science 283:393–397

Fruman DA, Mauvais-Jarvis F, Pollard DA, Yballe CM, Brazil D, Bronson RT, Kahn CR, Cantley LC (2000a) Hypoglycaemia, liver necrosis and perinatal death in mice lacking all isoforms of phosphoinositide 3-kinase p85 alpha. Nat Genet 26:379–382

Fruman DA, Satterthwaite AB, Witte ON (2000b) Xid-like phenotypes: a B cell signalosome takes shape. Immunity 13:1–3

Fu Z, Aronoff-Spencer E, Backer JM, Gerfen GJ (2003) The structure of the inter-SH2 domain of class IA phosphoinositide 3-kinase determined by site-directed spin labeling EPR and homology modeling. Proc Natl Acad Sci USA 100:3275–3280

Fukao T, Tanabe M, Terauchi Y, Ota T, Matsuda S, Asano T, Kadowaki T, Takeuchi T, Koyasu S (2002a) PI3K-mediated negative feedback regulation of IL-12 production in DCs. Nat Immunol 3:875–881

Fukao T, Yamada T, Tanabe M, Terauchi Y, Ota T, Takayama T, Asano T, Takeuchi T, Kadowaki T, Hata Ji J, Koyasu S (2002b) Selective loss of gastrointestinal mast cells and impaired immunity in PI3K-deficient mice. Nat Immunol 3:295–304

Garcia Z, Silio V, Marques M, Cortes I, Kumar A, Hernandez C, Checa AI, Serrano A, Carrera AC (2006) A PI3K activity-independent function of p85 regulatory subunit in control of mammalian cytokinesis. EMBO J 25:4740–4751

Geering B, Cutillas PR, Nock G, Gharbi SI, Vanhaesebroeck B (2007) Class IA phosphoinositide 3-kinases are obligate p85-p110 heterodimers. Proc Natl Acad Sci USA 104:7809–7814

Gupta S, Ramjaun AR, Haiko P, Wang Y, Warne PH, Nicke B, Nye E, Stamp G, Alitalo K, Downward J (2007) Binding of ras to phosphoinositide 3-kinase p110alpha is required for ras-driven tumorigenesis in mice. Cell 129:957–968

Gymnopoulos M, Elsliger MA, Vogt PK (2007) Rare cancer-specific mutations in PIK3CA show gain of function. Proc Natl Acad Sci USA 104:5569–5574

Hartley D, Meisner H, Corvera S (1995) Specific association of the beta isoform of the p85 subunit of phosphatidylinositol-3 kinase with the proto-oncogene c-cbl. J Biol Chem 270:18260–18263

Hawkins PT, Stephens LR, Suire S, Wilson M (2010) PI3K Signaling in Neutrophils. Curr Top Microbiol Immunol doi 10.1007/82_2010_40

Hess KL, Donahue AC, Ng KL, Moore TI, Oak J, Fruman DA (2004) Frontline: the p85alpha isoform of phosphoinositide 3-kinase is essential for a subset of B cell receptor-initiated signaling responses. Eur J Immunol 34:2968–2976

Hill KM, Huang Y, Yip SC, Yu J, Segall JE, Backer JM (2001) N-terminal domains of the class IA phosphoinositide 3-kinase regulatory subunit play a role in cytoskeletal but not mitogenic signaling. J Biol Chem 276:16374–16378

Huang CH, Mandelker D, Schmidt-Kittler O, Samuels Y, Velculescu VE, Kinzler KW, Vogelstein B, Gabelli SB, Amzel LM (2007) The structure of a human p110alpha/p85alpha complex elucidates the effects of oncogenic PI3Kalpha mutations. Science 318:1744–1748

Huddleston H, Tan B, Yang FC, White H, Wenning MJ, Orazi A, Yoder MC, Kapur R, Ingram DA (2003) Functional p85alpha gene is required for normal murine fetal erythropoiesis. Blood 102:142–145

Inukai K, Anai M, Van Breda E, Hosaka T, Katagiri H, Funaki M, Fukushima Y, Ogihara T, Yazaki Y, Kikuchi OY, Asano T (1996) A novel 55-kDa regulatory subunit for phosphatidylinositol 3-kinase structurally similar to p55PIK is generated by alternative splicing of the p85alpha gene. J Biol Chem 271:5317–5320

Jaiswal BS, Janakiraman V, Kljavin NM, Chaudhuri S, Stern HM, Wang W, Kan Z, Dbouk HA, Peters BA, Waring P, Dela Vega T, Kenski DM, Bowman KK, Lorenzo M, Li H, Wu J, Modrusan Z, Stinson J, Eby M, Yue P, Kaminker JS, de Sauvage FJ, Backer JM, Seshagiri S (2009) Somatic mutations in p85alpha promote tumorigenesis through class IA PI3K activation. Cancer cell 16:463–474

Janssen JW, Schleithoff L, Bartram CR, Schulz AS (1998) An oncogenic fusion product of the phosphatidylinositol 3-kinase p85beta subunit and HUMORF8, a putative deubiquitinating enzyme. Oncogene 16:1767–1772

Jimenez C, Jones DR, Rodriguez-Viciana P, Gonzalez-Garcia A, Leonardo E, Wennstrom S, von Kobbe C, Toran JL, R-Borlado L, Calvo V, Copin SG, Albar JP, Gaspar ML, Diez E, Marcos MA, Downward J, Martinez AC, Merida I, Carrera AC (1998) Identification and characterization of a new oncogene derived from the regulatory subunit of phosphoinositide 3-kinase. EMBO J 17:743–753

Jimenez C, Portela RA, Mellado M, Rodriguez-Frade JM, Collard J, Serrano A, Martinez AC, Avila J, Carrera AC (2000) Role of the PI3K regulatory subunit in the control of actin organization and cell migration. J Cell Biol 151:249–262

Jimenez C, Hernandez C, Pimentel B, Carrera AC (2002) The p85 regulatory subunit controls sequential activation of phosphoinositide 3-kinase by Tyr kinases and Ras. J Biol Chem 277:41556–41562

Kang S, Bader AG, Zhao L, Vogt PK (2005) Mutated PI 3-kinases: cancer targets on a silver platter. Cell cycle 4:578–581

Karkkainen MJ, Haiko P, Sainio K, Partanen J, Taipale J, Petrova TV, Jeltsch M, Jackson DG, Talikka M, Rauvala H, Betsholtz C, Alitalo K (2004) Vascular endothelial growth factor C is required for sprouting of the first lymphatic vessels from embryonic veins. Nat Immunol 5:74–80

Kharas MG, Deane JA, Wong S, O'Bosky KR, Rosenberg N, Witte ON, Fruman DA (2004) Phosphoinositide 3-kinase signaling is essential for ABL oncogene-mediated transformation of B-lineage cells. Blood 103:4268–4275

Kharas MG, Janes MR, Scarfone VM, Lilly MB, Knight ZA, Shokat KM, Fruman DA (2008) Ablation of PI3K blocks BCR-ABL leukemogenesis in mice, and a dual PI3K/mTOR inhibitor prevents expansion of human BCR-ABL+ leukemia cells. J Clin Invest 118:3038–3050

Kirwan JP, Varastehpour A, Jing M, Presley L, Shao J, Friedman JE, Catalano PM (2004) Reversal of insulin resistance postpartum is linked to enhanced skeletal muscle insulin signaling. J Clin Endocrinol Metab 89:4678–4684

Kok K, Geering B, Vanhaesebroeck B (2009) Regulation of phosphoinositide 3-kinase expression in health and disease. Trends Biochem Sci 34:115–127

Koyasu S (2003) The role of PI3K in immune cells. Nat Immunol 4:313–319

Lin X, Bohle AS, Dohrmann P, Leuschner I, Schulz A, Kremer B, Fandrich F (2001) Over-expression of phosphatidylinositol 3-kinase in human lung cancer. Langenbeck's Arch Surg 386:293–301

Lu-Kuo JM, Fruman DA, Joyal DM, Cantley LC, Katz HR (2000) Impaired kit- but not FcepsilonRI-initiated mast cell activation in the absence of phosphoinositide 3-kinase p85alpha gene products. J Biol Chem 275:6022–6029

Luo J, Cantley LC (2005) The negative regulation of phosphoinositide 3-kinase signaling by p85 and it's implication in cancer. Cell cycle 4:1309–1312

Luo J, Field SJ, Lee JY, Engelman JA, Cantley LC (2005a) The p85 regulatory subunit of phosphoinositide 3-kinase down-regulates IRS-1 signaling via the formation of a sequestration complex. J Cell Biol 170:455–464

Luo J, Sobkiw CL, Logsdon NM, Watt JM, Signoretti S, O'Connell F, Shin E, Shim Y, Pao L, Neel BG, Depinho RA, Loda M, Cantley LC (2005b) Modulation of epithelial neoplasia and lymphoid hyperplasia in PTEN+/− mice by the p85 regulatory subunits of phosphoinositide 3-kinase. Proc Natl Acad Sci USA 102:10238–10243

Luo J, Sobkiw CL, Hirshman MF, Logsdon MN, Li TQ, Goodyear LJ, Cantley LC (2006) Loss of class IA PI3K signaling in muscle leads to impaired muscle growth, insulin response, and hyperlipidemia. Cell Metab 3:355–366

Mauvais-Jarvis F, Ueki K, Fruman DA, Hirshman MF, Sakamoto K, Goodyear LJ, Iannacone M, Accili D, Cantley LC, Kahn CR (2002) Reduced expression of the murine p85alpha subunit of phosphoinositide 3-kinase improves insulin signaling and ameliorates diabetes. J Clin Invest 109:141–149

Miled N, Yan Y, Hon WC, Perisic O, Zvelebil M, Inbar Y, Schneidman-Duhovny D, Wolfson HJ, Backer JM, Williams RL (2007) Mechanism of two classes of cancer mutations in the phosphoinositide 3-kinase catalytic subunit. Science 317:239–242

Miyake K, Ogawa W, Matsumoto M, Nakamura T, Sakaue H, Kasuga M (2002) Hyperinsulinemia, glucose intolerance, and dyslipidemia induced by acute inhibition of phosphoinositide 3-kinase signaling in the liver. J Clin Invest 110:1483–1491

Mouta-Bellum C, Kirov A, Miceli-Libby L, Mancini ML, Petrova TV, Liaw L, Prudovsky I, Thorpe PE, Miura N, Cantley LC, Alitalo K, Fruman DA, Vary CP (2009) Organ-specific lymphangiectasia, arrested lymphatic sprouting, and maturation defects resulting from gene-targeting of the PI3K regulatory isoforms p85alpha, p55alpha, and p50alpha. Dev Dyn 238:2670–2679

Munugalavadla V, Sims EC, Borneo J, Chan RJ, Kapur R (2007) Genetic and pharmacologic evidence implicating the p85 alpha, but not p85 beta, regulatory subunit of PI3K and Rac2 GTPase in regulating oncogenic KIT-induced transformation in acute myeloid leukemia and systemic mastocytosis. Blood 110:1612–1620

Oak JS, Deane JA, Kharas MG, Luo J, Lane TE, Cantley LC, Fruman DA (2006) Sjogren's syndrome-like disease in mice with T cells lacking class 1A phosphoinositide-3-kinase. Proc Natl Acad Sci USA 103:16882–16887

Oak JS, Chen J, Peralta RQ, Deane JA, Fruman DA (2009) The p85β regulatory subunit of phosphoinositide 3-kinase has unique and redundant functions in B cells. Autoimmunity 42:447–458

Okkenhaug K, Vanhaesebroeck B (2001) New responsibilities for the PI3K regulatory subunit p85 alpha. Sci STKE 2001:PE1

Okkenhaug K, Vanhaesebroeck B (2003) PI3K in lymphocyte development, differentiation and activation. Nat Rev Immunol 3:317–330

Okkenhaug K, Patton DT, Bilancio A, Garcon F, Rowan WC, Vanhaesebroeck B (2006) The p110delta isoform of phosphoinositide 3-kinase controls clonal expansion and differentiation of Th cells. J Immunol 177:5122–5128

Park SW, Zhou Y, Lee J, Lu A, Sun C, Chung J, Ueki K, Ozcan U (2010) The regulatory subunits of PI3K, p85alpha and p85beta, interact with XBP-1 and increase its nuclear translocation. Nat Med 16:429–437

Parsons DW, Jones S, Zhang X, Lin JC, Leary RJ, Angenendt P, Mankoo P, Carter H, Siu IM, Gallia GL, Olivi A, McLendon R, Rasheed BA, Keir S, Nikolskaya T, Nikolsky Y, Busam DA, Tekleab H, Diaz LA Jr, Hartigan J, Smith DR, Strausberg RL, Marie SK, Shinjo SM, Yan H, Riggins GJ, Bigner DD, Karchin R, Papadopoulos N, Parmigiani G, Vogelstein B, Velculescu VE, Kinzler KW (2008) An integrated genomic analysis of human glioblastoma multiforme. Science 321:1807–1812

Philp AJ, Campbell IG, Leet C, Vincan E, Rockman SP, Whitehead RH, Thomas RJ, Phillips WA (2001) The phosphatidylinositol 3′-kinase p85alpha gene is an oncogene in human ovarian and colon tumors. Cancer Res 61:7426–7429

Pleiman CM, Hertz WM, Cambier JC (1994) Activation of phosphatidylinositol-3′ kinase by Src-family kinase SH3 binding to the p85 subunit. Science 263:1609–1612

Pons S, Asano T, Glasheen E, Miralpeix M, Zhang Y, Fisher TL, Myers MG Jr, Sun XJ, White MF (1995) The structure and function of p55PIK reveal a new regulatory subunit for phosphatidylinositol 3-kinase. Mol Cell Biol 15:4453–4465

Reynolds LF, Smyth LA, Norton T, Freshney N, Downward J, Kioussis D, Tybulewicz VL (2002) Vav1 transduces T cell receptor signals to the activation of phospholipase C-gamma1 via phosphoinositide 3-kinase-dependent and -independent pathways. J Exp Med 195:1103–1114

Rodriguez-Viciana P, Sabatier C, McCormick F (2004) Signaling specificity by Ras family GTPases is determined by the full spectrum of effectors they regulate. Mol Cell Biol 24:4943–4954

Rommel C, Camps M, Ji H (2007) PI3Kdelta and PI3Kgamma: partners in crime in inflammation in rheumatoid arthritis and beyond? Nat Rev Drug Discov 7:191–201

Rychahou PG, Murillo CA, Evers BM (2005) Targeted RNA interference of PI3K pathway components sensitizes colon cancer cells to TNF-related apoptosis-inducing ligand (TRAIL). Surgery 138:391–397

Shukla S, Maclennan GT, Hartman DJ, Fu P, Resnick MI, Gupta S (2007) Activation of PI3K-Akt signaling pathway promotes prostate cancer cell invasion. Int J Cancer 121:1424–1432

Soltoff SP, Cantley LC (1996) p120cbl is a cytosolic adapter protein that associates with phosphoinositide 3-kinase in response to epidermal growth factor in PC12 and other cells. J Biol Chem 271:563–567

Stahl U, Wenk J, Wagener F, Woenckhaus J, Gamerdinger U, Battmann A, Dreyer T (2004) Phosphoinositide 3-kinase (PI3-K) expression. Tumorigenesis of epithelial carcinoma of the mouth. Pathologe 25:31–37

Suzuki H, Terauchi Y, Fujiwara M, Aizawa S, Yazaki Y, Kadowaki T, Koyasu S (1999) Xid-like immunodeficiency in mice with disruption of the p85alpha subunit of phosphoinositide 3-kinase. Science 283:390–392

Suzuki H, Matsuda S, Terauchi Y, Fujiwara M, Ohteki T, Asano T, Behrens TW, Kouro T, Takatsu K, Kadowaki T, Koyasu S (2003) PI3K and Btk differentially regulate B cell antigen receptor-mediated signal transduction. Nat Immunol 4:280–286

Taniguchi CM, Tran TT, Kondo T, Luo J, Ueki K, Cantley LC, Kahn CR (2006a) Phosphoinositide 3-kinase regulatory subunit p85alpha suppresses insulin action via positive regulation of PTEN. Proc Natl Acad Sci USA 103:12093–12097

Taniguchi CM, Kondo T, Sajan M, Luo J, Bronson R, Asano T, Farese R, Cantley LC, Kahn CR (2006b) Divergent regulation of hepatic glucose and lipid metabolism by phosphoinositide 3-kinase via Akt and PKClambda/zeta. Cell Metab 3:343–353

Taniguchi CM, Aleman JO, Ueki K, Luo J, Asano T, Kaneto H, Stephanopoulos G, Cantley LC, Kahn CR (2007) The p85alpha regulatory subunit of phosphoinositide 3-kinase potentiates c-Jun N-terminal kinase-mediated insulin resistance. Mol Cell Biol 27:2830–2840

Terauchi Y, Tsuji Y, Satoh S, Minoura H, Murakami K, Okuno A, Inukai K, Asano T, Kaburagi Y, Ueki K, Nakajima H, Hanafusa T, Matsuzawa Y, Sekihara H, Yin Y, Barrett JC, Oda H, Ishikawa T, Akanuma Y, Komuro I, Suzuki M, Yamamura K, Kodama T, Suzuki H, Yamamura K, Kodama T, Suzuki H, Koyasu S, Aizawa S, Tobe K, Fukui Y, Yazaki Y, Kadowaki T (1999) Increased insulin sensitivity and hypoglycaemia in mice lacking the p85 alpha subunit of phosphoinositide 3-kinase. Nat Genet 21:230–235

The Cancer Genome Atlas Research Network (2008) Comprehensive genomic characterization defines human glioblastoma genes and core pathways. Nature 455:1061–1068

Ueki K, Yballe CM, Brachmann SM, Vicent D, Watt JM, Kahn CR, Cantley LC (2002a) Increased insulin sensitivity in mice lacking p85beta subunit of phosphoinositide 3-kinase. Proc Natl Acad Sci USA 99:419–424

Ueki K, Fruman DA, Brachmann SM, Tseng YH, Cantley LC, Kahn CR (2002b) Molecular balance between the regulatory and catalytic subunits of phosphoinositide 3-kinase regulates cell signaling and survival. Mol Cell Biol 22:965–977

Ueki K, Fruman DA, Yballe CM, Fasshauer M, Klein J, Asano T, Cantley LC, Kahn CR (2003) Positive and negative roles of p85 alpha and p85 beta regulatory subunits of phosphoinositide 3-kinase in insulin signaling. J Biol Chem 278:48453–48466

Vanhaesebroeck B, Leevers SJ, Ahmadi K, Timms J, Katso R, Driscoll PC, Woscholski R, Parker PJ, Waterfield MD (2001) Synthesis and function of 3-phosphorylated inositol lipids. Annu Rev Biochem 70:535–602

Vanhaesebroeck B, Ali K, Bilancio A, Geering B, Foukas LC (2005) Signalling by PI3K isoforms: insights from gene-targeted mice. Trends Biochem Sci 30:194–204

Vigorito E, Bardi G, Glassford J, Lam EW, Clayton E, Turner M (2004) Vav-dependent and vav-independent phosphatidylinositol 3-kinase activation in murine B cells determined by the nature of the stimulus. J Immunol 173:3209–3214

Walmsley MJ, Ooi SK, Reynolds LF, Smith SH, Ruf S, Mathiot A, Vanes L, Williams DA, Cancro MP, Tybulewicz VL (2003) Critical roles for Rac1 and Rac2 GTPases in B cell development and signaling. Science 302:459–462

Watanabe N, Nakajima H, Suzuki H, Oda A, Matsubara Y, Moroi M, Terauchi Y, Kadowaki T, Suzuki H, Koyasu S, Ikeda Y, Handa M (2003) Functional phenotype of phosphoinositide 3-kinase p85alpha-null platelets characterized by an impaired response to GP VI stimulation. Blood 102:541–548

Wetzker R, Rommel C (2004) Phosphoinositide 3-kinases as targets for therapeutic intervention. Curr Pharm Design 10:1915–1922

Winnay JN, Boucher J, Mori MA, Ueki K, Kahn CR (2010) A regulatory subunit of phosphoinositide 3-kinase increases the nuclear accumulation of X-box-binding protein-1 to modulate the unfolded protein response. Nat Med 16:438–445

Wu H, Shekar SC, Flinn RJ, El-Sibai M, Jaiswal BS, Sen KI, Janakiraman V, Seshagiri S, Gerfen GJ, Girvin ME, Backer JM (2009) Regulation of Class IA PI 3-kinases: C2 domain-iSH2 domain contacts inhibit p85/p110alpha and are disrupted in oncogenic p85 mutants. Proc Natl Acad Sci USA 106:20258–20263

Wymann MP, Zvelebil M, Laffargue M (2003) Phosphoinositide 3-kinase signalling – which way to target? Trends Pharmacol Sci 24:366–376

Yang FC, Chen S, Robling AG, Yu X, Nebesio TD, Yan J, Morgan T, Li X, Yuan J, Hock J, Ingram DA, Clapp DW (2006) Hyperactivation of p21ras and PI3K cooperate to alter murine and human neurofibromatosis type 1-haploinsufficient osteoclast functions. J Clin Invest 116:2880–2891

Yu J, Zhang Y, McIlroy J, Rordorf-Nikolic T, Orr GA, Backer JM (1998) Regulation of the p85/p110 phosphatidylinositol 3′-kinase: stabilization and inhibition of the p110alpha catalytic subunit by the p85 regulatory subunit. Mol Cell Biol 18:1379–1387

The Neurodevelopmental Implications of PI3K Signaling

Kathryn Waite and Britta J. Eickholt

Contents

Abstract Our understanding of the mechanisms involved in the formation of the complex arrangement of neurons and their interconnections within the brain has made significant progress in recent years. Current research has uncovered a network of intracellular signaling events that provide precise coordination of a diverse array of cellular responses, including trafficking events, cytoskeletal remodeling, gene transcription, and protein ubiquitination and translation. This chapter considers the specific cellular responses controlled by the phosphatidylinositol 3-kinase (PI3K) signaling pathway, which is instructive with regard to a number of important steps involved in the development of the brain. These range from the mediation of extrinsic signals – such as growth factors, axon guidance cues, and extracellular matrix components – to intrinsic effectors, such as downstream signaling components that act, for example, at the translation level. PI3K signaling is, consequently, at the heart of controlling neuronal migration and neuronal morphogenesis, as well

K. Waite and B.J. Eickholt (✉)
MRC Centre for Developmental Neurobiology, King's College London, New Hunt's House, London SE1 1UL, UK
e-mail: Britta.J.Eickholt@kcl.ac.uk

C. Rommel et al. (eds.). *Phosphoinositide 3-kinase in Health and Disease*, Volume 1
Current Topics in Microbiology and Immunology 346, DOI 10.1007/82_2010_82

as dendrite and synapse development. Many neurobehavioral disorders arise as a consequence of subtle developmental abnormalities. Unsurprisingly, therefore, aberrant PI3K signaling has been indicated by many studies to be a contributing factor to the pathophysiology of disorders such as schizophrenia and autism. In this chapter, we will focus on the specific, yet divergent, cellular processes that are achieved through PI3K signaling in neurons and are key to brain development.

1 PI3Ks: The Basics

Phosphatidylinositol 3-kinases (PI3Ks) are a group of enzymes capable of phosphorylating the hydroxyl group on position 3 of the inositol ring of inositol glycerophospholipids. These reactions create a number of different 3′ phosphoinositides (PI), leading to the activation of many intracellular signaling pathways through recruitment of intracellular signaling proteins to different membrane compartments. PI3Ks regulate functions as diverse as cell metabolism, survival and polarity, and vesicle trafficking. According to their specific substrate preference and sequence homology, the mammalian PI3Ks isoforms have been organized into the three classes I, II, and III (Vanhaesebroeck et al. 2001). Class I PI3Ks are divided into two subfamilies. The class IA subset of PI3Ks are coupled to receptor tyrosine kinases (RTKs) or intracellular adaptor proteins and exist as heterodimers composed of one of three p110 catalytic subunits – p110α, p110β or p110δ – which form complexes with one of the three regulatory subunits (Engelman et al. 2006). Class IB PI3Ks also form heterodimers, consisting of a regulatory p101 subunit and a p110γ catalytic subunit, but can – in contrast to the Class IA isoforms – trigger G protein-coupled receptor (GPCR) activation through association with heterotrimeric G proteins (Engelman et al. 2006). Interestingly, it was shown recently that the p110β PI3Ks also binds to heterotrimeric G proteins, suggesting that this isoform might integrate signals from RTKs as well as GPCRs (Guillermet-Guibert et al. 2008). Class II PI3Ks consist of a single p110-like catalytic subunit with three isoforms (PIK3C2α, PIK3C2β, and PIK3C2γ) that mediate phosphorylation of PI and, to a lesser extent, PI4P and PI4,5P2 (Engelman et al. 2006). It is currently believed that the Class II PI3K's main cellular function relates to cellular trafficking events, cell survival and receptor internalization, potentially through binding to clathrin or multidomain scaffolding proteins (Das et al. 2007). However, little is currently known regarding the neuron-specific roles of this PI3K. Similarly, the functions of class III PI3K Vps34 (Vacuole Protein-Sorting defective 34) in neurons is, as yet, not adequately understood. Vps34 phosphorylates PI to form PI3P, and is required for trafficking vesicles from the Golgi apparatus to the vacuole (Odorizzi et al. 2000). Because Vps34 is required for induction of cellular autophagy, it has been implicated in mechanisms underlying ischemia-induced stress responses in neurons (Funderburk et al. in press; Simonsen and Tooze 2009). Because of the limited understanding into the function of class II and class III PI3Ks during nervous system development, we will focus here on the contribution of class I PI3K during neuronal development and circuit formation, and explore the implications of aberrant class I PI3K signaling in the pathophysiology of schizophrenia and autism spectrum disorders (ASDs).

2 Isoform Specific Functions of Class I PI3K in the Nervous System

A function for Class I PI3Ks in the establishment of neuronal circuits in the nervous system can be postulated based on patterns of expression of specific isoforms. It appears that all of the Class I PI3K catalytic and regulatory subunits are expressed in both the developing and mature mammalian nervous system. For example, expressions of mRNAs for p85α, p50α, p55α, p85β, and p55γ are widespread in the adult brain, with high levels in the olfactory system, hypothalamus, the hippocampus, cortex, and cerebellum (Hörsch and Kahn 1999; Shin et al. 1998). mRNA expression of p55γ PI3K is limited to the cerebellum and anterior pituitary, with moderate levels in the olfactory bulb and hippocampus. Using antisera specific to the different PI3K regulatory subunits further indicated that p50α and p55γ may play important functions during neurogenesis and neuronal differentiation during cerebellar development (Trejo and Pons 2001). In the rat embryo, the p110α catalytic PI3K subunit is found throughout the central nervous system, with decreases in expression during postnatal development to lower expression in the adult (Ito et al. 1995). However, whilst p110α expression is not restricted to the developing nervous system, the p110δ PI3K appears to be particularly enriched in this tissue (Eickholt et al. 2007). Indeed, expression of p110α PI3K is particularly enriched in neurons at developmental stages that correspond to extension and guidance of neuronal processes (see Fig. 1).

In order to understand the functional significance of distinct isoforms to brain development it would be desirable to use gene-targeting studies in mice. To date, little data of this kind is available. In one example, deficiencies in the p85α regulatory PI3K subunit in p85α$^{-/-}$ mice results in losses of synapses and myelinated axons and affects normal learning behavior. Mice exhibit also restlessness, anxiety and motivation-deficit in water maze tests, as well as deficits in object recognition memory (Tohda et al. 2006, 2009). Restricted loss of the p85α PI3K in GnRH neurons demonstrate that p85α participates in the control of GnRH neuronal activity in male mice. This sex-specific phenotype raises the possibility that PI3K may be required in order to establish sex differences in GnRH neuronal function during development (Acosta-Martinez et al. 2009). Genetic inactivation of p110δ PI3K activity in neurons leads to a reduction in PI3K signaling and deficient axonal elongation under limiting growth conditions. In addition, mice with inactive p110δ show impaired axonal regeneration and functional recovery following a sciatic nerve crush injury, identifying that p110δ-mediated PI3K signaling is a crucial component for efficient axonal elongation (Eickholt et al. 2007).

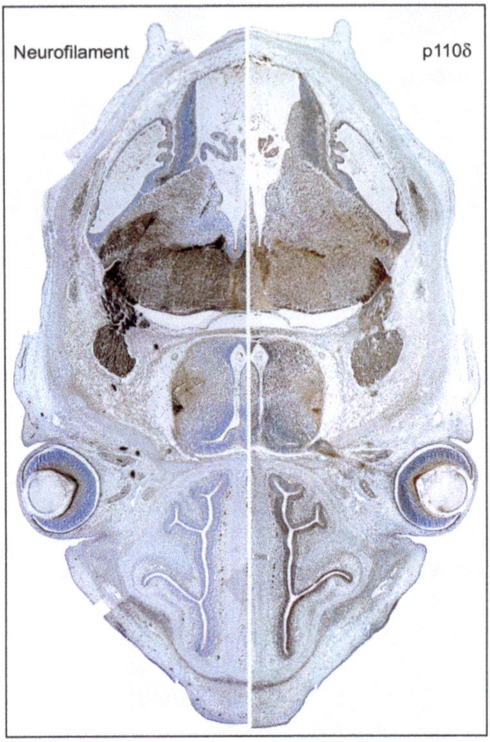

Fig. 1 Expression of p110δ PI3K in the developing brain. Horizontal sections through the mouse brain at E17 were labeled with antineurofilament (*left*) and anti-p110δ PI3K antibodies (*right*). (Eickholt and Vanhaesebroeck, unpublished data)

3 Upstream and Downstream of PI3K Signaling

PI3K regulatory subunits contain SH2 domains that are involved in mediating the recruitment of the p110 catalytic PI3K subunit to membrane-bound RTKs or intracellular adaptor proteins. By contrast, the p110γ PI3K (and potentially also p110β PI3K) convey to signaling mediated by heterotrimeric G proteins (Engelman et al. 2006). PI3K signaling regulation may, additionally, be achieved through upstream Ras activity, which explains the ability of activatory forms of Ras to trigger the PI3K pathway. Thus, PI3K signaling can be activated by a broad range of input signals, and these include growth factors, insulin, cytokines, components of the extracellular matrix and in GPCRs; all of which play important functions in neural development.

Class I PI3Ks activity results in the generation of phosphatidylinositol (3,4,5)-trisphosphate (PIP3), which is able to activate signaling molecules as well as regulators of the cytoskeleton at the cell membrane. One of the best studied downstream mediators of PI3K signaling in neural development is Akt. Activation of PI3K causes the translocation of Akt from the cytosol to the membrane, where it is activated by phosphoinositide dependent kinase 1 (PDK1) and a complex containing mammalian target of rapamycin (mTOR), Rictor and Sin1 (Jacinto et al. 2006;

Sarbassov et al. 2005). An important function of Akt is the phosphorylation and inactivation of the Hamartin/Tuberin protein complex (Dan et al. 2002), which is coded for by Tuberous Sclerosis Complex (TSC) 1 and 2, respectively. TSC1/2 is responsible for inhibition RHEB (Ras Homolog Enriched in Brain) and mTor (Inoki et al. 2002; Zhang et al. 2003). This PI3K regulated pathway is important for promoting protein translation and accelerating cell growth (Hay and Sonenberg 2004). Akt also negatively regulates Glycogen Synthase Kinase-3 (GSK-3), a multifunctional serine/threonine kinase that has been shown to regulate a number of neuronal responses, including microtubule dynamics (Geraldo and Gordon-Weeks 2009). Further downstream effectors of Akt important for brain development are members of the forkhead family of transcription regulators (FOXO) (Engelman et al. 2006). Akt phosphorylates FOXO proteins, an event that causes their inactivation. FOXOs have prominent roles in neural stem cell proliferation and renewal (Chen and Finkel 2009; Paik et al. 2009; Renault et al. 2009), and contributions of specific FOXOs have been indicated to play critical functions in the development of subsets of neuronal circuitries in humans (Konopka et al. 2009) (see Fig. 2).

In addition to Akt, several guanine nucleotide exchange factors for Rho family GTPases interact with PIP3 and have been shown to control PI3K induced cell shape changes and motility in a number of cell types, including neurons (Welch et al. 2003). A further important signaling output of PI3K activity is PI3,4P2. Following PI3K activation, PIP3 can be metabolized by SH2-domain containing inositol 5-phosphatases 1 and 2 (SHIP1 and SHIP2). This enzymatic activity generates PI3,4P2, which binds to sets of effector proteins that overlap with PIP3

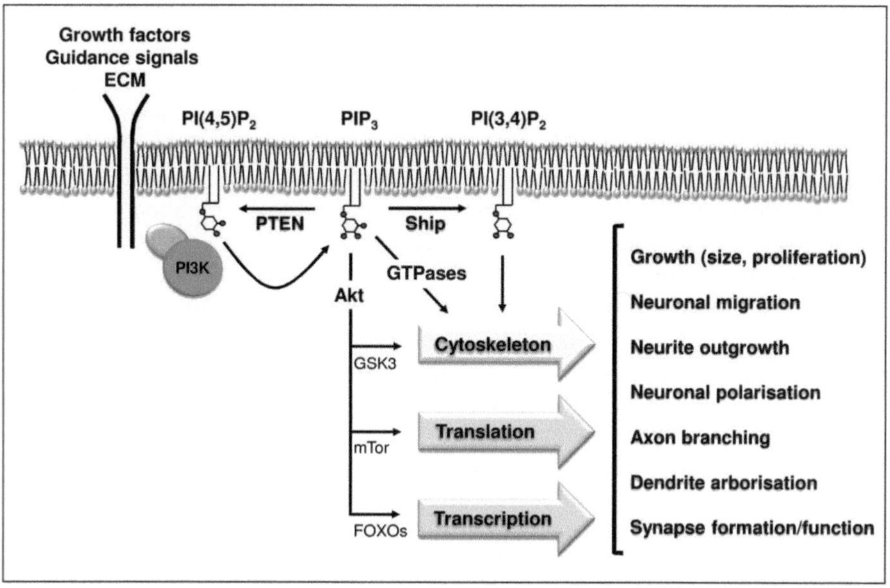

Fig. 2 Major PI3K signaling pathways controlling cellular processes in neuronal development

specific interactors (i.e., Akt) but also recruit others, such as the lamellipodin, which has been reported to bind selectively PI3,4P2, but not PIP3 (Krause et al. 2004) (see Fig. 2).

PI3K signaling is directly antagonized by the tumor suppressor PTEN (phosphatase and tensin homologue deleted on chromosome 10). PTEN was first identified in 1997 as a phosphatase and candidate tumor suppressor in brain, prostate, breast, and kidney tumors (Li et al. 1997; Steck et al. 1997). Germline mutations have been associated with a variety of syndromes, collectively known as the PTEN Hamartoma Tumor Syndrome (PHTS) (Orrico et al. 2009). Since then, a wealth of research has revealed additional functions for PTEN that are particularly important during brain development (van Diepen and Eickholt 2008). It is currently believed that PTEN's main role lies in the regulation of membranous phosphoinositides. However, PTEN has several other potential mechanisms of action including those relating to PTEN's protein phosphatase activities (Leslie et al. 2009). To date, the physiological significance of these roles, especially in neurons, remain largely unclear.

4 PI3K Signaling, Neuronal Migration and Cortical Development

Neuronal migration is a complex biological process, which plays a key role in physiological and pathological conditions. Following the period of proliferation, neural progenitors differentiate and form processes – neurites – through robust changes in the neuronal cytoskeleton. This process is integrated with the program of neuronal migration, when newly born neurons migrate towards their ultimate destination. In general, two modes of migration can be distinguished. During radial migration, neurons adopt characteristic bipolar morphologies and move along the radial fiber scaffold. In contrast, tangentially moving neurons are typically multipolar, with several leading processes orientated toward the direction of migration and one short trailing process. Both forms of neuronal migration incorporate mechanisms that depend on PI3K signaling. For example, the extracellular matrix protein Reelin – secreted by Cajal–Retzius cells in the marginal zone of the cortex – controls the radial migration of cortical neurons and directs the organization of neurons in the cortical plate in a PI3K-dependent fashion. Reelin binds to members of the low density lipoprotein receptor family, ApoER2 and VLDLR, and activates Src family kinases, as well as inducing tyrosine phosphorylation and activation of the intracellular adaptor protein Disabled-1 (Dab1) (Bock et al. 2003). Reelin also induces the PI3K regulatory subunit p85α to associate with Dab1, leading to the activation of Akt and inhibition of GSK-3 (Beffert et al. 2002). In an *in vitro* cortical slice assay, treatment with LY294002 has been shown to perturb the formation of an ordered cortical plate, uncovering the dependency of PI3K activity during early cortex development (Bock et al. 2003). While PI3K and Akt are required for the effects of Reelin during the organization of the cortical plate, this developmental process is independent of their downstream signaling partner mTor and GSK-3 (Jossin and Goffinet 2007). During radial migration of cortical neurons, Reelin induced PI3K signaling affects cytoskeletal in the leading processes of migrating neurons, involving LIMK1 dependent increases in n-cofilin phosphorylation at serine 3 (Chai et al. 2009). Although mice lacking *reelin*, *dab1*, or *apoer2* and *vldlr* display the typical *reeler* migration phenotypes (characterized by inversion of cortical layers, cerebellar dysplasia, and ataxia), n-cofilin regulation in this pathway seems to be selectively mediated by the Reelin receptor ApoER2 (Chai et al. 2009). Additional studies have also defined a pathway in which reelin binding to ApoER receptors activates the PI3K/Akt signaling, causing phosphorylation of BAD which helps to protect cells from cell death (Ohkubo et al. 2007).

In addition to Reelin, classical neurotrophins such as BDNF control the movement of neurons during cortical development. This is best exemplified by a functional implication of the BDNF tyrosine kinase receptor TrkB, which is expressed in migrating neurons in the cortical plate and tangentially moving interneurons (Ernfors et al. 1992). TrkB conditional knock-out mice show a delay in neuronal migration in the developing cerebral cortex, which required docking sites for both Shc and PLCγ within the intracellular domain of the TrkB receptor. This has led to the idea that PI3K signaling, in conjunction with a Ras/MAPK pathway, mediates this response downstream of TrkB (Medina et al. 2004). BDNF also accelerates tangential movements of neurons from the medial ganglionic eminence (MGE) into the developing cortex. Pharmacological inhibition of Trk receptor function using K252a, and TrkB-null mice, show a significant decrease neurons migrating tangentially in the embryonic cortex, which suggests that migration is regulated by TrkB signaling. BDNF (and also NT4) cause rapid activation of PI3K in MGE cells, and inhibition of PI3K (but not of MAP kinase or PLCγ) attenuates tangential migration (Polleux et al. 2002). These observations suggest that TrkB signaling, via PI3Ks, plays an important role in controlling tangential migration in the developing cerebral cortex. Additional upstream regulators of PI3K during cortical migration may involve HGF dependent Met receptor activation. Upon HGF stimulation it was demonstrated that concomitant activation of the Ras/ERK, PI3K/Akt and Rac1/p38 pathways is required to achieve the full capacity of cortical neurons to migrate *in vitro* (Segarra et al. 2006).

5 PI3K Signaling and Neuronal Morphogenesis

The establishment of proper neuronal morphology is vital for the assembly of functional neural circuits and it is regulated through the integration of a number of different intracellular signaling pathways. Amongst those, the PI3K pathway is particularly important for the segregation of neurite processes into axonal and dendritic compartments (Cosker and Eickholt 2007). During the early maturation stages in cultured hippocampal neurons, for example, asymmetric distributions of PIP3 is thought to result in increases in elongation and to initiate the specification of a neurite to adopt axonal characteristics (Menager et al. 2004; Shi et al. 2003). Restricted PIP3 accumulations then lead to the recruitment and activation of different downstream effectors in this compartment. In a positive feedback regulation, restricted PIP3 accumulations may be stabilized by Rho GTPases and the Par3/Par6/aPKC complex, which regulates the breaking of symmetry and persistence of PI3K activation required to implement "polarization" (Arimura and Kaibuchi 2005; Wiggin et al. 2005). Indeed, appropriate localization of the Par3/Par6/aPKC complex to the future axon relies on PI3K activity, but also incorporates Cdc42 (Schwamborn et al. 2006; Schwamborn and Puschel 2004; Shi et al. 2003). In addition to this feedback regulation, neuronal polarization involves the PI3K dependent regulation of ILK, Akt, and GSK-3, a signaling axis that targets microtubule dynamics through phosphorylation of collapsin response mediator protein 2 (CRMP-2) (Guo et al. 2007; Oinuma et al. 2007; Yoshimura et al. 2006). In this way, PI3K activity ensures, in the first instance, the persistent localization and activation of a positive feedback loop and, in the second instance, the coordination of cytoskeletal dynamics which collectively drive axon specification and elongation. The restricted accumulation of PI3K activity and signaling during the early stages of neuronal polarization may also be reinforced through sequestration of PTEN away from the membrane (Chadborn et al. 2006; Kreis et al. 2009), through physical interaction of the p85 PI3K subunit with Shootin1 (Toriyama et al. 2006), through increased Akt ubiquitination in the dendrite compartment (Yan et al. 2006) or enhanced local translation of CRMP2 in axons (Morita and Sobue 2009). Localized vesicular transport of HRas and the PIP3-interacting protein PIP3BP have also been shown to restrict accumulations of PI3K activity during the process of symmetry breaking, resulting in the formation of a single axon in the absence of extrinsic spatial cues (Fivaz et al. 2008; Horiguchi et al. 2006). Candidate extracellular mechanism involved in regulating PI3K signaling during the specification of axonal and dendritic compartments in the developing brain include IGF-1 (Sosa et al. 2006), and the axon guidance cues netrin (Adler et al. 2006) and, potentially, Semaphorins (Polleux et al. 2000).

Following polarization, neurons extend and elaborate their axonal and dendritic processes. During this course, neurons receive and integrate extracellular signals, which coordinate the cytoskeletal remodeling required for the guidance of their processes to appropriate target tissues. Axonal elongation is controlled by interactions between the specialized structure at the distal tip of an extending axon, the growth cone, and the local microenvironment. Different extracellular factors, including components of the ECM, axon guidance molecules and growth factors have been shown to alter axonal growth behavior through changes in PI3K signaling. It has long been recognized, for example, that neurotrophins increase axonal elongation, through activation of PI3K by regulating the growth cone cytoskeleton (Tucker 2002; Zhou et al. 2004). In hippocampal neurons, NGF induced PI3K activation accelerates axon elongation involving localized recruitment of TrkA by ganglioside sialidase PMGS (Da Silva et al. 2005). Similarly, NGF induced increases in axonal elongation in hippocampal neurons have been shown to result from inhibition of PTEN (Arevalo and Rodriguez-Tebar 2006). Conversely, inhibition of PI3K mediates growth cone collapse responses and attenuates axon elongation (Atwal et al. 2003; Chadborn et al. 2006; Eickholt 2002; Eickholt et al. 2007; Ito et al. 2006). The inhibitory axon guidance cues Semaphorin (Sema) 3A has been shown to suppress PI3K signaling concomitant with the activation of GSK-3, which depends on an intact PTEN phosphatase activity (Chadborn et al. 2006; Gallo 2008). PTEN is highly enriched in the axonal compartment and the central domain of sensory growth cones during axonal extension, where it colocalizes with microtubules and accumulates rapidly, during exposure with Sema3A, at the growth cone membrane. These findings highlight the importance of subcellular distributions of PTEN in the control of growth cone behavior (Chadborn et al. 2006), a mechanism potentially regulated by Myosin V transport (Kreis et al. 2009; van Diepen et al. 2009). Downstream of PI3K, Sema3A-induced axon retraction also involves activation of myosin II and its upstream regulator RhoA-kinase (Gallo 2006). Sema3A also inhibits ERM protein phosphorylation in growth cone filopodia through inactivation of PI3K, thereby contributing to the growth cone collapse response (Gallo 2008). During axon elongation, cytoskeletal dynamics within the growth cone are coordinated with the exocytosis of plasmalemmal precursor vesicles at the growth cone, a process regulated by IGF-1 activation of PI3K signaling. Indeed, IGF-1 has been shown to trigger translocation of the exocyst component exo70 to the plasma membrane, a mechanism involving the GTP-binding protein TC10. TC10 and exo70 function are both necessary for the addition of new membrane components and are essential for axon elongation stimulated by IGF-1 (Dupraz et al. 2009).

6 PI3K Signaling in the Control of Dendrite and Spine Development

The dendritic architecture of neurons determines the character of individual synaptic inputs and the functioning of a neuron within a circuit. Developmental programs that control the dendritic branching pattern are initiated relatively early during development and continue, long after neuronal migration is complete, well into postnatal stages. Genetic factors dictate to some extend the development of dendrites, yet the complex shape and arborization pattern of dendrites is additionally controlled by the influence of extrinsic signals, and these include growth factors and classical axon guidance molecules (Jan and Jan 2003). In many instances, a dependency on PI3K signaling in mediating these responses has been shown, mainly using hippocampal neuronal cultures as model systems. Reelin and CALEB/NGC (a EGF family member highly expressed in the nervous system), for example, influence both dendritic tree complexity through the PI3K/Akt/mTor signaling pathway (Brandt et al. 2007; Jossin and Goffinet 2007). This PI3K signaling axis is also important for the stimulatory effects of Sema4D on dendrite morphology (Vodrazka et al. 2009), suggesting that mTor may function, generally, as a convergence point for multiple signaling pathways in the control of dendrite growth and branching during development. Indeed, in the absence of extracellular signals, overactivation of PI3K signaling, achieved through constitutively active mutants of Ras, PI3K, Akt, or siRNA knockdown of PTEN, in hippocampal neurons, is sufficient in inducing growth and elaboration of dendrites (Jaworski et al. 2005).

The development of dendrites is often orchestrated with the establishment of dendritic spines, specialized structures that allow the fast transmission of information and form as protrusions on the dendritic surface. Dendritic spines form from initial filopodial processes that develop into mature spines. On the basis of a number of studies, PI3K signaling has been implicated to influence dendritic spine and synapse development. For example, activation of PI3K signaling causes increases in filopodia spine length. Conversely, pharmacological inhibition of this signaling pathway or overexpression of PTEN reduces both dendritic spine number and filopodia processes (Kumar et al. 2005). Downstream effectors of PI3K activity controlling spine morphology include the Akt/mTOR/p70S6K pathway controlling protein translation (Kumar et al. 2005). During this process, PIP3, indeed, appears to localize to dendritic filopodia and this distribution is enhanced by BDNF application. Interestingly, conditional loss of two downstream negative regulators of PI3K signaling, PTEN and Nf1 (neurofibromatosis type 1) enhance filopodial motility during synapse development (Luikart et al. 2008). Thus, intracellular control of dendritic filopodial dynamics during synapse formation converges on PI3Ks and levels of generated PIP3. As key regulators of both dendrite branching and spine development, it is therefore not surprising that the PI3K signaling pathway has emerged as key mediator in the establishment of functioning neuronal circuits. Consequently, subtle aberrations in this signaling pathway are likely to underlie the pathophysiology of neurodevelopmental disorders such as schizophrenia and autism (Kalkman 2006; Levitt and Campbell 2009).

7 PI3K Signaling in Neurodevelopmental Disorders

7.1 PI3K Signaling and Schizophrenia

Schizophrenia is a psychiatric disorder that affects 1% of the population. It is generally diagnosed at late adolescence or early adulthood, and characterized by a collection of symptoms typically described as positive (delusions, hallucinations, thought disorganization) and negative (impaired motivation, decreased emotional expression) (Andreasen 1995). Schizophrenia might be caused by genetic predisposition and/or environmental factors that influence normal brain development – including events during pregnancy, birth or childhood. It is thought that subtle developmental alterations in the brain increase the risk of developing schizophrenia in later life. At the anatomical level, schizophrenia is associated with an overall reduction in the volume of the cortex and hippocampus (Harrison 1999; Narr et al. 2005a, b) and atrophy of neurons in the cortex (Andersen and Pakkenberg 2003;

Glantz et al. 2006; Liu et al. 2007; Selemon et al. 1995). Decreases in both dendrite arborization and synapse number have also been reported to occur, mainly restricted to regions within the prefrontal cortex (Glantz and Lewis 2000; Kalus et al. 2000). Although the mechanisms underlying this disorder are largely unknown, there is increasing evidence that schizophrenia might develop – at least in part – as a consequence of aberrant PI3K signaling. For example, an association between schizophrenia and *AKT1* genetic variants was found in several samples and family cohorts (Emamian et al. 2004). However, negative findings have also been reported in some patients, provoking a debate about whether AKT1 is a potential schizophrenia susceptibility gene or not (Arguello and Gogos 2008; Liu et al. 2009; Norton et al. 2007; Pietilainen et al. 2009; Thiselton et al. 2008). In this context, however, it is interesting to note that that drugs used in the management of schizophrenia, such as the typical antipsychotic haloperidol as well as several atypical antipsychotics, can act as enhancers of PI3K signaling by increasing AKT or inactivating GSK-3 (Beaulieu et al. 2007).

A contribution in defects of the PI3K–Akt signaling pathway to the etiology of Schizophrenia may also occur as a consequence of aberrant function of Neuregulin-1 (NRG1) signaling. Genetic association study in an Icelandic population indicated that mutation of the *NRG1* gene raises the risk for schizophrenia (Stefansson et al. 2002), which has since been confirmed in other populations (Hong et al. 2004; Stefansson et al. 2003; Tosato et al. 2005). NRG1 triggers signaling by interaction with members of the ErbB RTK family, ErbB2, ErbB3 and ErbB4. Activated ErbB tyrosine kinase domains generate intracellular docking sites for the Grb2 and Shc adaptor proteins, which in turn activate the Raf–MEK–ERK and PI3K–Akt signaling pathways. NRG1 and ErbB receptors are expressed in a number of different regions in the developing brain, and, consequently, NRG1 has been shown to regulate a number of important neurodevelopmental events. It increases the proliferation of neuronal progenitors, controls neuronal migration and axon guidance, as well as glial development, synapse formation and function (Mei and Xiong 2008). Indeed, in many instances, neuronal responses have been shown to involve downstream activation of PI3K signaling (Croslan et al. 2008; Di Segni et al. 2005, 2006; Gambarotta et al. 2004; Li et al. 2001).

Another schizophrenia susceptible gene linked to PI3K signaling is *DTNBP1* (dystrobrevin-binding protein 1), which is located at one of the most promising susceptibility loci in schizophrenia linkage studies, chromosome 6p22.3 (Guo et al. 2009). *DTNBP1* encodes Dysbindin, a coiled-coil-containing protein involved in linking the cytoskeleton to the extracellular matrix, potentially providing molecular scaffolds for the recruitment of signaling proteins. In cortical neurons, the overexpression of dysbindin has been shown to increase phosphorylation of Akt, whilst siRNA silencing of dysbindin protein attenuates Akt phosphorylation (Numakawa et al. 2004). These results suggest that loss of dysbindin may play an important role in the pathogenesis of schizophrenia by altering PI3K signaling.

The *DISC* locus (Disrupted-in-schizophrenia) – a general risk factor for number psychiatric conditions including schizophrenia, bipolar disorder, major depression and autism – has also recently been linked to PI3K signaling (Chubb et al. 2008).

Two genes, *DISC1* and *DISC2*, encode a multifunctional scaffold protein and a putative noncoding RNA antisense to DISC1, respectively. It has been shown that DISC1 function is crucial for neurite outgrowth (Kamiya et al. 2006; Ozeki et al. 2003; Taya et al. 2007), is required for radial migration of cortical neurons and for proper arrangements of dendritic arbors of cortical neurons (Kamiya et al. 2005, 2006). In addition to this, DISC1 regulates cell cycle exit and neuronal differentiation (Mao et al. 2009; Ming and Song 2009), as well as synapse formation (Duan et al. 2007; Faulkner et al. 2008). DISC1 does so by interaction with a host of proteins, amongst which is Girdin (also known as KIAA1212). Girdin directly binds to Akt and enhances its kinase activity *in vitro* (Enomoto et al. 2005). Whilst DISC1 associates directly with Girdin and attenuates Akt activation, DISC1 removal permits Girdin to interact with and to enhance Akt signaling (Kim et al. 2009). In support of this, one can note that there are striking similarities that result as a consequence of DISC1- and PTEN-loss or, indeed, DISC1-loss and over-activation of PI3K signaling (Enomoto et al. 2009; Kim et al. 2009). In addition to the susceptible genes discussed here, there are many other schizophrenia risk factors, including environmental risk factors (such as steroid hormones and cannabis) as well as different growth factors (such as IGF-1 and BDNF), which are linked to affect neuronal morphogenesis and circuit formation through altering PI3K signaling (Gunnell and Holly 2004; Hashimoto et al. 2005). Thus, it seems that an optimally operating PI3K signaling system can potentially be seen as "protective" for individuals carrying other genetic predispositions associated with an increased risk to develop schizophrenia.

7.2 PI3K Signaling and Autism Spectrum Disorder

ASD is a genetically complex neurodevelopmental syndrome characterized by abnormalities in language development and social interaction, which coincide with restricted stereotypic and repetitive patterns of behavior and interests (Abrahams and Geschwind 2008; McCaffery and Deutsch 2005; Persico and Bourgeron 2006). ASD generally manifests itself before 3 years of age, but studies of toddlers at play have also shown signs of ASD as early as 14 months (Landa and Garrett-Mayer 2006). A subset of individuals with ASD show substantially accelerated brain growth during a critical postnatal period, leading to a statistical overrepresentation of individuals with enlarged brain volumes (macrocephaly) that may, however, disappear at later developmental stages (Courchesne and Pierce 2005; Fombonne et al. 1999; McCaffery and Deutsch 2005). The disturbance of early brain growth seen in ASD coincides with a time of critical events in neuronal circuit formation. Indeed, elaboration of neurite processes, synaptic connectivity and the extent of myelination have been shown, in different studies, to be affected in individuals with ASD (Herbert 2005; Herbert et al. 2004).

Deregulation of PI3K signaling is currently thought to be a contributing factor in the development of ASD. For example, a recent study that focused specifically on subjects with autism and severe macrocephaly, found germline mutations of PTEN in 20% of cases (Butler et al. 2005). Although subsequent studies only confirmed PTEN mutations as a rare cause of ASD, when ASD presents with macrocephaly, there is a high incidence of PTEN mutations. ASD and macrocephaly have also been observed together in patients suffering from two rare genetic disorders, Cowden Syndrome and Bannayan–Riley–Ruvalcaba Syndrome, which can both be caused by inherited mutations in the PTEN gene (Goffin et al. 2001; Zori et al. 1998). In this context, it is interesting to note that other related neurological disorders (which also harbor mutations in the PI3K pathway) are associated with ASD. For example, individuals with mutations in the *TSC1/2* complex – a direct substrate of Akt activity – develop tuberous sclerosis and, with a high incidence of 25–50%, also exhibit ASD (Asano et al. 2001; de Vries et al. 2005; Wiznitzer 2004).

The idea that accelerated brain growth and increased process extension, induced by aberrant PI3K/PTEN signaling, may underlie ASD is strongly supported by mouse models in which PTEN-loss is achieved in a tissue specific and temporal manner under the control of different promoters (van Diepen and Eickholt 2008). In the most compelling of these studies, $\mathrm{Pten}^{-/-}$ mutant mice were engineered to carry conditional deletions of PTEN under the control of neuron-specific enolase–$\mathrm{CRE}^{\mathrm{CK2}}$ (NSE–pten). This specific NSE–CRE line is distinct from other NSE–CRE transgenes, as it provides a spatially and temporally restricted pattern of CRE-activity, which commences at around birth and is largely restricted to sub-populations of forebrain neurons (Kwon et al. 2006a, b). NSE–$\mathrm{CRE}^{\mathrm{CK2}}$ mediated Pten-loss resulted in increases in p-Akt and p-S6 (a substrate of mTOR), and induced progressive changes in forebrain circuitries, generating ectopic dendrites and axonal tracts and increasing the number of synapses (Kwon et al. 2006a). The appearance of enlarged neuronal soma and macrocephaly, together with the development of behavioral deficits that include deficits in social interaction, seizures, and decreased learning, suggest that these mice may provide a direct causal link between PTEN deficiency and ASD (Kwon et al. 2006a). Significantly, systemic administration rapamycin (an mTOR inhibitor) was subsequently shown to reverse the majority of behavioral and structural pathologies associated with conditional PTEN deficiencies in this mouse model (Zhou et al. 2009). While ASD is unquestionably a complex genetic disorder, this work points towards the existence of neuronal circuits that might be key for the development of learning and social behavior. It also provides a compelling correlation between increased PI3K–Akt dependent mTOR signaling and autism-like symptoms in individuals bearing *PTEN* mutations.

8 Conclusions

A number of *in vitro* and *in vivo* studies have provided substantial evidence for an involvement of the PI3K signaling pathway during neuronal migration, the establishment of neuronal morphology and circuit formation. Increasingly, there are compelling links between deregulation of this signaling pathway and the pathogenesis of neurodevelopmental disorders such as schizophrenia and ASD. PI3Ks are key regulators of the neuronal cytoskeleton, but also evoke responses such as cell survival, proliferation and differentiation. PI3Ks are part of an increasingly complex signaling network, conveying information of a multitude of receptors to a number of different signal transduction pathways, allowing for a diverse range of cellular responses. Although not specifically discussed here, PI3K signaling functions beyond integration of neurons into functioning neuronal circuits, influences important aspects of synaptic plasticity (Arendt et al. 2010; Fraser et al. 2008; Howlett et al. 2008; Man et al. 2003; Peineau et al. 2007; Yang et al. 2001; Zhu et al. 2002). Future studies will be facilitated through the generation of more refined animal models that further control components of the PI3K pathway in space and time, and that may mimic more subtle developmental and postdevelopmental deregulation of this important signaling pathway.

Acknowledgments This work was funded by the Medical Research Council (Ref G0802289). We thank Carl Hobbs for the immunohistochemistry staining of p110δ PI3K in the developing brain.

References

Arendt KL, Royo M, Fernandez-Monreal M, Knafo S, Petrok CN, Martens JR, Esteban, JA (2010) PIP3 controls synaptic function by maintaining AMPA receptor clustering at the postsynaptic membrane. Nat Neurosci 13:36–44

Abrahams BS, Geschwind DH (2008) Advances in autism genetics: on the threshold of a new neurobiology. Nat Rev Genet 9:341–355

Acosta-Martinez M, Luo J, Elias C, Wolfe A, Levine JE (2009) Male-biased effects of gonadotropin-releasing hormone neuron-specific deletion of the phosphoinositide 3-kinase regulatory subunit p85{alpha} on the reproductive axis. Endocrinology 150:4203–4212

Adler CE, Fetter RD, Bargmann CI (2006) UNC-6/Netrin induces neuronal asymmetry and defines the site of axon formation. Nat Neurosci 9:511–518

Andersen BB, Pakkenberg B (2003) Stereological quantitation in cerebella from people with schizophrenia. Br J Psychiatry 182:354–361

Andreasen NC (1995) Symptoms, signs, and diagnosis of schizophrenia. Lancet 346:477–481

Arevalo MA, Rodriguez-Tebar A (2006) Activation of casein kinase II and inhibition of phosphatase and tensin homologue deleted on chromosome 10 phosphatase by nerve growth factor/p75NTR inhibit glycogen synthase kinase-3beta and stimulate axonal growth. Mol Biol Cell 17:3369–3377

Arguello PA, Gogos JA (2008) A signaling pathway AKTing up in schizophrenia. J Clin Invest 118:2018–2021

Arimura N, Kaibuchi K (2005) Key regulators in neuronal polarity. Neuron 48:881–884

Asano E, Chugani DC, Muzik O, Behen M, Janisse J, Rothermel R, Mangner TJ, Chakraborty PK, Chugani HT (2001) Autism in tuberous sclerosis complex is related to both cortical and subcortical dysfunction. Neurology 57:1269–1277

Atwal JK, Singh KK, Tessier-Lavigne M, Miller FD, Kaplan DR (2003) Semaphorin 3F antagonizes neurotrophin-induced phosphatidylinositol 3-kinase and mitogen-activated protein kinase kinase signaling: a mechanism for growth cone collapse. J Neurosci 23:7602–7609

Beaulieu JM, Gainetdinov RR, Caron MG (2007) The Akt-GSK-3 signaling cascade in the actions of dopamine. Trends Pharmacol Sci 28:166–172

Beffert U, Morfini G, Bock HH, Reyna H, Brady ST, Herz J (2002) Reelin-mediated signaling locally regulates protein kinase B/Akt and glycogen synthase kinase 3β. J Biol Chem 277:49958–49964

Bock HH, Jossin Y, Liu P, Förster E, May P, Goffinet AM, Herz J (2003) Phosphatidylinositol 3-Kinase interacts with the adaptor protein Dab1 in response to Reelin signaling and is required for normal cortical lamination. J Biol Chem 278:38772–38779

Brandt N, Franke K, Rasin MR, Baumgart J, Vogt J, Khrulev S, Hassel B, Pohl EE, Sestan N, Nitsch R et al (2007) The neural EGF family member CALEB/NGC mediates dendritic tree and spine complexity. EMBO J 26:2371–2386

Butler MG, Dasouki MJ, Zhou XP, Talebizadeh Z, Brown M, Takahashi TN, Miles JH, Wang CH, Stratton R, Pilarski R et al (2005) Subset of individuals with autism spectrum disorders and extreme macrocephaly associated with germline PTEN tumour suppressor gene mutations. J Med Genet 42:318–321

Chadborn NH, Ahmed AI, Holt MR, Prinjha R, Dunn GA, Jones GE, Eickholt BJ (2006) PTEN couples Sema3A signalling to growth cone collapse. J Cell Sci 119:951–957

Chai X, Forster E, Zhao S, Bock HH, Frotscher M (2009) Reelin stabilizes the actin cytoskeleton of neuronal processes by inducing n-cofilin phosphorylation at serine3. J Neurosci 29:288–299

Chen E, Finkel T (2009) Preview. The Tortoise, the hare, and the FoxO. Cell Stem Cell 5:451–452

Chubb JE, Bradshaw NJ, Soares DC, Porteous DJ, Millar JK (2008) The DISC locus in psychiatric illness. Mol Psychiatry 13:36–64

Cosker KE, Eickholt BJ (2007) Phosphoinositide 3-kinase signalling events controlling axonal morphogenesis. Biochem Soc Trans 35:207–210

Courchesne E, Pierce K (2005) Brain overgrowth in autism during a critical time in development: implications for frontal pyramidal neuron and interneuron development and connectivity. Int J Dev Neurosci 23:153–170

Croslan DR, Schoell MC, Ford GD, Pulliam JV, Gates A, Clement CM, Harris AE, Ford BD (2008) Neuroprotective effects of neuregulin-1 on B35 neuronal cells following ischemia. Brain Res 1210:39–47

Da Silva JS, Hasegawa T, Miyagi T, Dotti CG, Abad-Rodriguez J (2005) Asymmetric membrane ganglioside sialidase activity specifies axonal fate. Nat Neurosci 8:606–615

Dan HC, Sun M, Yang L, Feldman RI, Sui X-M, Ou CC, Nellist M, Yeung RS, Halley DJJ, Nicosia SV et al (2002) Phosphatidylinositol 3-kinase/Akt pathway regulates tuberous sclerosis tumor suppressor complex by phosphorylation of tuberin. J Biol Chem 277:35364–35370

Das M, Scappini E, Martin NP, Wong KA, Dunn S, Chen Y-J, Miller SLH, Domin J, O'Bryan JP (2007) Regulation of neuron survival through an intersectin-phosphoinositide 3′-kinase C2 {beta}-AKT pathway. Mol Cell Biol 27:7906–7917

de Vries P, Humphrey A, McCartney D, Prather P, Bolton P, Hunt A (2005) Consensus clinical guidelines for the assessment of cognitive and behavioural problems in Tuberous Sclerosis. Eur Child Adolesc Psychiatry 14:183–190

Di Segni A, Shaharabani E, Stein R, Pinkas-Kramarski R (2005) Neuregulins rescue PC12-ErbB-4 cells from cell death induced by beta-amyloid peptide: involvement of PI3K and PKC. J Mol Neurosci 26:57–69

Di Segni A, Farin K, Pinkas-Kramarski R (2006) ErbB4 activation inhibits MPP+-induced cell death in PC12-ErbB4 cells: involvement of PI3K and Erk signaling. J Mol Neurosci 29:257–267

Duan X, Chang JH, Ge S, Faulkner RL, Kim JY, Kitabatake Y, Liu XB, Yang CH, Jordan JD, Ma DK et al (2007) Disrupted-In-Schizophrenia 1 regulates integration of newly generated neurons in the adult brain. Cell 130:1146–1158

Dupraz S, Grassi D, Bernis ME, Sosa L, Bisbal M, Gastaldi L, Jausoro I, Caceres A, Pfenninger KH, Quiroga S (2009) The TC10-Exo70 complex is essential for membrane expansion and axonal specification in developing neurons. J Neurosci 29:13292–13301

Eickholt BJ (2002) An inactive pool of GSK-3 at the leading edge of growth cones is implicated in Semaphorin 3A signaling. J Cell Biol 157:211–217

Eickholt BJ, Ahmed AI, Davies M, Papakonstanti EA, Pearce W, Starkey ML, Bilancio A, Need AC, Smith AJ, Hall SM et al (2007) Control of axonal growth and regeneration of sensory neurons by the p110delta PI 3-kinase. PLoS One 2:e869
Emamian ES, Hall D, Birnbaum MJ, Karayiorgou M, Gogos JA (2004) Convergent evidence for impaired AKT1-GSK3[beta] signaling in schizophrenia. Nat Genet 36:131–137
Engelman JA, Luo J, Cantley LC (2006) The evolution of phosphatidylinositol 3-kinases as regulators of growth and metabolism. Nat Rev Genet 7:606–619
Enomoto A, Murakami H, Asai N, Morone N, Watanabe T, Kawai K, Murakumo Y, Usukura J, Kaibuchi K, Takahashi M (2005) Akt/PKB regulates actin organization and cell motility via Girdin/APE. Dev Cell 9:389–402
Enomoto A, Asai N, Namba T, Wang Y, Kato T, Tanaka M, Tatsumi H, Taya S, Tsuboi D, Kuroda K et al (2009) Roles of disrupted-in-schizophrenia 1-interacting protein girdin in postnatal development of the dentate gyrus. Neuron 63:774–787
Ernfors P, Merlio JP, Persson H (1992) Cells expressing mRNA for neurotrophins and their receptors during embryonic rat development. Eur J Neurosci 4:1140–1158
Faulkner RL, Jang MH, Liu XB, Duan X, Sailor KA, Kim JY, Ge S, Jones EG, Ming GL, Song H et al (2008) Development of hippocampal mossy fiber synaptic outputs by new neurons in the adult brain. Proc Natl Acad Sci USA 105:14157–14162
Fivaz M, Bandara S, Inoue T, Meyer T (2008) Robust neuronal symmetry breaking by Ras-triggered local positive feedback. Curr Biol 18:44–50
Fombonne E, Roge B, Claverie J, Courty S, Fremolle J (1999) Microcephaly and macrocephaly in autism. J Autism Dev Disord 29:113–119
Fraser MM, Bayazitov IT, Zakharenko SS, Baker SJ (2008) Phosphatase and tensin homolog, deleted on chromosome 10 deficiency in brain causes defects in synaptic structure, transmission and plasticity, and myelination abnormalities. Neuroscience 151:476–488
Funderburk SF, Wang QJ, Yue Z (In Press) The Beclin 1-VPS34 complex – at the crossroads of autophagy and beyond. Trends Cell Biol, Corrected Proof
Gallo G (2006) RhoA-kinase coordinates F-actin organization and myosin II activity during semaphorin-3A-induced axon retraction. J Cell Sci 119:3413–3423
Gallo G (2008) Semaphorin 3A inhibits ERM protein phosphorylation in growth cone filopodia through inactivation of PI3K. Dev Neurobiol 68:926–933
Gambarotta G, Garzotto D, Destro E, Mautino B, Giampietro C, Cutrupi S, Dati C, Cattaneo E, Fasolo A, Perroteau I (2004) ErbB4 expression in neural progenitor cells (ST14A) is necessary to mediate neuregulin-1beta1-induced migration. J Biol Chem 279:48808–48816
Geraldo S, Gordon-Weeks PR (2009) Cytoskeletal dynamics in growth-cone steering. J Cell Sci 122:3595–3604
Glantz LA, Lewis DA (2000) Decreased dendritic spine density on prefrontal cortical pyramidal neurons in schizophrenia. Arch Gen Psychiatry 57:65–73
Glantz LA, Gilmore JH, Lieberman JA, Jarskog LF (2006) Apoptotic mechanisms and the synaptic pathology of schizophrenia. Schizophr Res 81:47–63
Goffin A, Hoefsloot LH, Bosgoed E, Swillen A, Fryns JP (2001) PTEN mutation in a family with Cowden syndrome and autism. Am J Med Genet 105:521–524
Guillermet-Guibert J, Bjorklof K, Salpekar A, Gonella C, Ramadani F, Bilancio A, Meek S, Smith AJH, Okkenhaug K, Vanhaesebroeck B (2008) The p110β isoform of phosphoinositide 3-kinase signals downstream of G protein-coupled receptors and is functionally redundant with p110γ. Proc Natl Acad Sci USA 105:8292–8297
Gunnell D, Holly JMP (2004) Insulin-like growth factors, insulin resistance and schizophrenia. Br J Psychiatry 185:353–354
Guo W, Jiang H, Gray V, Dedhar S, Rao Y (2007) Role of the integrin-linked kinase (ILK) in determining neuronal polarity. Dev Biol 306:457–468
Guo AY, Sun J, Riley BP, Thiselton DL, Kendler KS, Zhao Z (2009) The dystrobrevin-binding protein 1 gene: features and networks. Mol Psychiatry 14:18–29
Harrison PJ (1999) Brains at risk of schizophrenia. Lancet 353:3–4

Hashimoto T, Bergen SE, Nguyen QL, Xu B, Monteggia LM, Pierri JN, Sun Z, Sampson AR, Lewis DA (2005) Relationship of brain-derived neurotrophic factor and its receptor TrkB to altered inhibitory prefrontal circuitry in schizophrenia. J Neurosci 25:372–383

Hay N, Sonenberg N (2004) Upstream and downstream of mTOR. Genes Dev 18:1926–1945

Herbert MR (2005) Large brains in autism: the challenge of pervasive abnormality. Neuroscientist 11:417–440

Herbert MR, Ziegler DA, Makris N, Filipek PA, Kemper TL, Normandin JJ, Sanders HA, Kennedy DN, Caviness VS Jr (2004) Localization of white matter volume increase in autism and developmental language disorder. Ann Neurol 55:530–540

Hong CJ, Huo SJ, Liao DL, Lee K, Wu JY, Tsai SJ (2004) Case-control and family-based association studies between the neuregulin 1 (Arg38Gln) polymorphism and schizophrenia. Neurosci Lett 366:158–161

Horiguchi K, Hanada T, Fukui Y, Chishti AH (2006) Transport of PIP3 by GAKIN, a kinesin-3 family protein, regulates neuronal cell polarity. J Cell Biol 174:425–436

Hörsch D, Kahn CR (1999) Region-specific mRNA expression of phosphatidylinositol 3-kinase regulatory isoforms in the central nervous system of C57BL/6J mice. J Comp Neurol 415:105–120

Howlett E, Lin CC, Lavery W, Stern M (2008) A PI3-kinase-mediated negative feedback regulates neuronal excitability. PLoS Genet 4:e1000277

Inoki K, Li Y, Zhu T, Wu J, Guan K-L (2002) TSC2 is phosphorylated and inhibited by Akt and suppresses mTOR signalling. Nat Cell Biol 4:648–657

Ito Y, Goto K, Kondo H (1995) Localization of mRNA for phosphatidylinositol 3-kinase in brain of developing and mature rats. Brain Res Mol Brain Res 34:149–153

Ito Y, Oinuma I, Katoh H, Kaibuchi K, Negishi M (2006) Sema4D/plexin-B1 activates GSK-3beta through R-Ras GAP activity, inducing growth cone collapse. EMBO Rep 7:704–709

Jacinto E, Facchinetti V, Liu D, Soto N, Wei S, Jung SY, Huang Q, Qin J, Su B (2006) SIN1/MIP1 maintains rictor-mTOR complex integrity and regulates Akt phosphorylation and substrate specificity. Cell 127:125–137

Jan YN, Jan LY (2003) The control of dendrite development. Neuron 40:229–242

Jaworski J, Spangler S, Seeburg DP, Hoogenraad CC, Sheng M (2005) Control of dendritic arborization by the phosphoinositide-3′-kinase-Akt-mammalian target of rapamycin pathway. J Neurosci 25:11300–11312

Jossin Y, Goffinet AM (2007) Reelin signals through phosphatidylinositol 3-kinase and Akt to control cortical development and through mTor to regulate dendritic growth. Mol Cell Biol 27:7113–7124

Kalkman HO (2006) The role of the phosphatidylinositide 3-kinase-protein kinase B pathway in schizophrenia. Pharmacol Ther 110:117–134

Kalus P, Muller TJ, Zuschratter W, Senitz D (2000) The dendritic architecture of prefrontal pyramidal neurons in schizophrenic patients. Neuroreport 11:3621–3625

Kamiya A, Kubo K, Tomoda T, Takaki M, Youn R, Ozeki Y, Sawamura N, Park U, Kudo C, Okawa M et al (2005) A schizophrenia-associated mutation of DISC1 perturbs cerebral cortex development. Nat Cell Biol 7:1167–1178

Kamiya A, Tomoda T, Chang J, Takaki M, Zhan C, Morita M, Cascio MB, Elashvili S, Koizumi H, Takanezawa Y et al (2006) DISC1-NDEL1/NUDEL protein interaction, an essential component for neurite outgrowth, is modulated by genetic variations of DISC1. Hum Mol Genet 15:3313–3323

Kim JY, Duan X, Liu CY, Jang MH, Guo JU, Pow-anpongkul N, Kang E, Song H, Ming GL (2009) DISC1 regulates new neuron development in the adult brain via modulation of AKT-mTOR signaling through KIAA1212. Neuron 63:761–773

Konopka G, Bomar JM, Winden K, Coppola G, Jonsson ZO, Gao F, Peng S, Preuss TM, Wohlschlegel JA, Geschwind DH (2009) Human-specific transcriptional regulation of CNS development genes by FOXP2. Nature 462:213–217

Krause M, Leslie JD, Stewart M, Lafuente EM, Valderrama F, Jagannathan R, Strasser GA, Rubinson DA, Liu H, Way M et al (2004) Lamellipodin, an Ena/VASP ligand, is implicated in the regulation of lamellipodial dynamics. Dev Cell 7:571–583

Kreis P, van Diepen MT, Eickholt BJ (2009) Regulation of PTEN in neurons by myosin-based transport mechanisms. Adv Enzyme Regul 50:119–124

Kumar V, Zhang MX, Swank MW, Kunz J, Wu GY (2005) Regulation of dendritic morphogenesis by Ras-PI3K-Akt-mTOR and Ras-MAPK signaling pathways. J Neurosci 25:11288–11299

Kwon CH, Luikart BW, Powell CM, Zhou J, Matheny SA, Zhang W, Li Y, Baker SJ, Parada LF (2006a) Pten regulates neuronal arborization and social interaction in mice. Neuron 50: 377–388

Kwon CH, Zhou J, Li Y, Kim KW, Hensley LL, Baker SJ, Parada LF (2006b) Neuron-specific enolase-cre mouse line with cre activity in specific neuronal populations. Genesis 44:130–135

Landa R, Garrett-Mayer E (2006) Development in infants with autism spectrum disorders: a prospective study. J Child Psychol Psychiatry 47:629–638

Leslie NR, Maccario H, Spinelli L, Davidson L (2009) The significance of PTEN's protein phosphatase activity. Adv Enzyme Regul 49:190–196

Levitt P, Campbell DB (2009) The genetic and neurobiologic compass points toward common signaling dysfunctions in autism spectrum disorders. J Clin Invest 119:747–754

Li J, Yen C, Liaw D, Podsypanina K, Bose S, Wang SI, Puc J, Miliaresis C, Rodgers L, McCombie R et al (1997) PTEN, a putative protein tyrosine phosphatase gene mutated in human brain, breast, and prostate cancer. Science 275:1943–1947

Li Y, Tennekoon GI, Birnbaum M, Marchionni MA, Rutkowski JL (2001) Neuregulin signaling through a PI3K/Akt/Bad pathway in Schwann cell survival. Mol Cell Neurosci 17:761–767

Liu L, Schulz SC, Lee S, Reutiman TJ, Fatemi SH (2007) Hippocampal CA1 pyramidal cell size is reduced in bipolar disorder. Cell Mol Neurobiol 27:351–358

Liu YC, Huang CL, Wu PL, Chang YC, Huang CH, Lane HY (2009) Lack of association between AKT1 variances versus clinical manifestations and social function in patients with schizophrenia. J Psychopharmacol 23:937–943

Luikart BW, Zhang W, Wayman GA, Kwon CH, Westbrook GL, Parada LF (2008) Neurotrophin-dependent dendritic filopodial motility: a convergence on PI3K signaling. J Neurosci 28:7006–7012

Man HY, Wang Q, Lu WY, Ju W, Ahmadian G, Liu L, D'Souza S, Wong TP, Taghibiglou C, Lu J et al (2003) Activation of PI3-kinase is required for AMPA receptor insertion during LTP of mEPSCs in cultured hippocampal neurons. Neuron 38:611–624

Mao Y, Ge X, Frank CL, Madison JM, Koehler AN, Doud MK, Tassa C, Berry EM, Soda T, Singh KK et al (2009) Disrupted in schizophrenia 1 regulates neuronal progenitor proliferation via modulation of GSK3beta/beta-catenin signaling. Cell 136:1017–1031

McCaffery P, Deutsch CK (2005) Macrocephaly and the control of brain growth in autistic disorders. Prog Neurobiol 77:38–56

Medina DL, Sciarretta C, Calella AM, von Bohlen und Halbach O, Unsicker K, Minichiello L (2004) TrkB regulates neocortex formation through the Shc/PLC[gamma]-mediated control of neuronal migration. EMBO J 23:3803–3814

Mei L, Xiong WC (2008) Neuregulin 1 in neural development, synaptic plasticity and schizophrenia. Nat Rev Neurosci 9:437–452

Menager C, Arimura N, Fukata Y, Kaibuchi K (2004) PIP3 is involved in neuronal polarization and axon formation. J Neurochem 89:109–118

Ming GL, Song H (2009) DISC1 partners with GSK3beta in neurogenesis. Cell 136:990–992

Morita T, Sobue K (2009) Specification of neuronal polarity regulated by local translation of CRMP2 and Tau via the mTOR-p70S6K pathway. J Biol Chem 284:27734–27745

Narr KL, Bilder RM, Toga AW, Woods RP, Rex DE, Szeszko PR, Robinson D, Sevy S, Gunduz-Bruce H, Wang YP et al (2005a) Mapping cortical thickness and gray matter concentration in first episode schizophrenia. Cereb Cortex 15:708–719

Narr KL, Toga AW, Szeszko P, Thompson PM, Woods RP, Robinson D, Sevy S, Wang Y, Schrock K, Bilder RM (2005b) Cortical thinning in cingulate and occipital cortices in first episode schizophrenia. Biol Psychiatry 58:32–40

Norton N, Williams HJ, Dwyer S, Carroll L, Peirce T, Moskvina V, Segurado R, Nikolov I, Williams NM, Ikeda M et al (2007) Association analysis of AKT1 and schizophrenia in a UK case control sample. Schizophr Res 93:58–65

Numakawa T, Yagasaki Y, Ishimoto T, Okada T, Suzuki T, Iwata N, Ozaki N, Taguchi T, Tatsumi M, Kamijima K et al (2004) Evidence of novel neuronal functions of dysbindin, a susceptibility gene for schizophrenia. Hum Mol Genet 13:2699–2708

Odorizzi G, Babst M, Emr SD (2000) Phosphoinositide signaling and the regulation of membrane trafficking in yeast. Trends Biochem Sci 25:229–235

Ohkubo N, Vitek MP, Morishima A, Suzuki Y, Miki T, Maeda N, Mitsuda N (2007) Reelin signals survival through Src-family kinases that inactivate BAD activity. J Neurochem 103:820–830

Oinuma I, Katoh H, Negishi M (2007) R-Ras controls axon specification upstream of glycogen synthase kinase-3beta through integrin-linked kinase. J Biol Chem 282:303–318

Orrico A, Galli L, Buoni S, Orsi A, Vonella G, Sorrentino V (2009) Novel PTEN mutations in neurodevelopmental disorders and macrocephaly. Clin Genet 75:195–198

Ozeki Y, Tomoda T, Kleiderlein J, Kamiya A, Bord L, Fujii K, Okawa M, Yamada N, Hatten ME, Snyder SH et al (2003) Disrupted-in-Schizophrenia-1 (DISC-1): mutant truncation prevents binding to NudE-like (NUDEL) and inhibits neurite outgrowth. Proc Natl Acad Sci USA 100:289–294

Paik JH, Ding Z, Narurkar R, Ramkissoon S, Muller F, Kamoun WS, Chae SS, Zheng H, Ying H, Mahoney J et al (2009) FoxOs cooperatively regulate diverse pathways governing neural stem cell homeostasis. Cell Stem Cell 5:540–553

Peineau S, Taghibiglou C, Bradley C, Wong TP, Liu L, Lu J, Lo E, Wu D, Saule E, Bouschet T et al (2007) LTP inhibits LTD in the hippocampus via regulation of GSK3beta. Neuron 53:703–717

Persico AM, Bourgeron T (2006) Searching for ways out of the autism maze: genetic, epigenetic and environmental clues. Trends Neurosci 29:349–358

Pietilainen OP, Paunio T, Loukola A, Tuulio-Henriksson A, Kieseppa T, Thompson P, Toga AW, van Erp TG, Silventoinen K, Soronen P et al (2009) Association of AKT1 with verbal learning, verbal memory, and regional cortical gray matter density in twins. Am J Med Genet B Neuropsychiatr Genet 150B:683–692

Polleux F, Morrow T, Ghosh A (2000) Semaphorin 3A is a chemoattractant for cortical apical dendrites. Nature 404:567–573

Polleux F, Whitford KL, Dijkhuizen PA, Vitalis T, Ghosh A (2002) Control of cortical interneuron migration by neurotrophins and PI3-kinase signaling. Development 129:3147–3160

Renault VM, Rafalski VA, Morgan AA, Salih DA, Brett JO, Webb AE, Villeda SA, Thekkat PU, Guillerey C, Denko NC et al (2009) FoxO3 regulates neural stem cell homeostasis. Cell Stem Cell 5:527–539

Sarbassov DD, Guertin DA, Ali SM, Sabatini DM (2005) Phosphorylation and regulation of Akt/PKB by the rictor-mTOR complex. Science 307:1098–1101

Schwamborn JC, Puschel AW (2004) The sequential activity of the GTPases Rap1B and Cdc42 determines neuronal polarity. Nat Neurosci 7:923–929

Schwamborn JC, Li Y, Puschel AW (2006) GTPases and the control of neuronal polarity. Methods Enzymol 406:715–727

Segarra J, Balenci L, Drenth T, Maina F, Lamballe F (2006) Combined signaling through ERK, PI3K/AKT, and RAC1/p38 is required for met-triggered cortical neuron migration. J Biol Chem 281:4771–4778

Selemon LD, Rajkowska G, Goldman-Rakic PS (1995) Abnormally high neuronal density in the schizophrenic cortex. A morphometric analysis of prefrontal area 9 and occipital area 17. Arch Gen Psychiatry 52:805–818, Discussion 819–820

Shi SH, Jan LY, Jan YN (2003) Hippocampal neuronal polarity specified by spatially localized mPar3/mPar6 and PI 3-kinase activity. Cell 112:63–75

Shin BC, Suzuki M, Inukai K, Anai M, Asano T, Takata K (1998) Multiple isoforms of the regulatory subunit for phosphatidylinositol 3-kinase (PI3-kinase) are expressed in neurons in the rat brain. Biochem Biophys Res Commun 246:313–319
Simonsen A, Tooze SA (2009) Coordination of membrane events during autophagy by multiple class III PI3-kinase complexes. J Cell Biol 186:773–782
Sosa L, Dupraz S, Laurino L, Bollati F, Bisbal M, Caceres A, Pfenninger KH, Quiroga S (2006) IGF-1 receptor is essential for the establishment of hippocampal neuronal polarity. Nat Neurosci 9:993–995
Steck PA, Pershouse MA, Jasser SA, Yung WK, Lin H, Ligon AH, Langford LA, Baumgard ML, Hattier T, Davis T et al (1997) Identification of a candidate tumour suppressor gene, MMAC1, at chromosome 10q23.3 that is mutated in multiple advanced cancers. Nat Genet 15:356–362
Stefansson H, Sigurdsson E, Steinthorsdottir V, Bjornsdottir S, Sigmundsson T, Ghosh S, Brynjolfsson J, Gunnarsdottir S, Ivarsson O, Chou TT et al (2002) Neuregulin 1 and susceptibility to schizophrenia. Am J Hum Genet 71:877–892
Stefansson H, Sarginson J, Kong A, Yates P, Steinthorsdottir V, Gudfinnsson E, Gunnarsdottir S, Walker N, Petursson H, Crombie C et al (2003) Association of neuregulin 1 with schizophrenia confirmed in a Scottish population. Am J Hum Genet 72:83–87
Taya S, Shinoda T, Tsuboi D, Asaki J, Nagai K, Hikita T, Kuroda S, Kuroda K, Shimizu M, Hirotsune S et al (2007) DISC1 regulates the transport of the NUDEL/LIS1/14-3-3epsilon complex through kinesin-1. J Neurosci 27:15–26
Thiselton DL, Vladimirov VI, Kuo PH, McClay J, Wormley B, Fanous A, O'Neill FA, Walsh D, Van den Oord EJ, Kendler KS et al (2008) AKT1 is associated with schizophrenia across multiple symptom dimensions in the Irish study of high density schizophrenia families. Biol Psychiatry 63:449–457
Tohda C, Nakanishi R, Kadowaki M (2006) Learning deficits and agenesis of synapses and myelinated axons in phosphoinositide-3 kinase-deficient mice. Neurosignals 15:293–306
Tohda C, Nakanishi R, Kadowaki M (2009) Hyperactivity, memory deficit and anxiety-related behaviors in mice lacking the p85alpha subunit of phosphoinositide-3 kinase. Brain Dev 31:69–74
Toriyama M, Shimada T, Kim KB, Mitsuba M, Nomura E, Katsuta K, Sakumura Y, Roepstorff P, Inagaki N (2006) Shootin1: a protein involved in the organization of an asymmetric signal for neuronal polarization. J Cell Biol 175:147–157
Tosato S, Dazzan P, Collier D (2005) Association between the neuregulin 1 gene and schizophrenia: a systematic review. Schizophr Bull 31:613–617
Trejo JL, Pons S (2001) Phosphatidylinositol-3-OH kinase regulatory subunits are differentially expressed during development of the rat cerebellum. J Neurobiol 47:39–50
Tucker KL (2002) Neurotrophins and the control of axonal outgrowth. Panminerva Med 44:325–333
van Diepen MT, Eickholt BJ (2008) Function of PTEN during the formation and maintenance of neuronal circuits in the brain. Dev Neurosci 30:59–64
van Diepen MT, Parsons M, Downes CP, Leslie NR, Hindges R, Eickholt BJ (2009) MyosinV controls PTEN function and neuronal cell size. Nat Cell Biol 11:1191–1196
Vanhaesebroeck B, Leevers SJ, Ahmadi K, Timms J, Katso R, Driscoll PC, Woscholski R, Parker PJ, Waterfield MD (2001) Synthesis and function of 3-phosphorylated inositol lipids. Annu Rev Biochem 70:535–602
Vodrazka P, Korostylev A, Hirschberg A, Swiercz JM, Worzfeld T, Deng S, Fazzari P, Tamagnone L, Offermanns S, Kuner R (2009) The semaphorin 4D-plexin-B signalling complex regulates dendritic and axonal complexity in developing neurons via diverse pathways. Eur J Neurosci 30:1193–1208
Welch HC, Coadwell WJ, Stephens LR, Hawkins PT (2003) Phosphoinositide 3-kinase-dependent activation of Rac. FEBS Lett 546:93–97
Wiggin GR, Fawcett JP, Pawson T (2005) Polarity proteins in axon specification and synaptogenesis. Dev Cell 8:803–816

Wiznitzer M (2004) Autism and tuberous sclerosis. J Child Neurol 19:675–679
Yan D, Guo L, Wang Y (2006) Requirement of dendritic Akt degradation by the ubiquitin-proteasome system for neuronal polarity. J Cell Biol 174:415–424
Yang F, He X, Feng L, Mizuno K, Liu XW, Russell J, Xiong WC, Lu B (2001) PI-3 kinase and IP3 are both necessary and sufficient to mediate NT3-induced synaptic potentiation. Nat Neurosci 4:19–28
Yoshimura T, Arimura N, Kawano Y, Kawabata S, Wang S, Kaibuchi K (2006) Ras regulates neuronal polarity via the PI3-kinase/Akt/GSK-3beta/CRMP-2 pathway. Biochem Biophys Res Commun 340:62–68
Zhang H, Cicchetti G, Onda H, Koon HB, Asrican K, Bajraszewski N, Vazquez F, Carpenter CL, Kwiatkowski DJ (2003) Loss of Tsc1/Tsc2 activates mTOR and disrupts PI3K-Akt signaling through downregulation of PDGFR. J Clin Invest 112:1223–1233
Zhou FQ, Zhou J, Dedhar S, Wu YH, Snider WD (2004) NGF-induced axon growth is mediated by localized inactivation of GSK-3beta and functions of the microtubule plus end binding protein APC. Neuron 42:897–912
Zhou J, Blundell J, Ogawa S, Kwon CH, Zhang W, Sinton C, Powell CM, Parada LF (2009) Pharmacological inhibition of mTORC1 suppresses anatomical, cellular, and behavioral abnormalities in neural-specific Pten knock-out mice. J Neurosci 29:1773–1783
Zhu JJ, Qin Y, Zhao M, Van Aelst L, Malinow R (2002) Ras and Rap control AMPA receptor trafficking during synaptic plasticity. Cell 110:443–455
Zori RT, Marsh DJ, Graham GE, Marliss EB, Eng C (1998) Germline PTEN mutation in a family with Cowden syndrome and Bannayan-Riley-Ruvalcaba syndrome. Am J Med Genet 80:399–402

[illegible]

PI3 Kinase Regulation of Skeletal Muscle Hypertrophy and Atrophy

David J. Glass

Contents

Abstract Activation of the PI3 kinase pathway can induce skeletal muscle hypertrophy, defined as an increase in skeletal muscle mass. In mammals, skeletal muscle hypertrophy occurs as a result of an increase in the size, as opposed to the number, of pre-existing skeletal muscle fibers. This pathway's effects on skeletal muscle have been implicated most prominently downstream of Insulin-like growth factor 1 signaling. IGF-1's pro-hypertrophy activity comes predominantly through its ability to activate the Phosphoinositide 3-kinase (PI3K)/Akt signaling pathway. Akt is a serine-threonine protein kinase that can induce protein synthesis and block the transcriptional upregulation of key mediators of skeletal muscle atrophy, the E3 ubiquitin ligases MuRF1 and MAFbx (also called Atrogin-1), by phosphorylating and thereby inhibiting the nuclear translocation of the FOXO (also called "forkhead") family of transcription factors. Once phosphorylated by Akt, the FOXOs are excluded from the nucleus, and upregulation of *MuRF1* and *MAFbx* is blocked.

D.J. Glass
Novartis Institute for Biomedical Research, 100 Technology Square, Cambridge, MA 02139
e-mail: david.glass@novartis.com

C. Rommel et al. (eds.). *Phosphoinositide 3-kinase in Health and Disease*, Volume 1
Current Topics in Microbiology and Immunology 346, DOI 10.1007/82_2010_78

MuRF1 and MAFbx mediate atrophy by ubiquitinating particular protein substrates, causing them to undergo degradation by the proteasome. MuRF1's substrates include several components of the sarcomeric thick filament, including Myosin Heavy Chain (MyHC). Thus, by blocking MuRF1 activation, IGF-1 helps prevent the breakdown of the thick filament under atrophy conditions.

IGF1/PI3K/Akt signaling also can dominantly inhibit the effects of a secreted protein called "myostatin," which is a member of the TGFβ family of proteins. Deletion or inhibition of myostatin causes an increase in skeletal muscle size, because myostatin acts both to inhibit myoblast differentiation and to block the Akt pathway. Thus by blocking myostatin, PI3K/Akt activation stimulates differentiation and protein synthesis by this distinct mechanism. Myostatin induces the phosphorylation and activation of the transcription factors of Smad2 and Smad3, downstream of the ActRII (Activin Receptor type II)/Alk (Activin Receptor-like kinase) receptor complex. Other TGFβ-like molecules can also block differentiation, including TGF-b1, GDF-11, activinA, BMP-2 and BMP-7. As mentioned, myostatin also downregulates the Akt/mTOR/p70S6 protein synthesis pathway, which mediates both differentiation in myoblasts and hypertrophy in myotubes. Blockade of the Akt/mTOR pathway, using siRNA to RAPTOR, a component of "TORC1" (TOR signaling Complex 1), increases myostatin-induced phosphorylation of Smad2; this establishes a "feed-forward mechanism," because myostatin can downregulates TORC1, and this downregulation in turn amplifies myostatin signaling. Blockade of RAPTOR also facilitates myostatin's inhibition of muscle differentiation. When added to post-differentiated myotubes, myostatin causes a decrease in their diameter – however, this does not happen through the normal "atrophy pathway." Rather than causing upregulation of the E3 ubiquitin ligases MuRF1 and MAFbx, previously shown to mediate skeletal muscle atrophy, myostatin decreases expression of these atrophy markers in differentiated myotubes, as well as other genes normally upregulated during differentiation, such as MyoD and myogenin. These findings show that myostatin signaling acts by blocking genes induced during differentiation, even in a myotube, as opposed to activating the distinct "atrophy program."

1 Protein Synthesis Pathways Downstream of PI3K

Stimulation of the Phosphatidylinositol-3 Kinase (PI3K)/Akt pathway results in the downstream activation of targets that induce protein synthesis (Bodine et al. 2001b; Rommel et al. 2001) (Fig. 1). Weightbearing or anabolic exercise leads to activation of the PI3K/Akt pathway by directly inducing muscle expression of IGF-1 (DeVol et al. 1990; Yan et al. 1993), which is sufficient to induce hypertrophy of skeletal muscle (Vandenburgh et al. 1991), as was demonstrated in transgenic mice in which IGF-1 is overexpressed in skeletal muscle (Coleman et al. 1995; Musaro et al. 2001). Activation of Akt is sufficient to induce hypertrophy *in vivo*, as was shown by the production of transgenic mice in which a mutant, constitutively active, form of Akt is conditionally expressed in adult skeletal muscle (Izumiya et al. 2008; Lai et al. 2004; Mammucari et al. 2007). In that setting, acute activation of Akt, for 2–3 weeks, is sufficient to induce a doubling in the size of skeletal muscle; this increase

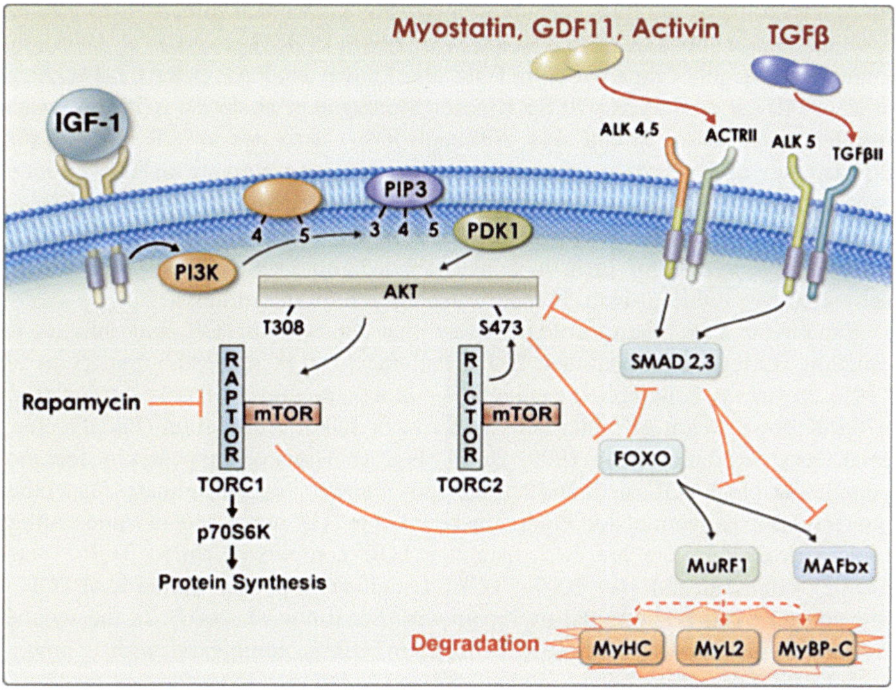

Fig. 1 Hypertrophy signaling dominantly regulates atrophy and myostatin signaling. On the *left*, PI3k/Akt hypertrophy signaling is demonstrated. Activation of the PI3k/Akt pathway with IGF-1 induces skeletal muscle hypertrophy via the PI3k/Akt pathway, which upregulates protein synthesis downstream of TORC1 signaling. Myostatin's activates the transcription factors Smad2 and Smad3, which block muscle differentiation, including the E3 ligases MuRF1 and MAFbx, which are normally upregulated during muscle atrophy. This distinguishes the effects of myostatin with "typical atrophy." Smad2 and 3 activation also lead to myostatin's inhibitory effects on Akt. Substrates of MuRF1 include Myosin Heavy Chain

occurs via an increase in the average cross-sectional area of individual muscle fibers, caused by an increase in TORC1/p70S6K protein synthesis pathways (Lai et al. 2004). Conversely, in settings of skeletal muscle atrophy, Akt activation is downregulated (Sugita et al. 2005). Inhibiting PI3Kinase ablates Akt signaling downstream of IGF-1 activation, which is sufficient to block IGF-1's pro-hypertrophy effects (Rommel et al. 2001). This demonstrated that PI3 kinase is required for IGF-1-induced hypertrophy in skeletal muscle.

2 Hypertrophy Mediators Downstream of PI3K and Akt: The Akt/TORC1/p70S6K and Akt/GSK3 Pathways

Genetic experiments performed in Drosophila originally helped to define the pathway, which included PI3K and Akt that can control cell size (Fig. 1). This is the pathway that is recruited by IGF-1 in mammals. In mammals, IGF-1 induces phosphorylation of the IGF-1 receptor, a tyrosine kinase that directly activates the insulin receptor substrate (IRS) (Bohni et al. 1999), PI3K (Leevers et al. 1996), the mammalian target of rapamycin (mTOR, also known as FRAP or RAFT-1) (Zhang et al. 2000), as well as of p70 S6 Kinase (Montagne et al. 1999) (p70S6K), each resulted in decreases in cell size. Although IGF-1 activates mTOR and p70S6K downstream of PI3K/Akt activation, amino acids can activate mTOR directly, causing a subsequent stimulation of p70S6K activity (Burnett et al. 1998; Hara et al. 1998). Thus, mTOR appears to have an important and central function in integrating a variety of growth signals, from simple nutritional stimulation to activation by protein growth factors, resulting in protein synthesis.

Rapamycin is a pharmacologic agent that binds to mTOR and inhibits its function (Lorenz and Heitman 1995; Pallafacchina et al. 2002; Sabers et al. 1995). *In vitro*, when applied to myotube cultures, rapamycin blocks activation of p70S6K downstream of either activated Akt or IGF-1 stimulation (Pallafacchina et al. 2002; Rommel et al. 1999, 2001) (Fig. 1). However, rapamycin does not completely block IGF-mediated hypertrophy *in vitro*, which indicates that other pathways downstream of Akt but independent of mTOR play a role in some settings of hypertrophy. There are two distinct mTOR complexes, called TORC1 and TORC2 (Bhaskar and Hay 2007). TORC1 includes a protein called RAPTOR – this complex can be inhibited by rapamycin (Schalm et al. 2003). In the second, rapamycin-independent complex, TORC2, mTOR is complexed with a protein called RICTOR, which has been characterized as being able to phosphorylate Akt on serine 473, in a positive feedback mechanism (Sarbassov et al. 2005).

In addition to stimulating p70S6 kinase, activation of mTOR via TORC1 inhibits PHAS-1 (also known as 4E-BP), which is a negative regulator of the protein initiation factor eIF-4E (Hara et al. 1997; Proud 2004). PHAS-1 can directly bind raptor (Hara et al. 2002; Kim do et al. 2002). Mutations in PHAS-1 that inhibit interaction with raptor also inhibit mTOR-mediated phosphorylation of PHAS-1 (Choi et al. 2003). Finally, overexpression of raptor can enhance the phosphorylation of PHAS-1 by mTOR *in vitro* (Choi et al. 2003; Schalm et al. 2003). mTOR binds PHAS-1 by a TOR signaling (TOS) motif; this same motif is found in p70S6K (Schalm et al. 2003), demonstrating both a mechanism for mTOR's interaction with its downstream signaling components, and the possibility that there may be some selective hierarchy in signaling (because the same motif binds mTOR, one might wonder whether there is competition for the same mTOR interaction site).

In summary, mTOR can increase protein synthesis by modulating two distinct pathways, the p70S6K pathway and the PHAS-1 pathway (Fig. 1).

3 A Second Hypertrophy Mediator Downstream of PI3K and Akt: GSK3β

Glycogen synthase kinase 3 beta, GSK3β, is a distinct substrate of Akt that can modulate hypertrophy. Phosphorylation by Akt inhibits GSK3β activity (Cross et al. 1995). Expression of a dominant-negative kinase inactive form of GSK3β induces significant hypertrophy in myotubes (Rommel et al. 2001), as does inhibition of GSK3 with Lithium (Vyas et al. 2002). GSK3β blocks protein translation initiated by the eIF2B protein (Hardt and Sadoshima 2002), and blocks differentiation induced by a transcription factor called NFAT (Rommel et al. 2001) (van der Velden et al. 2008). Inhibition or loss of GSK-3β protein activity results in enhanced myotube formation and muscle-specific gene expression during differentiation (van der Velden et al. 2008). In addition, GSK-3β inhibition restores myogenic differentiation following calcineurin blockade, which further implicates the involvement of the transcription factor NFAT in myoblast differentiation downstream of IGF-1 activation (van der Velden et al. 2008). Furthermore, GSK-3β-deficient myoblasts demonstrated enhanced nuclear translocation of NFATc3, and elevated NFAT-sensitive promoter transactivation (van der Velden et al. 2008). IGF-1-mediated inhibition of GSK-3β is, therefore, a distinct mechanism for inducing hypertrophy and promoting myoblast differentiation.

4 Skeletal Muscle Atrophy Occurs via Induction of Distinct E3 Ubiquitin Ligases, Whose Expression Can Be Inhibited by the PI3K/Akt Pathway

Skeletal muscle atrophy occurs in a variety of settings, including disuse, denervation, cachexia, renal failure, and burns. There is a distinct set of genes which are inversely regulated under IGF-1-induced hypertrophy conditions vs. Dexamethasone (DEX)-induced atrophy (Latres et al. 2005); these include the gene MAFbx (Bodine et al. 2001a) (for Muscle Atrophy F-box; also called Atrogin-1 (Gomes et al. 2001)). The second gene, MuRF1 (Bodine et al. 2001a) (for Muscle Ring Finger1) is significantly upregulated under atrophy conditions (Bodine et al. 2001a). Both MuRF1 and MAFbx/Atrogin were shown to encode E3 ubiquitin ligases (Bodine et al. 2001a). Expression of *MuRF1* and *MAFbx* is stimulated in thirteen distinct models of skeletal muscle atrophy (Bodine et al. 2001a; Dehoux et al. 2003; Deruisseau et al. 2004; Gomes et al. 2001; Li et al. 2003). Mice which are null for MuRF1 (MuRF1−/−) and mice which are null for MAFbx (MAFbx−/−) appear phenotypically normal. However, under atrophy conditions, significantly less muscle mass is lost in either MuRF1−/− or MAFbx−/− animals in comparison with control littermates (Bodine et al. 2001a).

MuRF1 encodes a protein that contains three domains: a RING-finger domain (Borden and Freemont 1996), which is required for ubiquitin ligase activity (Kamura et al. 1999); a "B-box," whose function is unclear; and a "coiled-coil domain," which may be required for the formation of heterodimers between MuRF1 and a related protein, MuRF2 (Centner et al. 2001). Proteins that have these three domains have been called "RBCC" proteins (for RING, B-BOX, Coiled-Coil domain) (Saurin et al. 1996), or "TRIM" proteins (for tripartite motif) (Reymond et al. 2001). MuRF1 has been demonstrated to have ubiquitin ligase activity which depends on the presence of the RING domain for that activity (Bodine et al. 2001a).

MuRF1 has been shown to bind to the myofibrillar protein titin, at the M line (Centner et al. 2001; McElhinny et al. 2002; Pizon et al. 2002).

MuRF1 and Myosin Heavy Chain (MyHc) physically interact – as demonstrated by immune precipitation of epitope-tagged MuRF1 protein, which co-immunoprecipitated Myosin Heavy Chain protein; this finding led to the discovery that MyHc was a substrate of MuRF1 (Clarke et al. 2007). Subsequently, it was shown that several other proteins in the thick filament of muscle were also degraded by MuRF1, including Myosin light chain and Myosin binding protein C (Cohen et al. 2009).

MAFbx/Atrogin-1 contains an F-box domain, a characteristic motif seen in a family of E3 ubiquitin ligases called SCFs (for Skp1, Cullin, F-box) (Jackson and Eldridge 2002). F-box containing E3 ligases usually bind a substrate only after that substrate has first been post-translationally modified, for example by phopshorylation (Jackson and Eldridge 2002). This suggests the possibility of a signaling pathway in which a potential substrate is first phosphorylated as a response to an atrophy-induced stimulus, and then degraded via MAFbx.

Substrates have been suggested for MAFbx in skeletal muscle including MyoD (Tintignac et al. 2005) and eIF3f (Csibi et al. 2009; Lagirand-Cantaloube et al. 2008). eIF3f is a translation initiation factor; it is ubiquitinated and degraded in a MAFbx-dependent manner in myotubes (Lagirand-Cantaloube et al. 2008); upregulation of MAFbx causes breakdown of eIF3f (Lagirand-Cantaloube et al. 2008), giving a mechanism for MAFbx to control protein synthesis in addition to protein breakdown (Lagirand-Cantaloube et al. 2008).

In cardiac muscle, it was shown that while MAFbx has no effect on Akt activation in response to IGF-1 or insulin challenge in cardiomyocytes, nevertheless MAFbx can repress Akt-dependent hypertrophy by activating the Forkhead transcription factors through a distinct type of ubiquitination – ubiquitination using Lysine 63, which perturbs transcriptional activity (in this case that of the FOXO transcription factors) rather than inducing proteasomal degradation (Li 2007); FOXO activitation was shown to be required to activate the atrophy transcriptional program (Sandri et al. 2004; Stitt et al. 2004), as will be discussed further. Because FOXO proteins regulate MAFbx expression in skeletal and cardiac muscle, these findings indicated the presence of a feed-forward mechanism in which MAFbx is activated by, and in turn coactivates, FOXO3a and FOXO1 (Li 2007), making it clear why IGF-1's ability to inhibit FOXO via activation of Akt is necessary to inhibit uncontrolled atrophy in skeletal muscle.

5 IGF-1/PI3K/Akt Inhibition of FOXO Transcription Factors Blocks Upregulation of MuRF1 and MAFbx

Studies of differentiated myotube cultures demonstrated that treatment of myotubes with the cachectic glucocorticoid dexamethasone promotes enhanced protein breakdown and increased expression of genes broadly involved in the ubiquitin-proteasome proteolytic pathway (Du et al. 2000; Hong and Forsberg 1995;

Wang et al. 1998). *In vitro* treatment of myotubes with dexamethasone induces atrophy, accompanied by the specific increased expression of *MAFbx* and *MuRF1* (Sandri et al. 2004; Stitt et al. 2004). The upregulation of *MAFbx* and *MuRF1* was antagonized by simultaneous treatment with IGF-1 (Sacheck et al. 2004; Sandri et al. 2004; Stitt et al. 2004), acting through the PI3K/Akt pathway (Sandri et al. 2004; Stitt et al. 2004); this finding demonstrated a novel role for Akt – in addition to stimulating skeletal muscle hypertrophy, Akt stimulation could dominantly inhibit the induction of atrophy signaling (Fig. 1). Similarly, *MuRF1* and *MAFbx* were activated in a separate model of atrophy, diabetes, and here too IGF-1 blocked the transcriptional upregulation (Lee et al. 2004). Genetic activation of Akt was shown to be sufficient to block the atrophy-associated increases in *MAFbx* and *MuRF1* transcription (Stitt et al. 2004). The mechanism by which Akt inhibited *MAFbx* and *MuRF1* upregulation was demonstrated to involve the FOXO family of transcription factors (Lee et al. 2004; Sandri et al. 2004; Stitt et al. 2004). In myotubes, FOXO transcription factors are excluded from the nucleus when phosphorylated by Akt, and translocate to the nucleus upon dephosphorylation. The translocation and activity of FOXO transcription factors is required for upregulation of MuRF1 and MAFbx – in the case of FOXO3, activation was demonstrated to be sufficient to induce atrophy (Sandri et al. 2004), a finding which was subsequently supported by the transgenic expression of FOXO1, which resulted in atrophic phenotype (Kamei et al. 2004).

6 Akt Regulation of Myostatin

In addition to IGF-1, other secreted proteins have been demonstrated to perturb skeletal muscle size. Myostatin, also called GDF-8, is a TGF-β family member, which is a negative regulator of muscle mass (Lee et al. 2004). Myostatin's effect was demonstrated in studies with mice which were made null for the myostatin gene(McPherron et al. 1997), and also by correlating increases in muscle mass that were observed in strains of cattle with a loss of myostatin (Grobet et al. 1997; Kambadur et al. 1997; McPherron and Lee 1997); the loss of myostatin resulted in a more than doubling in muscle mass. It has been suggested that other TGF-β superfamily molecules, distinct from myostatin, play a role in modulating skeletal muscle size, because Myostatin$^{-/-}$ mice that are mated with mice that are transgenic for follistatin (TG^{follistatin}), which is capable of inhibiting not only myostatin, but also its close relative GDF-11, and other TGF-β molecules such as activins, resulted in an even greater increase in muscle size (33).

In vitro studies with myostatin have been performed on rodent cells. In these studies, it has been shown that myostatin can block the differentiation of myoblasts from myotubes (Langley et al. 2002; McFarlane et al. 2006; Rios et al. 2004; Yang et al. 2007). Experiments both *in vitro* and *in vivo* have demonstrated that myostatin signals first bind to the type II activin receptor, IIb, which then allows for interaction with type I receptors, ALK4 or ALK5 (Tsuchida et al. 2008). The binding of myostatin to these receptor complexes results in the phosphorylation and activation of the transcription factors, Smad2 and Smad3, which translocate to the nucleus upon phosphorylation (Rebbapragada et al. 2003). In a study of myostatin and other TGF-b molecules on human skeletal myoblasts (HuSkMC) and myotubes, HuSkMCs respond to myostatin at physiologic concentrations, 0.1–300 ng/ml, resulting in a decrease in fusion index, myotube diameter, creatine kinase (CK) activity, and expression of MyoD and myogenin (Trendelenburg et al. 2009). It was previously demonstrated that follistatin, a more general inhibitor of TGFβ molecules, could induce an additive increase in muscle mass when combined with myostatin(Lee 2007). A range of other TGFβ molecules are shown to be able to block muscle differentiation, including the more distantly related activins, and BMP-2 (Trendelenburg et al. 2009). Myostatin inhibits activation of Akt, in both myoblasts and myotubes (Trendelenburg et al. 2009). It was recently reported that muscle-specific ablation of TORC1 (by ablating RAPTOR) results in a dystrophic phenotype (Bentzinger et al. 2008). Inhibition of RAPTOR, and thus TORC1, does not by itself block muscle differentiation, but does contribute to myostatin's inhibitory effects, by resulting in an increase in myostatin-induced Smad phosphorylation, establishing a feed-forward mechanism: myostatin activates Smad2, which inhibits Akt, inhibiting TORC1, which in turn potentiates myostatin's activation of Smad2. These findings are outlined in Fig. 1.

Addition of IGF-1 dominantly blocks the effects of myostatin, when applied to either myoblasts or myotubes (Trendelenburg et al. 2009). The precise intersection between the two pathways may be multifold, but it is clear that Akt is a particular nexus, and that IGF-1 can rescue the activation of the PI3K/Akt pathway that is blunted by myostatin.

The demonstration that IGF-1 can dominantly overcome myostatin inhibition via this pathway adds to the rationale for treatment regimens that activate the PI3K/Akt pathway in clinical settings where myostatin is active.

7 Conclusion

A considerable amount of recent progress has been made in the understanding of signaling pathways that mediate skeletal muscle hypertrophy and atrophy. Whereas it was appreciated many years ago that hypertrophy comes about through an increase in the rate of protein synthesis, and atrophy through an increase in protein degradataion, only now can specific signaling pathways be drawn, since the particular molecular mediators of hypertrophy and atrophy in skeletal muscle have only recently been determined. Furthermore, it is only through recent studies that it is understood that hypertrophy pathways can be dominant over atrophy mediators. These findings help to give hope that novel drug targets may be found to block skeletal muscle atrophy, seen in a variety of clinical conditions, from the cachexia of AIDS, sepsis, and cancer, to the gradual loss of muscle mass observed during normal aging.

Acknowledgments Thanks to Drs. M. Fishman, B. Richardson, A. Mackenzie, as well as the rest of the Novartis Community, for their enthusiastic support and input. For this work, studies performed in large part by A.U. Trendelenburg, B. Clarke, and E. Latres, in particular, respectively, were referred.

References

Bentzinger C, Romanino K, Cloëtta D, Lin S, Mascarenhas J, Oliveri F, Xia J, Casanova E, Costa C, Brink M et al (2008) Skeletal muscle-specific ablation of raptor, but not of rictor, causes metabolic changes and results in muscle dystrophy. Cell Metab 8:411–424

Bhaskar PT, Hay N (2007) The two TORCs and Akt. Dev Cell 12:487–502

Bodine SC, Latres E, Baumhueter S, Lai VK, Nunez L, Clarke BA, Poueymirou WT, Panaro FJ, Na E, Dharmarajan K et al (2001a) Identification of ubiquitin ligases required for skeletal muscle atrophy. Science 294:1704–1708

Bodine SC, Stitt TN, Gonzalez M, Kline WO, Stover GL, Bauerlein R, Zlotchenko E, Scrimgeour A, Lawrence JC, Glass DJ et al (2001b) Akt/mTOR pathway is a crucial regulator of skeletal muscle hypertrophy and can prevent muscle atrophy *in vivo*. Nat Cell Biol 3:1014–1019

Bohni R, Riesgo-Escovar J, Oldham S, Brogiolo W, Stocker H, Andruss BF, Beckingham K, Hafen E (1999) Autonomous control of cell and organ size by CHICO, a Drosophila homolog of vertebrate IRS1-4. Cell 97:865–875

Borden KL, Freemont PS (1996) The RING finger domain: a recent example of a sequence-structure family. Curr Opin Struct Biol 6:396–401

Burnett PE, Barrow RK, Cohen NA, Snyder SH, Sabatini DM (1998) RAFT1 phosphorylation of the translational regulators p70 S6 kinase and 4E-BP1. Proc Natl Acad Sci USA 95: 1432–1437

Centner T, Yano J, Kimura E, McElhinny AS, Pelin K, Witt CC, Bang ML, Trombitas K, Granzier H, Gregorio CC et al (2001) Identification of muscle specific ring finger proteins as potential regulators of the titin kinase domain. J Mol Biol 306:717–726

Choi KM, McMahon LP, Lawrence JC Jr (2003) Two motifs in the translational repressor PHAS-I required for efficient phosphorylation by mTOR and recognition by raptor. J Biol Chem 278:19667–19673

Clarke BA, Drujan D, Willis MS, Murphy LO, Corpina RA, Burova E, Rakhilin SV, Stitt TN, Patterson C, Latres E et al (2007) The E3 Ligase MuRF1 degrades myosin heavy chain protein in dexamethasone-treated skeletal muscle. Cell Metab 6:376–385

Cohen S, Brault JJ, Gygi SP, Glass DJ, Valenzuela DM, Gartner C, Latres E, Goldberg AL (2009) During muscle atrophy, thick, but not thin, filament components are degraded by MuRF1-dependent ubiquitylation. J Cell Biol 185:1083–1095

Coleman ME, DeMayo F, Yin KC, Lee HM, Geske R, Montgomery C, Schwartz RJ (1995) Myogenic vector expression of insulin-like growth factor I stimulates muscle cell differentiation and myofiber hypertrophy in transgenic mice. J Biol Chem 270:12109–12116

Cross DA, Alessi DR, Cohen P, Andjelkovich M, Hemmings BA (1995) Inhibition of glycogen synthase kinase-3 by insulin mediated by protein kinase B. Nature 378:785–789

Csibi A, Leibovitch M-P, Cornille K, Tintignac LA, Leibovitch SA (2009) MAFbx/Atrogin-1 controls the activity of the initiation factor eIF3-f in skeletal muscle atrophy by targeting multiple C-terminal lysines. J Biol Chem 284:4413–4421

Dehoux MJM, van Beneden RP, Fernandez-Celemin L, Lause PL, Thissen J-PM (2003) Induction of MafBx and Murf ubiquitin ligase mRNAs in rat skeletal muscle after LPS injection. FEBS Lett 544:214–217

Deruisseau KC, Kavazis AN, Deering MA, Falk DJ, Van Gammeren D, Yimlamai T, Ordway GA, Powers SK (2004) Mechanical ventilation induces alterations of the ubiquitin-proteasome pathway in the diaphragm. J Appl Physiol 98:1314–1321

DeVol DL, Rotwein P, Sadow JL, Novakofski J, Bechtel PJ (1990) Activation of insulin-like growth factor gene expression during work-induced skeletal muscle growth. Am J Physiol 259: E89–E95

Du J, Mitch WE, Wang X, Price SR (2000) Glucocorticoids induce proteasome C3 subunit expression in L6 muscle cells by opposing the suppression of its transcription by NF-kappa B. J Biol Chem 275:19661–19666

Gomes MD, Lecker SH, Jagoe RT, Navon A, Goldberg AL (2001) Atrogin-1, a muscle-specific F-box protein highly expressed during muscle atrophy. Proc Natl Acad Sci U S A 98: 14440–14445

Grobet L, Martin LJ, Poncelet D, Pirottin D, Brouwers B, Riquet J, Schoeberlein A, Dunner S, Menissier F, Massabanda J et al (1997) A deletion in the bovine myostatin gene causes the double-muscled phenotype in cattle. Nat Genet 17:71–74

Hara K, Yonezawa K, Kozlowski MT, Sugimoto T, Andrabi K, Weng QP, Kasuga M, Nishimoto I, Avruch J (1997) Regulation of eIF-4E BP1 phosphorylation by mTOR. J Biol Chem 272:26457–26463

Hara K, Yonezawa K, Weng QP, Kozlowski MT, Belham C, Avruch J (1998) Amino acid sufficiency and mTOR regulate p70 S6 kinase and eIF-4E BP1 through a common effector mechanism [published erratum appears in J Biol Chem 1998 Aug 21;273(34):22160]. J Biol Chem 273:14484–14494

Hara K, Maruki Y, Long X, Yoshino K, Oshiro N, Hidayat S, Tokunaga C, Avruch J, Yonezawa K (2002) Raptor, a binding partner of target of rapamycin (TOR), mediates TOR action. Cell 110:177–189

Hardt SE, Sadoshima J (2002) Glycogen synthase kinase-3beta: a novel regulator of cardiac hypertrophy and development. Circ Res 90:1055–1063

Hong DH, Forsberg NE (1995) Effects of dexamethasone on protein degradation and protease gene expression in rat L8 myotube cultures. Mol Cell Endocrinol 108:199–209

Izumiya Y, Hopkins T, Morris C, Sato K, Zeng L, Viereck J, Hamilton JA, Ouchi N, LeBrasseur NK, Walsh K (2008) Fast/glycolytic muscle fiber growth reduces fat mass and improves metabolic parameters in obese mice. Cell Metab 7:159–172

Jackson PK, Eldridge AG (2002) The SCF ubiquitin ligase: an extended look. Mol Cell 9:923–925

Kambadur R, Sharma M, Smith TP, Bass JJ (1997) Mutations in myostatin (GDF8) in double-muscled Belgian Blue and Piedmontese cattle. Genome Res 7:910–916

Kamei Y, Miura S, Suzuk M, Kai Y, Mizukami J, Taniguchi T, Mochida K, Hata T, Matsuda J, Aburatani H et al (2004) Skeletal muscle FOXO1 (FKHR) transgenic mice have less skeletal muscle mass, down-regulated Type I (slow twitch/red muscle) fiber genes, and impaired glycemic control. J Biol Chem 279:41114–411123

Kamura T, Koepp DM, Conrad MN, Skowyra D, Moreland RJ, Iliopoulos O, Lane WS, Kaelin WG Jr, Elledge SJ, Conaway RC et al (1999) Rbx1, a component of the VHL tumor suppressor complex and SCF ubiquitin ligase [see comments]. Science 284:657–661

Kim do H, Sarbassov dos D, Ali SM, King JE, Latek RR, Erdjument-Bromage H, Tempst P, Sabatini DM (2002) mTOR interacts with raptor to form a nutrient-sensitive complex that signals to the cell growth machinery. Cell 110:163–175

Lagirand-Cantaloube J, Offner N, Csibi A, Leibovitch MP, Batonnet-Pichon S, Tintignac LA, Segura CT, Leibovitch SA (2008) The initiation factor eIF3-f is a major target for Atrogin1/MAFbx function in skeletal muscle atrophy. EMBO J 27:1266–1276

Lai K-MV, Gonzalez M, Poueymirou WT, Kline WO, Na E, Zlotchenko E, Stitt TN, Economides AN, Yancopoulos GD, Glass DJ (2004) Conditional activation of Akt in adult skeletal muscle induces rapid hypertrophy. Mol Cell Biol 24:9295–9304

Langley B, Thomas M, Bishop A, Sharma M, Gilmour S, Kambadur R (2002) Myostatin inhibits myoblast differentiation by down-regulating MyoD expression. J Biol Chem 277:49831–49840

Latres E, Amini AR, Amini AA, Griffiths J, Martin FJ, Wei Y, Lin HC, Yancopoulos GD, Glass DJ (2005) Insulin-like growth factor-1 (IGF-1) inversely regulates atrophy-induced genes via the phosphatidylinositol 3-kinase/akt/mammalian target of rapamycin (PI3K/Akt/mTOR) pathway. J Biol Chem 280:2737–2744

Lee S-J (2007) Quadrupling muscle mass in mice by targeting TGFbeta signaling pathways. PLoS One 2:e789

Lee SW, Dai G, Hu Z, Wang X, Du J, Mitch WE (2004) Regulation of muscle protein degradation: coordinated control of apoptotic and ubiquitin-proteasome systems by phosphatidylinositol 3 kinase. J Am Soc Nephrol 15:1537–1545

Leevers SJ, Weinkove D, MacDougall LK, Hafen E, Waterfield MD (1996) The Drosophila phosphoinositide 3-kinase Dp110 promotes cell growth. EMBO J 15:6584–6594

Li H-H (2007) Atrogin-1 inhibits Akt-dependent cardiac hypertrophy in mice via ubiquitin-dependent coactivation of Forkhead proteins. J Clin Invest 117:3211–3223

Li Y-P, Chen Y, Li AS, Reid MB (2003) Hydrogen peroxide stimulates ubiquitin conjugating activity and expression of genes for specific E2 and E3 proteins in skeletal muscle myotubes. Am J Physiol Cell Physiol 285:C806–C812

Lorenz MC, Heitman J (1995) TOR mutations confer rapamycin resistance by preventing interaction with FKBP12-rapamycin. J Biol Chem 270:27531–27537

Mammucari C, Milan G, Romanello V, Masiero E, Rudolf R, Del Piccolo P, Burden SJ, Di Lisi R, Sandri C, Zhao J et al (2007) FoxO3 controls autophagy in skeletal muscle *in vivo*. Cell Metab 6:458–471

McElhinny AS, Kakinuma K, Sorimachi H, Labeit S, Gregorio CC (2002) Muscle-specific RING finger-1 interacts with titin to regulate sarcomeric M-line and thick filament structure and may have nuclear functions via its interaction with glucocorticoid modulatory element binding protein-1. J Cell Biol 157:125–136

McFarlane C, Plummer E, Thomas M, Hennebry A, Ashby M, Ling N, Smith H, Sharma M, Kambadur R (2006) Myostatin induces cachexia by activating the ubiquitin proteolytic system through an NF-?B-independent, FoxO1-dependent mechanism. J Cell Physiol 209:501–514

McPherron AC, Lee SJ (1997) Double muscling in cattle due to mutations in the myostatin gene. Proc Natl Acad Sci U S A 94:12457–12461

McPherron AC, Lawler AM, Lee SJ (1997) Regulation of skeletal muscle mass in mice by a new TGF-beta superfamily member. Nature 387:83–90

Montagne J, Stewart MJ, Stocker H, Hafen E, Kozma SC, Thomas G (1999) Drosophila S6 kinase: a regulator of cell size [see comments]. Science 285:2126–2129

Musaro A, McCullagh K, Paul A, Houghton L, Dobrowolny G, Molinaro M, Barton ER, Sweeney HL, Rosenthal N (2001) Localized Igf-1 transgene expression sustains hypertrophy and regeneration in senescent skeletal muscle. Nat Genet 27:195–200

Pallafacchina G, Calabria E, Serrano AL, Kalhovde JM, Schiaffino S (2002) A protein kinase B-dependent and rapamycin- sensitive pathway controls skeletal muscle growth but not fiber type specification. Proc Natl Acad Sci U S A 25:25

Pizon V, Iakovenko A, van der Ven PFM, Kelly R, Fatu C, Furst DO, Karsenti E, Gautel M (2002) Transient association of titin and myosin with microtubules in nascent myofibrils directed by the MURF2 RING-finger protein. J Cell Sci 115:4469–4482

Proud CG (2004) mTOR-mediated regulation of translation factors by amino acids. Biochem Biophys Res Commun 313:429–436

Rebbapragada A, Benchabane H, Wrana JL, Celeste AJ, Attisano L (2003) Myostatin signals through a transforming growth factor {beta}-like signaling pathway to block adipogenesis. Mol Cell Biol 23:7230–7242

Reymond A, Meroni G, Fantozzi A, Merla G, Cairo S, Luzi L, Riganelli D, Zanaria E, Messali S, Cainarca S et al (2001) The tripartite motif family identifies cell compartments. EMBO J 20:2140–2151

Rios R, Fernandez-Nocelos S, Carneiro I, Arce VM, Devesa J (2004) Differential response to exogenous and endogenous myostatin in myoblasts suggests that myostatin acts as an autocrine factor *in vivo*. Endocrinology 145:2795–2803

Rommel C, Clarke BA, Zimmermann S, Nunez L, Rossman R, Reid K, Moelling K, Yancopoulos GD, Glass DJ (1999) Differentiation stage-specific inhibition of the raf-MEK-ERK pathway by Akt. Science 286:1738–1741

Rommel C, Bodine SC, Clarke BA, Rossman R, Nunez L, Stitt TN, Yancopoulos GD, Glass DJ (2001) Mediation of IGF-1-induced skeletal myotube hypertrophy by PI(3)K/Akt/mTOR and PI(3)K/Akt/GSK3 pathways. Nat Cell Biol 3:1009–1013

Sabers CJ, Martin MM, Brunn GJ, Williams JM, Dumont FJ, Wiederrecht G, Abraham RT (1995) Isolation of a protein target of the FKBP12-rapamycin complex in mammalian cells. J Biol Chem 270:815–822

Sacheck JM, Ohtsuka A, McLary SC, Goldberg AL (2004) IGF-1 stimulates muscle growth by suppressing protein breakdown and expression of atrophy-related ubiquitin-ligases, atrogin-1 and MuRF1. Am J Physiol Endocrinol Metab 287:E591–E601

Sandri M, Sandri C, Gilbert A, Skurk C, Calabria E, Picard A, Walsh K, Schiaffino S, Lecker SH, Goldberg AL (2004) Foxo transcription factors induce the atrophy-related ubiquitin ligase atrogin-1 and cause skeletal muscle atrophy. Cell 117:399–412

Sarbassov DD, Guertin DA, Ali SM, Sabatini DM (2005) Phosphorylation and regulation of Akt/PKB by the rictor-mTOR complex. Science 307:1098–1101

Saurin AJ, Borden KL, Boddy MN, Freemont PS (1996) Does this have a familiar RING? Trends Biochem Sci 21:208–214

Schalm SS, Fingar DC, Sabatini DM, Blenis J (2003) TOS motif-mediated raptor binding regulates 4E-BP1 multisite phosphorylation and function. Curr Biol 13:797–806

Stitt TN, Drujan D, Clarke BA, Panaro FJ, Timofeyva Y, Kline WO, Gonzalez M, Yancopoulos GD, Glass DJ (2004) The IGF-1/PI3K/Akt pathway prevents expression of muscle atrophy-induced ubiquitin ligases by inhibiting FOXO transcription factors. Mol Cell 14:395–403

Sugita H, Kaneki M, Sugita M, Yasukawa T, Yasuhara S, Martyn JA (2005) Burn injury impairs insulin-stimulated Akt/PKB activation in skeletal muscle. Am J Physiol Endocrinol Metab 288:E585–E591

Tintignac LA, Lagirand J, Batonnet S, Sirri V, Leibovitch MP, Leibovitch SA (2005) Degradation of MyoD mediated by the SCF (MAFbx) ubiquitin ligase. J Biol Chem 280:2847–2856

Trendelenburg AU, Meyer A, Rohner D, Boyle J, Hatakeyama S, Glass DJ (2009) Myostatin reduces Akt/TORC1/p70S6K signaling, inhibiting myoblast differentiation and myotube size. Am J Physiol Cell Physiol 296:C1258–C1270

Tsuchida K, Nakatani M, Uezumi A, Murakami T, Cui X (2008) Signal transduction pathway through activin receptors as a therapeutic target of musculoskeletal diseases and cancer. Endocr J 55:11–21

van der Velden JLJ, Schols AMWJ, Willems J, Kelders MCJM, Langen RCJ (2008) Glycogen synthase kinase 3 suppresses myogenic differentiation through negative regulation of NFATc3. J Biol Chem 283:358–366

Vandenburgh HH, Karlisch P, Shansky J, Feldstein R (1991) Insulin and IGF-I induce pronounced hypertrophy of skeletal myofibers in tissue culture. Am J Physiol 260:C475–C484

Vyas DR, Spangenburg EE, Abraha TW, Childs TE, Booth FW (2002) GSK-3beta negatively regulates skeletal myotube hypertrophy. Am J Physiol Cell Physiol 283:C545–C551

Wang L, Luo GJ, Wang JJ, Hasselgren PO (1998) Dexamethasone stimulates proteasome- and calcium-dependent proteolysis in cultured L6 myotubes. Shock 10:298–306

Yan Z, Biggs RB, Booth FW (1993) Insulin-like growth factor immunoreactivity increases in muscle after acute eccentric contractions. J Appl Physiol 74:410–414

Yang W, Zhang Y, Li Y, Wu Z, Zhu D (2007) Myostatin induces cyclin D1 degradation to cause cell cycle arrest through a phosphatidylinositol 3-kinase/AKT/GSK-3beta pathway and is antagonized by insulin-like growth factor 1. J Biol Chem 282:3799–3808

Zhang H, Stallock JP, Ng JC, Reinhard C, Neufeld TP (2000) Regulation of cellular growth by the drosophila target of rapamycin dTOR [In Process Citation]. Genes Dev 14:2712–2724

Taking PI3Kδ and PI3Kγ One Step Ahead: Dual Active PI3Kδ/γ Inhibitors for the Treatment of Immune-Mediated Inflammatory Diseases

Christian Rommel

Contents

Abstract The multiple roles of PI3Kδ (p110δ) and PI3Kγ (p110γ) in various immune cells have encouraged the development of small-molecule inhibitors of both PI3K isoforms, alone or in combination, for the treatment of immune-mediated inflamatory diseases. Recent findings suggest a previously unrecognized interdependent cooperativity between p110δ and p110γ, if not all class I PI3K isoforms, expressed and activated in leukocytes and endothelium. For example, the activity of p110δ and p110γ in combination appears to be necessary for mediating efficient (velocity and direction) trafficking of immune competent cells to sites of inflammation. This chapter will focus on the emerging evidence of the dynamic interplay of p110δ and p110γ supporting the hypothesis that dual-blockade of both, p110δ and p110γ, presents a unique therapeutic opportunity in that pharmacological inhibition of the two PI3K isoforms simultaneously may yield

C. Rommel
Intellikine Inc, 10931 North Torrey Pines Road, La Jolla, CA 92037, USA
e-mail: christian@intellikine.com

C. Rommel et al. (eds.). *Phosphoinositide 3-kinase in Health and Disease*, Volume 1
Current Topics in Microbiology and Immunology 346, DOI 10.1007/82_2010_79

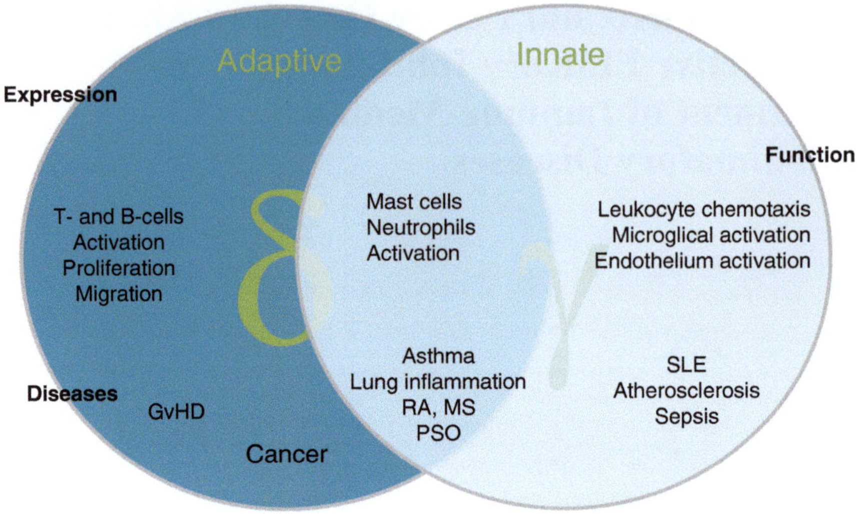

Fig. 1 The PI3K pathway stimulates numerous processes that promote cell proliferation, growth, motility, metabolism and survival. Inappropriate activation of these processes contributes to the development and progression of immune-mediated inflammatory disorders (Fruman and Bismuth 2009). The isoforms p110δ and p110γ are mainly expressed in cells of the immune system and contributes to innate and adaptive immunity. p110δ and p110γ regulates diverse immune cell functions, including: Suppression of B-cell activation and function (δ); Suppression of T-lymphocyte proliferation (δ); T-cell trafficking (δ); Th1-Th2 differentiation and Treg function (δ); Inhibition of neutrophil (leukocyte) chemotaxis (δ/γ); Inhibition of mast cell activation (δ/γ); Intact macrophage phagocytosis (δ/γ); Endothelium activation (δ/γ); Microglial activation (γ). Isoform-specific p110δ and p110γ inhibitors are expected to have therapeutic effects in these diseases without interfering with general PI3K signaling critical to the normal function of other cellular systems (Rommel et al. 2007). *GvHD* graft versus host disease, *RA* rheumatoid arthritis, *MS* multiple sclerosis, *PSO* psoriasis, *SLE* systemic lupus erythematosus

superior clinical results in the treatment of a variety of complex immune-mediated inflammatory diseases (Fig. 1).

1 p110δ and p110γ Function in Neutrophils

Neutrophils not only represent the first line of defense guard against microbial exposure but they are also involved in many immune pathologies in which the functions of adaptive immune cells are thought to be important, such as psoriasis, SLE, chronic obstructive pulmonary disease, and chronic severe asthma. In rheumatoid arthritis, numerous neutrophils accumulate in the inflamed joint, where they contribute to cartilage erosion by releasing proteolytic enzymes and toxic oxidative products. Pre-exposure of neutrophils to TNFα or LPS strongly augments the generation of reactive oxygen species (ROS) induced by subsequent stimuli such as chemoattractants. A marked impairment of ROS production by cytokine-primed neutrophils, in response to fMLP, has been shown in the absence of p110γ (Andrews et al. 2007). Similarly, pharmacological inhibition of p110γ with selective inhibitors shows that this isoform has the pivotal role in the initiation of a biphasic pathway of ROS production, induced by fMLP, in TNFα-primed human neutrophils (Condliffe et al. 2005). The first phase is mediated solely by p110γ activity while the second phase of ROS production is mediated by p110δ. It is, therefore, possible that both p110γ and p110δ functions are required for the organization and activation of the PHOX (phagocyte NADPH oxidase) complex, which is involved in ROS production, whereby under certain conditions p110γ might function as a gatekeeper ensuring maximal response. These findings by the Hawkins/Stephens labs have pioneered the principal concept of an interdependent dynamic play of p110δ and p110γ in immune cells managing adequate functional responses to multiple inflammatory stimuli. Notably, p110γ does not have a crucial role in respiratory burst in neutrophils responding to serum-opsonized zymosan suggesting that p110γ might not be important for direct phagocyte–pathogen interactions (Hirsch et al. 2000). Future work is needed to investigate the role of p110δ/γ (or all class I PI3Ks) in phagocyte–pathogen interactions.

Complement component C5a binds C5a receptor (C5aR) and facilitates leukocyte chemotaxis and release of inflammatory mediators. The receptor for C5a plays a critical role in numerous inflammatory conditions. Increased complement activation and excessive production of C5a have been implicated in the pathogenesis of several inflammatory diseases, and considerable effort has gone into developing C5a and/or C5aR antagonists (Mackay 2008). Konrad et al. used a mouse model of passive reverse lung Arthus reaction to define the requirements of PI3K isoform-selective signaling for Fcgamma receptor (FcγR) and the C5aR in the acute inflammatory response to IgG immune complexes (Konrad et al. 2008). Genetic deletion of p110γ abrogated C5aR signaling crucial for FcγR-mediated activation of lung macrophages. Thus, in p110γ-deficient mice, IgG immune complex-induced FcγR regulation, cytokine release, and neutrophil recruitment were abrogated. Notably, however, C5a production occurred normally in p110γ-deficient mice but was impaired in p110δ-deficient mice. Consequently, the absence of p110δ led to resistance to acute immune complex-mediated lung injury. These results demonstrate that both, p110γ and p110δ coordinate the inflammatory actions of C5aR and FcγR and define p110δ as a novel and essential element of FcγR signaling in the generation of C5a in immune complex disease.

2 p110δ and p110γ Function in Mast Cells

Mast cells play a central role in the initiation, and possibly "periodic" maintenance, of inflammatory responses associated with allergic and inflammatory disorders. Receptor-mediated mast cell growth, differentiation, homing to their target tissues, survival, and activation, are controlled, predominately but not exclusively, by PI3K-driven pathways (Kim et al. 2008). Not only are mast cells strategically localized along the microvasculature in tissues, but also their multifunctional capacity to release, under some circumstances independently of degranulation, vasoactive mediators, such as TNFα, VEGF, tryptase, histamine and serotonin, together with various proinflammatory cytokines and chemokines, means they are important immune-modulatory cells in acute and chronic inflammation. Many mediators, including SCF, and immune complexes containing antigen and IgE or IgG, affect mast-cell differentiation and activation. Activated SCF receptors (cKIT), the high-affinity Fc receptors – all generate PtdIns P_3 (Andrews et al. 2007; Condliffe et al. 2005; Hirsch et al. 2000) and PtdIns P_2 (Andrews et al. 2007; Condliffe et al. 2005) signals through PI3Ks (Kok et al. 2009; Wymann et al. 2003). A role for class I PI3K in mast-cell activation was also shown by nonisoform-specific PI3K inhibitors, which completely blocked antigen-triggered mast-cell degranulation (Duan 2005). However, which of the class I PI3K isoforms (and when) are engaged after ligand-specific mast-cell stimulation only became clear very recently. Without any doubts, both p110γ and p110δ are important in mast-cell function. p110δ is required for SCF-induced and, in part, IL-3-induced mast-cell proliferation, adhesion, and migration (Ali et al. 2004). Mast-cell degranulation and cytokine release in response to antigen–IgE immune complexes also depends on p110δ, to a certain extent. Consequently, p110δ -deficient mice are largely protected from passive cutaneous anaphylaxis. Interestingly, p110δ-deficient mice are also resistant to cutaneous anaphylaxis (edema formation following intradermal administration of adenosine) and passive systemic anaphylactic shock (Ali et al. 2008). Although p110γ is not directly activated by the high affinity Fcepsilon receptor ($Fc_\varepsilon RI$), it assists in the amplification of some mast-cell function through secondary autocrine and paracrine loops, involving adenosine-mediated, and probably also chemokine-mediated, pertussis-toxin-sensitive GPCR activation (Laffargue et al. 2002). In this scenario, speculatively though, p110δ activated by $Fc_\varepsilon RI$ and p110γ subsequently activated by a second independent GPCR-stimuli might coordinate, temporally and spatially, to achieve optimum mast-cell function. The initial phase of degranulation that is triggered by antigen–IgE immune complexes and mediated by p110δ provides a limited release of GPCR agonists, such as adenosine, that trigger autocrine and paracrine positive-feedback loops, which lead to p110γ activation and greater granule release. Therefore, it seems conceivable that, at least in this scenario, p110δ functions upstream of p110γ and that the two enzymes operate episodically at two distinct but crucial levels of the same signaling cascade in mast cells (Rommel et al. 2007; Wymann et al. 2003; Rückle et al. 2006). In this way, these two PI3K isoforms may provide a signaling platform that integrates distinct and multiple extracellular stimuli during towards comprehensive mast-cell activation. It is intriguing that this model of p110δ and p110γ dynamic interplay is analogous to cytokine-primed ROS production of neutrophils, as reviewed above. Nevertheless, the relative contribution of these two PI3K isoforms in IgE-dependent allergic responses *in vivo* remains controversial. A side-by-side comparative analysis of the role of p110γ and p110δ in mast cell function, using genetic approaches and newly developed isoform-selective pharmacologic inhibitors, confirmed that both PI3K isoforms play an important role in $Fc_{\varepsilon}RI$-activated mast cell degranulation *in vitro*. However, under *in vivo* conditions, only p110δ was found to be required for an optimal immune complex-dependent hypersensitivity response in mice (Ali et al. 2008). Although these observations identify p110δ as the high-ranked therapeutic target among PI3K isoforms, considering all the current results together, going after p110δ and p110γ appears to be a valid strategy ensuring maximal impact on allergy- and mast cell-related diseases that are often complex and eminently strong reactions in acute phase settings (Rommel et al. 2007; Rückle et al. 2006).

3 p110δ and p110γ Function in NK Cells

The Colucci lab determined the relative contribution of p110γ and p110δ to NK cell trafficking in mice model systems (Saudemont et al. 2009). An elegant combination of single-cell imaging analysis of transgenic cells reporting on PI3K activity in real time and small molecule inhibitors of p110γ and p110δ revealed that the tyrosine-kinase coupled p110δ is linked to GPCR signaling and, depending on the GPCR, may even be preferentially activated over p110γ. By using p110γ- or p110δ-deficient mice, they showed that both isoforms were essential for NK cell chemotaxis to CXCL12 and to CCL3 and, *in vivo*, for normal NK cell migration to the inflamed peritoneum. However, only p110δ was indispensable for chemotaxis to S1P and CXCL10 and for NK cell distribution throughout lymphoid and nonlymphoid tissues and for extravasation to tumors. p110δ can act downstream of GPCRs in NK cells and has a key, nonredundant, role as a regulator of NK cell trafficking. Complementary study by Tassi et al. showed that p110γ and p110δ, both have a crucial role in the development and functions of murine NK cells (Tassi et al. 2007). p110γ deficiency prevented full NK cell maturation, whereas concurrent absence of p110γ and p110δ exacerbated this defect, resulting in a very small population of immature NK cells in the bone marrow and periphery. Effective NK cell-mediated cytotoxicity required the combined activity of both isoforms.

4 p110δ and p110γ Function in Lymphocytes

The establishment of T cell-mediated inflammation requires the migration of primed T lymphocytes from the blood stream and their retention in antigenic sites. While naive T lymphocyte recirculation in the lymph and blood is constitutively regulated and occurs in the absence of inflammation, the recruitment of primed T cells to nonlymphoid tissue and their retention at the site are enhanced by various inflammatory signals, including TCR engagement by antigen-displaying endothelium and resident antigen-presenting cells. Two groups, Berg and Cantrell, showed independently that T cell receptor-induced p110δ activity is required for T cell localization to antigenic tissue in mice (Jarmin et al. 2008; Sinclair et al. 2008). T lymphocytes from p110δ-deficient mice displayed normal constitutive trafficking and migratory responses to nonspecific stimuli. However, these cells where impaired in their responsiveness to TCR-induced migration and failed to localize efficiently to antigenic tissue. Antigen-induced T cell trafficking and subsequent inflammation was abrogated by selective pharmacological inhibition of p110δ. There is somewhat a role of p110γ in this biology of adaptive immunity, however more distantly related to the process of antigen trafficking of activated T cells. It has also been reported that, p110γ might be dispensable for constitutive migration of naive CD8 T cells and subsequent activation and differentiation into effector CD8 T cells, but does play a central role in the migration of effector CD8 T cells into inflammatory sites (Martin et al. 2008). Moreover, the Hirsch lab focused on dendritic cell biology and showed that those specialized cells when obtained from p110γ-deficient mice showed a reduced ability to respond to chemokines *in vitro* and *in vivo* to cruise to draining lymph nodes under inflammatory conditions (Del Prete et al. 2004). p110γ-deficient mice had a selective defect in the number of skin Langerhans cells and in lymph nodes. Furthermore, p110γ-deficient mice showed some deficiency in mounting contact hypersensitivity and delayed-type hypersensitivity reactions. This defect was directly related to the reduced ability of antigen-loaded dendritic cells to migrate from the periphery to draining lymph nodes. Thus, p110γ plays a nonredundant role in dendritic cell trafficking and in the activation of specific immunity. B cell and T cell chemotactic responses are both sensitive to the general PI3K inhibitors, wortmannin and LY294002. Several groups set out to explore whether B cell responses were affected in mice either lacking p110δ or p110γ (Okkenhaug et al. 2007; Okkenhaug and Vanhaesebroeck 2003; Reif et al. 2004). T cells lacking p110γ, but not B cells, show reduced chemotactic responses to the lymphoid chemokines, CCL19, CCL21, and CXCL12. B cells deficient in p110δ showed diminished chemotactic responses, particularly to CXCL13. Adoptive transfer experiments with wortmannin-treated wild-type B cells and with p110δ-deficient B cells revealed reduced homing to Peyer's patches and splenic white pulp cords. Finally, a recent report descriebd a possible role of p110γ facilitating an efficient induction of CXCR3 on activated T cells (Barbi et al. 2008). These observations establish that p110γ and p110δ function in lymphocyte chemotaxis, and show differential roles for PI3K family members in B and T cell migration. The multiple roles of p110δ and p110γ in both antigenic lymphocyte trafficking and dendritic cell migration for the process of antigen presentation may help to explain the efficacy of PanPI3K and dual-active p110δ/γ inhibitors in preventing specific immune responses *in vivo*.

Contrary to p110δ, little (too little) is known about the role of p110γ in the regulation of B cell activation and/or function. p110δ has a crucial role in B cells by either mediating BCR signaling, proliferation/survival, antibody production or antigen presentation. Or in other words, p110δ is a "mission critical" signaling enzyme for B cells (Okkenhaug et al. 2002; Bilancio et al. 2006; Durand et al. 2009).

5 p110δ and p110γ Function in Inflammatory Leukocyte Recruitment

PI3K plays a fundamental role in regulating neutrophil recruitment into sites of inflammation but the individual roles of p110δ and p110γ isoforms remains controversial (Hirsch et al. 2000; Sasaki et al. 2000; Li et al. 2000). The role of p110δ in neutrophil directional movement, *in vitro* and *in vivo*, was first described in two subsequent reports from scientist from former ICOS (Sadhu et al. 2003; Puri et al. 2004). The lab of Teixeira evaluated side-by-side the role of p110γ and p110δ for neutrophil influx induced by either exogenous administration or by inducible endogenous generation of the chemokine CXCL1 (Pinho 2007). The induction of neutrophil recruitment into the pleural cavity or the tibia-femoral joint induced by the injection of CXCL1 was not significantly different in p110γ-deficient mice nor by treatment with IC87114, a p110δ specific inhibitor or with AS605240, a relatively potent and specific p110γ small molecule inhibitor. CXCL1-induced neutrophil recruitment was impaired by the administration of IC87114 when combined either with the p110γ inhibitor or to the p110γ-deficient mice. Ag challenge of immunized mice induced CXCR2-dependent neutrophil recruitment that was blocked only by the combination of pharmacological inhibition of p110δ in a p110γ-deficient background. Nevertheless, neutrophil recruitment to bronchoalveolar lavage induced by exogenously added or endogenous production of CXCL1 was prevented solely by the lack of p110γ activity; whereas the penetration and accumulation of neutrophils into the lung tissues was only significantly inhibited in p110γ-deficient mice when treated with IC87114. In the same setting, neutrophil recruitment induced by exogenous administration of the chemoattractants C5a or fMLP appeared to rely rather on p110γ. These data demonstrated to the first time that there is a tissue- and stimulus-dependent role of p110γ and p110δ for neutrophil recruitment *in vivo* induced by different inflammatory mediators (Puri et al. 2005). Moreover, this study also posed the eminent question how to target inflammatory recruitment by either p110δ or p110γ or rather considering an inhibitor that blocks both isoforms ("two birds with one stone")?

Additional studies by Puri and coworkers came to similar conclusions yet by exploring an alternative experimental model system (Liu et al. 2007). To reveal the role of p110γ in leukocyte recruitment, they used intravital microscopy to directly visualize leukocyte rolling, adhesion, and emigration in postcapillary venules in p110γ-deficient mice. Lack of p110γ had no significant effects on leukocyte rolling flux or rolling velocity and moderate effects on adhesion in response to CXCL2 or CXCL1. Of note, leukocyte emigration was severely impaired in p110γ-deficient mice in an early (1–2 h) response to both chemokines (similar to the original reports of Wymann and coworkers (Hirsch et al. 2000). Applying chimeric approach of mice receiving bone marrow transplants revealed that this early response was entirely dependent upon p110γ in neutrophils but not parenchymal cells (e.g., endothelium). More prolonged responses to

CXCL1 (for 4–5 h) or TNFα (6–8 h) were rather p110γ independent and largely dependent on p110δ. These results reveal that leukocyte tissue emigration response to a subset of inflammatory chemokines is entirely dependent upon PI3K, more accurately, p110γ or p110δ, but these are temporally coordinated events *in vivo*.

The above described studies pinpoint to a time- and context-dependent interplay of p110δ and p110γ in inflammatory leukocyte recruitment. This emerging concept has been further supported by another study. Puri et al. investigated the importance of p110γ in the endothelium relative to its leukocyte counterpart in facilitating neutrophil interactions in an inflammatory milieu (Puri et al. 2005). Again, by using the bone marrow chimera approach, reconstitution of p110γ function in neutrophils of p110-deficient mice, did not prevent a significant reduction in accumulation of these cells in an acute lung injury model. The authors proposed a perturbation in selectin-mediated adhesion as manifested by marked reduction in wild-type neutrophil attachment to and increase in rolling velocities on p110γ-deficient microvessels *in vivo* in response to TNFα as the underlying mechanism for this observation. These in vivo-based observations are supported by the *in vitro* studies led by Stephens and Hawkins (Ferguson et al. 2007). Namely, the dependence of chemokinesis (rather than strict chemotaxis) on p110γ activity is context dependent, both with respect to the state of priming of the neutrophils and the type of surface on which they are migrating. They proposed a model in which p110γ regulates neutrophil movement, such as chemokine-induced polarization, integrin-based adhesion and the accumulation of polymerized (F)-actin at the leading-edge of the chemoattractant directed plasma membrane (Ferguson et al. 2007). Nevertheless, the alteration of adhesion *in vivo* was further augmented by a deficiency in p110δ, suggesting that the activity of both catalytic subunits is required for efficient capture of neutrophils by cytokine-stimulated endothelium. Indeed, E-selectin-mediated adhesion in p110γ-deficient mice was nearly completely impaired, but no defect in nuclear factor kappa B (NF-kappaB)-induced gene expression was observed (Liu et al. 2007). It also needs to be remembered that p110γ although enriched in leukocytes is expressed and functional in other cells including endothelial cells as mapped by those *in vivo* studies.

6 p110δ and p110γ in Inflammatory Arthritis

Rheumatoid arthritis is generally considered to be an immune-mediated disease. Microscopically, rheumatoid joints are characterized by inflammatory infiltrates in the synovium and the synovial fluid, by pannus formation and, ultimately, by cartilage and bone erosion. Detailed studies in animal models that recapitulate many of the features of rheumatoid arthritis in humans have proposed the involvement of multiple steps in the pathogenesis of the disease, with interplay between the adaptive and innate immune systems. The activities of T cells, B cells, dendritic cells, mast cells, macrophages and neutrophils have all been shown to have important roles and to affect at least one or several aspects of the disease. Hong Ji et al. at Serono (now Merck-Serono) showed for the first time that p110γ mice are largely protected in mouse models of rheumatoid arthritis and that this protection correlated with defective neutrophil migration, further validating p110γ as a potential therapeutic target for inflammatory diseases (Camps et al. 2005). An orally active and selective p110γ small molecule inhibitor suppressed the progression of joint inflammation and damage in antigen- and antibody-mediated mouse models of rheumatoid arthritis.

Using an antibody-model (K/BxN serum transfer model of arthritis; also used by Ji et al.), in which mast cells, neutrophils participate micromanaged by immune-complexes, varies chemokines (e.g., CXCL2 and LTB4), Randis et al. showed that either genetic deletion or selective inhibition of p110δ diminishes joint erosion to a level comparable to p110γ (Randis et al. 2008). Moreover, the induction and progression of joint destruction was profoundly reduced in the absence of both PI3K isoforms and correlated with a limited ability of neutrophils to migrate into tissue in response to LTB (Condliffe et al. 2005). In vitro and *in vivo* chemokines such as MIP1α or IL8 rely on p110γ while LTB (Condliffe et al. 2005) utilizes p110δ for neutrophil recruitment. Similarly to the Teixeira report, the interplay between p110γ and p110δ is not all-encompassing, as some chemoattractant-induced neutrophil extravasation processes are predominately contingent on p110γ while others favor p110δ.

These results taken together not only demonstrate that either blockade of p110δ or p110γ has potential therapeutic value in the treatment of complex inflammatory conditions, but also provide inspiring evidence that dual inhibition of these two PI3K isoforms may yield superior clinical results (Rommel et al. 2007).

7 p110δ and p110γ in Allergic and Inflammatory Lung Disease

The PI3K pathway plays a crucial role in the expression and activation of inflammatory mediators, inflammatory cell recruitment, immune cell function, tissue remodeling, and also corticosteroid insensitivity in chronic inflammatory respiratory diseases. Airway inflammation involves recruitment and activation of inflammatory cells and changes in epithelial and fibroblast cells of the lung; commonly portrayed by an excessive expression of chemokines, cytokines, growth factors, enzymes, reactive oxygen, proteases and cell adhesion molecules. Invasion of T-lymphocytes, eosinophils, macrophages/monocytes, and mast cells into the airway wall is a hallmark of this immune-mediated pathology. Chronic inflammation often leads to structural changes in the airway, such as increased thickness of airway smooth muscle, neo-angiogenesis, increased number of mucus-secreting cells, sub-epithelial fibrosis, collectively referred to as airway remodeling (Ito et al. 2007). The therapeutic potential of PI3K for the treatment of airway inflammation was first described by using the LY294002 compound (Duan 2005). Intratracheal administration of LY294002 ameliorated OVA-induced increases in total cell counts, eosinophil counts, and levels of CCL11 and IL-5, IL-13 in bronchoalveolar lavage fluid. In addition, blockade of all class I PI3Ks by LY294002 repressed OVA-induced tissue eosinophilia.

Gene-deficient mutant mice and first generation isoform-selective inhibitors for p110δ and p110γ begun to shed light on the role of individual PI3K isoforms in preclinical models of immune-mediated inflammatory diseases (Nashed et al. 2007). In fact, p110δ inhibitor IC87114 (highly selective, moderately potent, suitable for certain *in vivo* studies) significantly reduced the serum levels of OVA-specific IgE and leukotriene C_4 release into the airspace, OVA-induced lung tissue eosinophilia, airway mucus production, and inflammation score, and importantly, OVA-induced increase in expression of IL-4, IL-5, IL-13, ICAM-1, VCAM-1, CCL5, and CCL11 (Lee et al. 2006). How does p110γ "perform" in preclinical models of allergic and inflammatory lung disease? Finan's team at Novartis research reported that allergic airway hyperresponsiveness, inflammation and remodeling do not develop in p110γ-deficient mice (Thomas et al. 2005, 2009). Similarly, Takeda et al. showed that in OVA-sensitized and OVA-challenged p110γ-deficient mice, levels of airway inflammation and airway remodeling were significantly decreased (Takeda et al. 2009). No significant differences were detected in serum OVA-specific IgE and IgG1 levels and CD4/CD8 balance in bronchoalveolar lavage fluid. To determine in which phase of allergic responses p110γ plays a role, splenocytes from OVA-sensitized wild type mice or p110γ-deficient mice were transferred to control mice of either genotype. Comparable increased levels of eosinophils were induced in both wild type mice but not in both p110γ-deficient recipient mice. Hence, p110γ might be involved in allergic airway inflammation by regulating the challenge:effector phase of allergic responses. Lim et al. examined the role of p110γ to airway remodeling under conditions of chronic allergen challenge (Lim et al. 2009). Wild-type mice sensitized to ovalbumin and chronically challenged with OVA for 1 month showed significantly reduced numbers of bronchoalveolar lavage and peribronchial eosinophils whereas wild-type mice developed significantly increased levels of eosinophilic inflammation and airway remodeling. The reduced eosinophil recruitment to the airway in p110γ-deficient mice challenged with OVA was associated with significantly reduced numbers of TGFβ1 positive peribronchial cells, reduced numbers of pSmad 2/3+ airway epithelial cells, and pSmad 2/3 positive peribronchial cells, as well as significantly reduced levels of peribronchial fibrosis. The group of Teixeira performed once again a set of meticulous experiments *in vivo* (Pinho et al. 2005). At first, the role of p110γ was evaluated in respect to both, migration and survival of eosinophils in a model of allergic pleurisy in mice. Eosinophil accumulation in p110γ-deficient mice was inhibited at 48 h, as compared with wild-type mice but not at earlier time-points (6 and 24 h). Although the dynamic range (early vs. delayed response) is somewhat contrary to results observed in models of mice peritonitis (Hirsch et al. 2000; Camps et al. 2005), subsequent experiments with adoptive transfer of bone marrow showed that p110γ in eosinophils but not in nonbone marrow-derived cells was required for their accumulation. Treatment with PI3K inhibitors 24 h after antigen challenge markedly cleared the accumulated eosinophils, an effect associated with inhibition of Akt phosphorylation and an increased number of apoptotic events. These data suggest an important role of p110γ for the maintenance of eosinophilic inflammation *in vivo*, whereas other isoforms, likely p110δ (but also p110γ) may be relevant for the recruitment process. It appears again that the multi-step process of *in vivo* leukocyte recruitment to sites of inflammation (e.g., exiting the vascular system and transitioning into tissue-based chemotaxis requires (at least) both p110δ and p110γ or in other words, it always "takes two to tango."

Closing the loop: From isoform nonselective PanPI3K via rather isoform-selective inhibitors of p110δ and p110γ to the arrival of first generation dual-active p110δ/γ inhibitors. Targegen's research team led by Rich Soll showed in a well-designed study that dual inhibition of p110δ and p110γ by the small molecule inhibitor TG100-115 offers clinical-relevant efficacy in rodent models of acute and chronic lung inflammation (Doukas et al. 2009). In a battery of preclinical studies, the potential of TG100-115 was evaluated as a form of anti-inflammatory therapy for respiratory diseases such as asthma and COPD. Aerosolized TG100-115 markedly reduced the pulmonary eosinophilia, IL-13 and mucin accumulation in a murine asthma model. In order to mimic pulmonary neutrophilia characteristic of COPD, mice were exposed to either intranasal LPS or inhaled smoke. TG100-115 ameliorated these inflammatory patterns, most notably in the smoke model, where interventional therapy overcame the steroid-resistant nature of the pulmonary inflammation. These studies support the hypothesis that dual-active p110δ/γ inhibitors, given either systemically or topically, provides a potential therapeutic strategy for the treatment of airway inflammation.

A recent report credited p110δ with a clinically highly relevant role namely, inhibition of p110δ restored glucocorticoid function in smoking-induced airway inflammation in mice (Marwick et al. 2009). There is an increasing prevalence of resistance to chronic glucocorticoid therapy in severe asthma and COPD. Recent studies have shown that histone deacetylase activity, which is critical to glucocorticoid function, is altered by oxidant stress and may be implicated in the development of glucocorticoid insensitivity. It is widely accepted that there is a dual role of PI3K in oxidative stress by either ROS-dependent activation of PI3K or PI3K-dependent ROS induction – or in other words, a relationship in reverse order between ROS and PI3K signaling. Marwick et al. set out to determine the role of p110γ and p110δ in the development of cigarette smoke-induced glucocorticoid insensitivity (2009). Wild-type, p110γ- and p110δ-deficient mice were used in a model of cigarette smoke-induced glucocorticoid insensitivity. Cigarette smoke exposure in mice increased tyrosine nitration of histone deacetylase 2 in the lung, correlating with reduced enzymatic activity and an abbreviated glucocorticoid response. Histone deacetylase 2 activity and the effects of glucocorticoids were restored in p110δ- but not p110γ-deficient mice, correlating with reduced histone deacetylase 2 tyrosine nitration. These data show that therapeutic inhibition of p110δ may restore glucocorticoid function in oxidative stress-induced glucocorticoid insensitivity.

8 p110δ and p110γ (p110δ/γ) Deficient Mice

Mice lacking both the p110γ and p110δ isoforms are viable and fertile with regular growth rates and behavior, yet displayed severe impairment of thymocyte development (Ji et al. 2007; Webb et al. 2005; Swat et al. 2006). The Turner lab showed the block in T cell development occurs at the beta-selection checkpoint, and reduction of pre-TCR-induced signaling resulted in a lack of proliferative expansion and increased apoptosis. The survival defect in p110δ/γ-deficient thymocytes was associated with decreased levels of the antiapoptotic protein Bcl2. These findings were further supported by a very similar report of the Diacovo lab concluding that class I PI3Ks work in concert to protect developing thymocytes from apoptosis. Ji et al. drilled deeper into the biology further phenotyping pathological consequences of the double mutant mice (Ji et al. 2007). Although mice lacking p110δ and p110γ displayed T-cell lymphopenia, the mice developed T-cell and eosinophil infiltration of mucosal organs, elevated IgE levels, and a skewing toward Th2 immune responses. Side by side comparison of small-molecule selective inhibitors for both isoforms, demonstrated that in mature T cells, p110δ, but not p110γ, controls Th1 and Th2 cytokine secretion. Thus, the pathology in the p110δ/γ double mutant mice is likely to be secondary to a developmental block in the thymus that potentially leads to lymphopenia-associated inflammatory responses. In summary, mice lacking p110δ and p110γ function developed Th2-like eosinophilic inflammation in multiple mucosal organs. On the contrary, combined treatment or the application of dual-active p110δ/γ selective inhibitors in matured fully developed vertebrates (including human patients), in whom the maintenance of the peripheral cell pool occurs primarily via thymic-independent pathways, lymphopenia-induced pathology is unlikely to occur. Several studies have demonstrated that genetic or pharmacologic selective inhibition of p110δ leads to markedly enhanced B-cell switch to IgE and increased IgE levels *in vivo*, despite reduced type 2 cytokine production. The molecular profile (e.g., antigen specificity and affinity) and the physiological consequences of PI3Kδ-triggered elevation of serum IgE remains unknown and deserves prudent evaluation (Ji et al. 2007; Zhang et al. 2008; Omori and Rickert 2007) (Fig. 2).

9 Conclusions

First generation kinase inhibitors often lack one or all of the following properties: *in vivo* potency, ideal degree of selectivity or pharmaceutical profile supporting robust and desired dose:exposure:specificity relations. Cutting corners by just improving one over the other assets has proven to only create fallouts. LY294002 and wortmannin, the historically famous first generation small molecule PI3K inhibitors, have been invaluable tools in the research-wide efforts of the understanding of PI3K signaling in health and disease (Vlahos et al. 1994; Arcaro and

Fig. 2 This *funny* picture (Horsham, UK, 2003 and inspired by G Larrson) perhaps illustrates the challenges (or in other words the unmet clarity) of which PI3K isoform (and smart combination thereof) needs to be inhibited when, where and for how long...? This cartoon is dedicated to my respected colleague and sympathetic competitor, and the other way around, Dr. Pete Finan

Wymann 1993; Walker et al. 2000). LY294002 is a reasonably potent small molecule kinase inhibitor, yet lacks class I PI3K isoform selectivity and carries a number of kinome off-target activities. LY294002 also has limited pharmaceutical value, for example poor pharmacokinetic properties hampering its utility for chronic treatment *in vivo*. Wortmannin, on the other hand, is a rather remarkably potent, allosteric and mechanistically irreversible inhibitor of all PI3K isoforms and carries less kinome liabilities. The chemical instability of wortmannin, however, limits and complicates its use for *in vivo* pharmacology studies. Both molecules, LY294002 and Wortmannin, have undoubtedly been instrumental in elucidating the cell biology directed by PI3Ks. However, they have also generated some false expectations due to lack of specificity. The latter was put to the test upon arrival of genetically targeted mutant mice and first generation of isoform-selective inhibitors (Rückle et al. 2006). The dissection of isoform narrowed function of all class I PI3Ks has generated new excitement while set forth new directions. PI3K inhibitors that lack isoform selectivity have basically established two major results: (1) efficacy (meaningful impact on a number of physiological and pathophysiological processes *in vitro* and *in vivo*) and (2) limited tolerability (a narrow window between pharmacological doses needed to drive the desired biological effect and dose-limiting unwanted toxicities). These findings have primed a fundamental question (a universal and widespread question in drug development) whether more selective inhibitors will be comparably efficacious while being significantly more tolerated. Or in other words – does safety comes at the expense of efficacy? Although this question needs to be refined in respect to individual isoforms and their specific expressions, and functions, in certain disease processes (e.g., life-threatening diseases such as cancer vs. life-quality disease such as allergy) – it remains the key question for those of us dedicated to the discovery and development of drugs targeting PI3K-dependent signaling for the treatment of human disease. Unsurprisingly, the path to resolve this matter has been proven to be cumbersome, tedious and time consuming. Of note, most resource holders and decision makers in the world of the pharmaceutical industry have little tolerance and patience for long and costly research processes habitually associated to unpredictable outcomes. Opposing the force has been the remarkable collaborative spirit, the passionate interest and exquisite know-how in drug discovery of the entire PI3K academic community (with several outstanding scientific and charismatic leaders) which has been instrumental to the success of PI3K drug development with several programs currently in clinical evaluation. Another 20+ year journey from purifying an enzyme, cloning a gene, creating a knock-out mice, collecting X-ray structures, designing inhibitors, developing drugs into the clinic putting the original hypothesis to the final test. First generation drugs are often "learning tools" and will be outperformed by better drugs and knowledge. All this can only arrive from clinical experience. Thus, science and drug development of PI3K is nearly just to set in motion again.

Surprisingly (or perhaps less though?), the chemical space of isoform-selective class I PI3K inhibitors was originally pioneered, and dominated for a long period of time by biotech-spirited groups such as Thrombogenix, ICOS, Serono, PIramed, Targegen rather than traditional pharma companies. Categorically, large pharma (alone or in combination) is catching up, although the biotech's of nowadays such as Calistoga, Cellzome and Intellikine remain competitive in the race for clinical (commercial) viable PI3K designer drugs, to say the least (Knight and Shokat 2007a).

Mouse genetic engineering is one technology that is uniquely positioned to address the biological role and relevance of PI3K isoform selectivity. Several genetic mutant (knock-in, knock-out) mice for nearly every class I PI3K isoform (including corresponding regulatory and accessory proteins) have been successfully generated (Vanhaesebroeck et al. 2004; Knight and Shokat 2007b). Clearly, these genetic tools have provided a huge advantage. There are, however, a few serious caveats to be kept in mind when applying the power of genetic-based approaches for target validation (Hooft van Huijsduijnen and Rommel 2006). Knocking off the activity of PI3K isoforms by mouse genetic engineering often leads to situations where either the protein (gene) has been entirely removed (knock-out) and/or its activity is completely and constitutively blocked (knock-out, knock-in) during and post development in every cell, tissue and organ. Pharmacological inhibition is fundamentally different by being often incomplete (neither complete nor selective), with limited and usually reversible exposure to organism-wide cells, tissue, organs and physiological compartments, acute and temporally as well as reversible (in most cases). Thus, between a LY-mice and a p110δ-deficient mice is a significant gap hampering a straight away understanding of the value and validation of targeting PI3Ks selectively for the treatment of inflammation (e.g., p110γ) or solid tumor (e.g., p110α oncogene). A powerful illustration of the gap between the application of either molecular or pharmacological experimental strategies to dissect PI3K signaling (not limited to) was presented to the field by the discovery of highly specific and independent scaffold-based and kinase-based functions of p110γ in cardiovascular physiology in mice (Hirsch et al. 2009).

The PI3K isoforms p110δ and p110γ perhaps represent an ideal model for kicking off the concept of targeting two isoforms as opposed to solely one or all class I PI3K isoforms (PanPI3K) in order to achieve clinical relevant therapeutic efficacy while limiting toxicities. Yet, PI3Kβ and PI3Kγ might also be hooked up with one and another. Shaun Jackson's laboratory had originally identified a key role for PI3Kβ isoform in regulating the formation and stability of integrin:adhesion bonds, necessary for shear activation of platelets. The team also developed an isoform-selective PI3Kβ inhibitor which eliminated occlusive thrombus formation but did not prolong bleeding time *in vivo*. These studies defined PI3Kβ as an important new target for antithrombotic therapy (Jackson et al. 2005). However, it might be the interplay between PI3Kβ and PI3Kγ allowing continuous signaling via PI3K isoforms required for platelet ADP receptor function in dynamic thrombus stabilization (Cosemans et al. 2006). Another such example of direct or indirect interdependence is the role of p110α and β in oncogenic signaling either downstream of growth factor tyrosine kinase signaling or PTEN tumor suppressor or in form of oncogenic gain-off-function mutations within PIK3CA and PIK3R1 (Engelman 2009; Liu et al. 2009). Thus, more evidence is coming along suggesting that individual class I PI3K isoforms might work in combinatorial mode to drive biological output which also varies depending on cell type and upstream signal input.

PI3K-depdendent signaling is highly complex in cells of the immune system due to a uniform expression and activity of all four class I PI3K catalytic isoforms, together with numerous regulatory subunits. Nevertheless, p110δ and p110γ are both enriched in cells of the hematological system with multiple crucial roles in various immune cells. Some of these functions might be redundant (implying compensation) whereas are complementary (suggesting additive or temporarily separated effects). Indeed, there is now an emerging appreciation of some interdependence between p110δ and p110γ in the regulation of certain immune functions of leukocytes, lymphocytes as well as endothelium. In general, most immune-mediated inflammatory diseases such as rheumatoid arthritis require dysregulation of both adaptive and innate immune cell functions. p110δ is primarily crucial for T and B cell function, for example Ag-triggered T cell trafficking or B cell proliferation and survival, whereas p110γ is a pivotal signaling enzyme downstream of a number of proinflammatory chemokines facilitating leukocyte recruitment (and survival and activation?) to (and at) sites of inflammation and tissue damage. Dual-active p110δ/γ inhibitors offer the opportunity of a parallel action against both adaptive and innate immune responses. More specifically, comprehensive impact of mast cell degranulation or neutrophil ROS production in response to immune-complex or chemokine stimuli was only seen when both isoforms were inhibited simultaneously. The concept of dual targeting is further supported by the fact that neutrophils respond to most chemokines via p110γ while to some leukotrienes via p110δ, and both families of chemoattractant cytokines are present and active at site of inflammation drawing in neutrophils and other leukocytes. The experiments by the labs of Puri, Teixeira and Doukas have convincingly shown that concomitant blockade of p110δ and p110γ is most effective in inflammatory cell recruitment of joint or airway inflammation. The TG100-115 study has successfully established the first proof-of-concept that dual-active p110δ/γ inhibitors are in range and show superior efficacy in preclinical models of inflammation. These encouraging and pioneering results were revealed by mouse genetics and first generation inhibitors. Next generation mono- and dual-active p110δ/γ inhibitors are already in the run generating exciting preclinical data and poised to enter into the clinic in the not too distant future.

If there was an agreement that dual-active p110δ/γ inhibitors present an attractive therapeutic strategy then consequently two noticeable questions turn up, namely: What diseases are most effectively suited by drugs of such a molecular profile? And, what safety concerns can be anticipated?

Regrettably, once again the answer is not straight forward. p110δ and p110γ have been successfully validated in numerous preclinical models of allergic, autoimmune and inflammatory disease and beyond. p110γ blockade led to impressive efficacy in preclinical models of atherosclerosis (Fougerat et al. 2008; Chang et al. 2007), whereas acute severe allergic responses or certain B cell malignancies may be sensitive to p110δ inhibition (Bilancio et al. 2006; Giese et al. CAL101). The real issue is not the lack of disease phenotypes, but rather concerns about the relevance of such studies and their design for the purpose to validate therapeutic strategies (Knight and Shokat 2007b; Hooft van Huijsduijnen and Rommel 2006). Numerous preclinical models of arthritis and multiple sclerosis have been established in rodents, but none of these, however, are truly mimicking the human disease (Firestein 2009).

At a higher level, therapeutic mode of action of p110δ inhibitors will not be limited to the B cell function expanding its activity against some specialized functions of T cells, neutrophils, and mast cells. The hypothesis that small molecule inhibitors of p110δ or other targets such as JAK are representing the small molecule alternatives to previously established protein-based therapeutics is indeed exciting and attractive. The potential therapeutic value of dialing-in a p110γ action into a p110δ small molecule inhibitor is the proven pan-chemokine wide inhibitory activity of p110γ antagonists. Consequently, diseases such as rheumatoid arthritis, multiple sclerosisupus but also psoriasis as well as acute allergic or chronic inflammatory disease of the lung are high ranked indications for tagging onto the product profile of an orally active, potent and selective dual-active p110δ/γ small molecule kinase inhibitor. The challenge for near term clinical development is to design creatively proof-of-concept studies to demonstrate on-target biological activity in humans and the therapeutic validation of p110δ/γ immune-signaling inhibition in patients with such immune mediated disorders. It is probably save to say that small molecule p110δ, p110γ or p110δ/γ inhibitors will differentiate from currently approved anti-inflammatory drugs, such as methotrexate or SAIDs, by the principal mode and mechanism of action. Unquestionable, such drugs will likely show effects on normal, host-defense functions of the immune system (Zhang et al. 2008; Maus et al. 2007). Such potentially problematic adverse events need to be anticipated, monitored and managed in patients as long as they have proven to be measurable and reversible in preceding preclinical safety and toxicology studies. One faces the "catch 22" situation when trying to develop meaningful anti-inflammatory drugs free of minimal immuno-suppressive penalty – which, nevertheless, remains the ultimate goal.

As the people in Tuscany say: "In order to make a fair and democratic decision one needs more than two people, but three are too many. . .." This nicely resembles the challenge in drug development when trying to agree on the right profile between super-selective and eventually dirty compounds. The PI3K field has now a precious in-depth knowledge and an unparalleled battery of tools at hand ranging from numerous genetic mutant mice, several crystal structures, broad-spectrum as well as highly selective inhibitors, human genetic data and the access to the invaluable insights from the emerging clinical data. There have been some painful, yet insightful lessons from the research and development efforts on the family of p38 MAP kinases, chemokine receptors and others (Johnson et al. 2004). The commitment has to be to respectfully embark on such experiences and to set forth appropriate future directions. The PI3K/mTOR pathway is a rich opportunity for drug development and (in my mind) will deliver a set of distinct therapeutics against several dreadful human diseases. As this book attests, this is an exciting and privileged time for the entire field and research community, both in academia and industry.

Acknowledgments I thank my former Serono team (Montse Camps, Hong Ji, Christian Pasquali, Jeff Shaw, Thomas Ruckle and Kaspar Schwarz et al.), and all collaborators (e.g., Bart Vanhaesebroeck, Emilio Hirsch, Matthias Wymann, Klaus Okkenhaug, David Fruman, Ana Carrera, Glen Prestwich, Reinhard Wetzker, Roger Williams, Len Stephens, and Phil Hawkins, Mauro Teixeira, Gary S. Firestein, David Boyle and Charles Mackay for their permanent and invaluable impact on my professional as well as personal path. I also like to thank Kevan Shokat, Yi Liu, Pingda Ren, Mike Martin, Ed Ulm, Pam Klein and Troy Wilson for the "Intellikine opportunity." Finally, I do would like to thank my lifetime mentors George Yancopoulos (big molecules) and Matthias Kaspar Schwarz (small molecules) for having taught me how to think and apply science for drug discovery and development. I am very grateful to my colleagues and friends Peter Vogt and Bart Vanhaesebroeck for the idea and vision of doing this book together.

References

Ali K, Bilancio A, Thomas M, Pearce W, Gilfillan AM, Tkaczyk C, Kuehn N, Gray A, Giddings J, Peskett E, Fox R, Bruce I, Walker C, Sawyer C, Okkenhaug K, Finan P, Vanhaesebroeck B (2004) Essential role for the p110delta phosphoinositide 3-kinase in the allergic response. Nature 431(7011):1007–1011

Ali K, Camps M, Pearce WP, Ji H, Rückle T, Kuehn N, Pasquali C, Chabert C, Rommel C, Vanhaesebroeck B (2008) Isoform-specific functions of phosphoinositide 3-kinases: p110 delta but not p110 gamma promotes optimal allergic responses *in vivo*. J Immunol 180 (4):2538–2544

Andrews S, Stephens LR, Hawkins PT (2007) PI3K class IB pathway in neutrophils. Sci STKE 2007(407):cm3

Arcaro A, Wymann MP (1993) Wortmannin is a potent phosphatidylinositol 3-kinase inhibitor. Biochem J 296:297–301

Barbi J, Cummings HE, Lu B, Oghumu S, Rückle T, Rommel C, Lafuse W, Whitacre CC (2008) p110g facilitating an efficient induction of CXCR3 on activated T cells. Satoskar AR Blood 112(8):3048–3051

Bilancio A, Okkenhaug K, Camps M, Emery JL, Ruckle T, Rommel C, Vanhaesebroeck B (2006) Key role of the p110delta isoform of PI3K in B-cell antigen and IL-4 receptor signaling: comparative analysis of genetic and pharmacologic interference with p110delta function in B cells. Blood 107(2):642–650

Camps M, Rückle T, Ji H, Ardissone V, Rintelen F, Shaw J, Ferrandi C, Chabert C, Gillieron C, Françon B, Martin T, Gretener D, Perrin D, Leroy D, Vitte PA, Hirsch E, Wymann MP, Cirillo R, Schwarz MK, Rommel C (2005) Blockade of PI3Kgamma suppresses joint inflammation and damage in mouse models of rheumatoid arthritis. Nat Med 11(9):936–943

Chang JD, Sukhova GK, Libby P, Schvartz E, Lichtenstein AH, Field SJ, Kennedy C, Madhavarapu S, Luo J, Wu D, Cantley LC (2007) Deletion of the phosphoinositide 3-kinase p110gamma gene attenuates murine atherosclerosis. Proc Natl Acad Sci USA 104(19):8077–8082

Condliffe AM, Davidson K, Anderson KE, Ellson CD, Crabbe T, Okkenhaug K, Vanhaesebroeck B, Turner M, Webb L, Wymann MP, Hirsch E, Ruckle T, Camps M, Rommel C, Jackson SP, Chilvers ER, Stephens LR, Hawkins PT (2005) Sequential activation of class IB and class IA PI3K is important for the primed respiratory burst of human but not murine neutrophils. Blood 106(4):1432–1440

Cosemans JM, Munnix IC, Wetzker R, Heller R, Jackson SP, Heemskerk JW (2006) Continuous signaling via PI3K isoforms β and γ is required for platelet ADP receptor function in dynamic thrombus stabilization. Blood 108(9):3045–3052

Del Prete A, Vermi W, Dander E, Otero K, Barberis L, Luini W, Bernasconi S, Sironi M, Santoro A, Garlanda C, Facchetti F, Wymann MP, Vecchi A, Hirsch E, Mantovani A, Sozzani S (2004) Defective dendritic cell migration and activation of adaptive immunity in PI3Kgamma-deficient mice. EMBO J 23(17):3505–3515

Doukas J, Eide L, Stebbins K, Racanelli-Layton A, Dellamary L, Martin M, Dneprovskaia E, Noronha G, Soll R, Wrasidlo W, Acevedo LM, Cheresh DA (2009) Aerosolized phosphoinositide 3-kinase gamma/delta inhibitor TG100–115 [3-[2, 4-diamino-6-(3-hydroxyphenyl)-pteridin-7-yl]phenol] as a therapeutic candidate for asthma and chronic obstructive pulmonary disease. J Pharmacol Exp Ther 328(3):758–765

Duan W (2005) Aguinaldo Datiles AM, Leung BP, Vlahos CJ, Wong WS. An anti-inflammatory role for a phosphoinositide 3-kinase inhibitor LY294002 in a mouse asthma model. Int Immunopharmacol 5(3):495–502

Durand CA, Hartvigsen K, Fogelstrand L, Kim S, Iritani S, Vanhaesebroeck B, Witztum JL, Puri KD, Gold MR (2009) Phosphoinositide 3-kinase p110delta regulates natural antibody production, marginal zone and B-1 B cell function, and autoantibody responses. J Immunol 183 (9):5673–5684

Engelman JA (2009) Targeting PI3K signalling in cancer: opportunities, challenges and limitations. Nat Rev Cancer 9(8):550–562

Ferguson GJ, Milne L, Kulkarni S, Sasaki T, Walker S, Andrews S, Crabbe T, Finan P, Jones G, Jackson S, Camps M, Rommel C, Wymann M, Hirsch E, Hawkins P, Stephens L (2007) PI(3) Kgamma has an important context-dependent role in neutrophil chemokinesis. Nat Cell Biol 9 (1):86–91

Firestein GS (2009) Rheumatoid arthritis in a mouse? Nat Clin Pract Rheumatol 5(1):1

Fougerat A, Gayral S, Gourdy P, Schambourg A, Rückle T, Schwarz MK, Rommel C, Hirsch E, Arnal JF, Salles JP, Perret B, Breton-Douillon M, Wymann MP, Laffargue M (2008) Genetic and pharmacological targeting of phosphoinositide 3-kinase-gamma reduces atherosclerosis and favors plaque stability by modulating inflammatory processes. Circulation 117(10):1310–1317

Fruman DA, Bismuth G (2009) Fine tuning the immune response with PI3K. Immunol Rev 228(1): 253–272

Giese N CAL101 ref (Barcelona. ASH)

Hirsch E, Katanaev VL, Garlanda C, Azzolino O, Pirola L, Silengo L, Sozzani S, Mantovani A, Altruda F, Wymann MP (2000) Central role for G protein-coupled phosphoinositide 3-kinase gamma in inflammation. Science 287(5455):1049–1053

Hirsch E, Braccini L, Ciraolo E, Morello F, Perino A (2009) Twice upon a time: PI3K's secret double life exposed. Trends Biochem Sci 34(5):244–248

Hooft van Huijsduijnen R, Rommel C (2006) Decompartmentalizing target validation-thinking outside the pipeline boxes. J Mol Med 84(10):802–813

Ito K, Caramori G, Adcock IM (2007) Therapeutic potential of phosphatidylinositol 3-kinase inhibitors in inflammatory respiratory disease. J Pharmacol Exp Ther 321(1):1–8

Jackson SP, Schoenwaelder SM, Goncalves I, Nesbitt WS, Yap CL, Wright CE, Kenche V, Anderson KE, Dopheide SM, Yuan Y, Sturgeon SA, Prabaharan H, Thompson PE, Smith GD, Shepherd PR, Daniele N, Kulkarni S, Abbott B, Saylik D, Jones C, Lu L, Giuliano S, Hughan SC, Angus JA, Robertson AD, Salem HH (2005) PI 3-kinase p110beta: a new target for antithrombotic therapy. Nat Med 11(5):507–514

Jarmin SJ, David R, Ma L, Chai JG, Dewchand H, Takesono A, Ridley AJ, Okkenhaug K, Marelli-Berg FM (2008) T cell receptor-induced phosphoinositide-3-kinase p110delta activity is required for T cell localization to antigenic tissue in mice. J Clin Invest 118(3): 1154–1164

Ji H, Rintelen F, Waltzinger C (2007) Bertschy Meier D, Bilancio A, Pearce W, Hirsch E, Wymann MP, Rückle T, Camps M, Vanhaesebroeck B, Okkenhaug K, Rommel C. Inactivation of PI3Kgamma and PI3Kdelta distorts T-cell development and causes multiple organ inflammation. Blood 110(8):2940–2947

Johnson Z, Power CA, Weiss C, Rintelen F, Ji H, Ruckle T, Camps M, Wells TN, Schwarz MK, Proudfoot AE, Rommel C (2004) Chemokine inhibition–why, when, where, which and how? Biochem Soc Trans 32(Pt 2):366–377

Kim MS, Rådinger M, Gilfillan AM (2008) The multiple roles of phosphoinositide 3-kinase in mast cell biology. Trends Immunol 29(10):493–501

Knight ZA, Shokat KM (2007a) Chemically targeting the PI3K family. Biochem Soc Trans 35 (pt 2):245–249

Knight ZA, Shokat KM (2007b) Chemical genetics: where genetics and pharmacology meet. Cell 128(3):425–430

Kok K, Geering B, Vanhaesebroeck B (2009) Regulation of phosphoinositide 3-kinase expression in health and disease. Trends Biochem Sci 34(3):115–127

Konrad S, Ali SR, Wiege K, Syed SN, Engling L, Piekorz RP, Hirsch E, Nürnberg B, Schmidt RE, Gessner JE (2008) Phosphoinositide 3-kinases gamma and delta, linkers of coordinate C5a receptor-Fcgamma receptor activation and immune complex-induced inflammation. J Biol Chem 283(48):33296–33303

Laffargue M, Calvez R, Finan P, Trifilieff A, Barbier M, Altruda F, Hirsch E, Wymann MP (2002) Phosphoinositide 3-kinase gamma is an essential amplifier of mast cell function. Immunity 16(3):441–451

Lee KS, Lee HK, Hayflick JS, Lee YC, Puri KD (2006) Inhibition of phosphoinositide 3-kinase delta attenuates allergic airway inflammation and hyperresponsiveness in murine asthma model. FASEB J 20(3):455–465

Li Z, Jiang H, Xie W, Zhang Z, Smrcka AV, Wu D (2000) Roles of PLC-beta2 and -beta3 and PI3Kgamma in chemoattractant-mediated signal transduction. Science 287(5455): 1046–1049

Lim DH, Cho JY, Song DJ, Lee SY, Miller M, Broide DH (2009) PI3K gamma-deficient mice have reduced levels of allergen-induced eosinophilic inflammation and airway remodeling. Am J Physiol Lung Cell Mol Physiol 296(2):L210–L219

Liu L, Puri KD, Penninger JM, Kubes P (2007) Leukocyte PI3Kgamma and PI3Kdelta have temporally distinct roles for leukocyte recruitment *in vivo*. Blood 110(4):1191–1198

Liu P, Cheng H, Roberts TM, Zhao JJ (2009) Targeting the phosphoinositide 3-kinase pathway in cancer. Nat Rev Drug Discov 8(8):627–644

Mackay CR (2008) Moving targets: cell migration inhibitors as new anti-inflammatory therapies. Nat Immunol 9(9):988–998

Martin AL, Schwartz MD, Jameson SC, Shimizu Y (2008) Selective regulation of CD8 effector T cell migration by the p110 gamma isoform of phosphatidylinositol 3-kinase. J Immunol 180(4):2081–2088

Marwick JA, Caramori G, Stevenson CS, Casolari P, Jazrawi E, Barnes PJ, Ito K, Adcock IM, Kirkham PA, Papi A (2009) Inhibition of PI3Kdelta restores glucocorticoid function in smoking-induced airway inflammation in mice. Am J Respir Crit Care Med 179(7):542–548

Maus UA, Backi M, Winter C, Srivastava M, Schwarz MK, Rückle T, Paton JC, Briles D, Mack M, Welte T, Maus R, Bohle RM, Seeger W, Rommel C, Hirsch E, Lohmeyer J, Preissner KT (2007) Importance of phosphoinositide 3-kinase gamma in the host defense against pneumococcal infection. Am J Respir Crit Care Med 175(9):958–966

Nashed BF, Zhang T, Al-Alwan M, Srinivasan G, Halayko AJ, Okkenhaug K, Vanhaesebroeck B, Hayglass KT, Marshall AJ (2007) Role of the phosphoinositide 3-kinase p110delta in generation of type 2 cytokine responses and allergic airway inflammation. Eur J Immunol 37(2): 416–424

Okkenhaug K, Vanhaesebroeck B (2003) PI3K in lymphocyte development, differentiation and activation. Nat Rev Immunol 3(4):317–330

Okkenhaug K, Bilancio A, Farjot G, Priddle H, Sancho S, Peskett E, Pearce W, Meek SE, Salpekar A, Waterfield MD, Smith AJ, Vanhaesebroeck B (2002) Impaired B and T cell antigen receptor signaling in p110delta PI 3-kinase mutant mice. Science 297(5583):1031–1034

Okkenhaug K, Ali K, Vanhaesebroeck B (2007) Antigen receptor signalling: a distinctive role for the p110delta isoform of PI3K. Trends Immunol 28(2):80–87

Omori SA, Rickert RC (2007) Phosphatidylinositol 3-kinase (PI3K) signaling and regulation of the antibody response. Cell Cycle 6(4):397–402

Pinho V (2007) de Castro Russo R, Amaral FA, de Sousa LP, Barsante MM, de Souza DG, Alves-Filho JC, Cara DC, Hayflick JS, Rommel C, Ruckle T, Rossi AG, Teixeira MM. Tissue- and stimulus-dependent role of phosphatidylinositol 3-kinase isoforms for neutrophil recruitment induced by chemoattractants *in vivo*. J Immunol 179(11):7891–7898

Pinho V, Souza DG, Barsante MM, Hamer FP, De Freitas MS, Rossi AG, Teixeira MM (2005) Phosphoinositide-3 kinases critically regulate the recruitment and survival of eosinophils *in vivo*: importance for the resolution of allergic inflammation. J Leukoc Biol 77(5): 800–810

Puri KD, Doggett TA, Douangpanya J, Hou Y, Tino WT, Wilson T, Graf T, Clayton E, Turner M, Hayflick JS, Diacovo TG (2004) Mechanisms and implications of phosphoinositide 3-kinase delta in promoting neutrophil trafficking into inflamed tissue. Blood 103(9):3448–3456

Puri KD, Doggett TA, Huang CY, Douangpanya J, Hayflick JS, Turner M, Penninger J, Diacovo TG (2005) The role of endothelial PI3Kgamma activity in neutrophil trafficking. Blood 106 (1):150–157

Randis TM, Puri KD, Zhou H, Diacovo TG (2008) Role of PI3Kdelta and PI3Kgamma in inflammatory arthritis and tissue localization of neutrophils. Eur J Immunol 38(5):1215–1224

Reif K, Okkenhaug K, Sasaki T, Penninger JM, Vanhaesebroeck B, Cyster JG (2004) Cutting edge: differential roles for phosphoinositide 3-kinases, p110gamma and p110delta, in lymphocyte chemotaxis and homing. J Immunol 173(4):2236

Rommel C, Camps M, Ji H (2007) PI3K delta and PI3K gamma: partners in crime in inflammation in rheumatoid arthritis and beyond? Nat Rev Immunol 7(3):191–201

Rückle T, Schwarz MK, Rommel C (2006) PI3Kgamma inhibition: towards an 'aspirin of the 21st century'? Nat Rev Drug Discov 5(11):903–918

Sadhu C, Masinovsky B, Dick K, Sowell CG, Staunton DE (2003) Essential role of phosphoinositide 3-kinase delta in neutrophil directional movement. J Immunol 170(5):2647–2654

Sasaki T, Irie-Sasaki J, Jones RG, Oliveira-dos-Santos AJ, Stanford WL, Bolon B, Wakeham A, Itie A, Bouchard D, Kozieradzki I, Joza N, Mak TW, Ohashi PS, Suzuki A, Penninger JM (2000) Function of PI3Kgamma in thymocyte development, T cell activation, and neutrophil migration. Science 287(5455):1040–1046

Saudemont A, Garçon F, Yadi H, Roche-Molina M, Kim N, Segonds-Pichon A, Martín-Fontecha A, Okkenhaug K, Colucci F (2009) p110gamma and p110delta isoforms of phosphoinositide 3-kinase differentially regulate natural killer cell migration in health and disease. Proc Natl Acad Sci USA 106(14):5795–5800

Sinclair LV, Finlay D, Feijoo C, Cornish GH, Gray A, Ager A, Okkenhaug K, Hagenbeek TJ, Spits H, Cantrell DA (2008) Phosphatidylinositol-3-OH kinase and nutrient-sensing mTOR pathways control T lymphocyte trafficking. Nat Immunol 9(5):513–521

Swat W, Montgrain V, Doggett TA, Douangpanya J, Puri K, Vermi W, Diacovo TG (2006) Essential role of PI3Kdelta and PI3Kgamma in thymocyte survival. Blood 107(6):2415–2422

Takeda M, Ito W, Tanabe M, Ueki S, Kato H, Kihara J, Tanigai T, Chiba T, Yamaguchi K, Kayaba H, Imai Y, Okuyama K, Ohno I, Sasaki T, Chihara J (2009) Allergic airway hyperresponsiveness, inflammation, and remodeling do not develop in phosphoinositide 3-kinase gamma-deficient mice. T J Allergy Clin Immunol 123(4):805–812

Tassi I, Cella M, Gilfillan S, Turnbull I, Diacovo TG, Penninger JM, Colonna M (2007) p110gamma and p110delta phosphoinositide 3-kinase signaling pathways synergize to control development and functions of murine NK cells. Immunity 27(2):214–227

Thomas MJ, Smith A, Head DH, Milne L, Nicholls A, Pearce W, Vanhaesebroeck B, Wymann MP, Hirsch E, Trifilieff A, Walker C, Finan P, Westwick J (2005) Airway inflammation: chemokine-induced neutrophilia and the class I phosphoinositide 3-kinases. Eur J Immunol 35(4):1283–1291

Thomas M, Edwards MJ, Sawicka E, Duggan N, Hirsch E, Wymann MP, Owen C, Trifilieff A, Walker C, Westwick J, Finan P (2009) Essential role of phosphoinositide 3-kinase gamma in eosinophil chemotaxis within acute pulmonary inflammation. Immunology 126(3):413–422

Vanhaesebroeck B, Rohn JL, Waterfield MD (2004) Gene targeting: attention to detail. Cell 118 (3):274–276

Vlahos CJ et al (1994) A specific inhibitor of phosphatidylinositol 3 kinase, 2-(4-morpholinyl)-8-phenyl-4H-1-benzopyran-4-one (LY294002). J Biol Chem 269(7):5241–5248

Walker EH, Pacold E, Perisic O, Stephens L, Hawkins PT, Wymann MP, Williams RL (2000) Structural Determinants of Phosphoinositide 3-Kinase Inhibition by Wortmannin, LY294002, Quercetin, Myricetin, and Staurosporine. Mol Cell 6:909–919

Webb LM, Vigorito E, Wymann MP, Hirsch E, Turner M (2005) Cutting edge: T cell development requires the combined activities of the p110gamma and p110delta catalytic isoforms of phosphatidylinositol 3-kinase. J Immunol 175(5):2783–2787

Wymann MP, Björklöf K, Calvez R, Finan P, Thomast M, Trifilieff A, Barbier M, Altruda F, Hirsch E, Laffargue M (2003) Phosphoinositide 3-kinase gamma: a key modulator in inflammation and allergy. Biochem Soc Trans 31(Pt 1):275–280

Zhang TT, Okkenhaug K, Nashed BF, Puri KD, Knight ZA, Shokat KM, Vanhaesebroeck B, Marshall AJ (2008) Genetic or pharmaceutical blockade of p110delta phosphoinositide 3-kinase enhances IgE production. J Allergy Clin Immunol 122(4):811–819

Index

Contents of Volume II

MIX
Papier aus verantwortungsvollen Quellen
Paper from responsible sources
FSC® C105338

If you have any concerns about our products,
you can contact us on
ProductSafety@springernature.com

In case Publisher is established outside the EU,
the EU authorized representative is:
Springer Nature Customer Service Center GmbH
Europaplatz 3, 69115 Heidelberg, Germany

Printed by Libri Plureos GmbH
in Hamburg, Germany